Lecture Notes in Physics

Volume 945

The Lecture Notes in Physics

The series Lecture Notes in Physics (LNP), founded in 1969, reports new developments in physics research and teaching-quickly and informally, but with a high quality and the explicit aim to summarize and communicate current knowledge in an accessible way. Books published in this series are conceived as bridging material between advanced graduate textbooks and the forefront of research and to serve three purposes:

- to be a compact and modern up-to-date source of reference on a well-defined topic
- to serve as an accessible introduction to the field to postgraduate students and nonspecialist researchers from related areas
- to be a source of advanced teaching material for specialized seminars, courses and schools

Both monographs and multi-author volumes will be considered for publication. Edited volumes should, however, consist of a very limited number of contributions only. Proceedings will not be considered for LNP.

Volumes published in LNP are disseminated both in print and in electronic formats, the electronic archive being available at springerlink.com. The series content is indexed, abstracted and referenced by many abstracting and information services, bibliographic networks, subscription agencies, library networks, and consortia.

Proposals should be sent to a member of the Editorial Board, or directly to the managing editor at Springer:

Christian Caron
Springer Heidelberg
Physics Editorial Department I
Tiergartenstrasse 17
69121 Heidelberg/Germany
christian.caron@springer.com

More information about this series at http://www.springer.com/series/5304

Helmut Satz

Extreme States of Matter in Strong Interaction Physics

An Introduction

Second Edition

 Springer

Helmut Satz
Universität Bielefeld Fakultät für Physik
Bielefeld, Germany

ISSN 0075-8450 ISSN 1616-6361 (electronic)
Lecture Notes in Physics
ISBN 978-3-319-71893-4 ISBN 978-3-319-71894-1 (eBook)
https://doi.org/10.1007/978-3-319-71894-1

Library of Congress Control Number: 2017962299

Printed on acid-free paper

This Springer imprint is published by Springer Nature
The registered company is Springer International Publishing AG
The registered company address is: Gewerbestrasse 11, 6330 Cham, Switzerland

Für Karin

Preface

It was quite naturally a pleasure for me that the resonance of this little book was sufficient to have the publishers propose a second edition. The past 5 years have indeed brought considerable progress both in phenomenology and in experiment, although there still remain some quite fundamental unsolved issues. Nevertheless, the advances of the recent years certainly merit an appropriate update.

Besides minor updates throughout, the main modifications concern

- The calibration of quarkonium production in nuclear collisions
- The relation between strangeness suppression and deconfinement
- The search for critical behavior in multihadron production moments

These led to significant revisions of Chaps. 10 and 11. In addition, I have added a new Chap. 13 on fluids and flow. The role of a hydrodynamic evolution of the medium produced in high-energy collisions has become more and more pronounced over the years, and so such a chapter seems relevant. There is, of course, a certain danger involved in this: combining (local) equilibrium thermodynamics with hydrodynamics introduces ever more model dependence and hence tends to weaken the direct relation between nuclear collision data and statistical QCD. It is to be hoped that future theoretical work can accommodate this problem.

Thanks go once more to D. Miller for helpful comments. Furthermore, I am very happy that R. Baier, P. Huovinen, and R. Vogt have looked over the new chapter and corrected quite a number of misleading formulations. Whatever remains wrong is certainly my doing.

Bielefeld, Germany
August 2017

Helmut Satz

Preface to the First Edition

Over the past 50 years, the thermodynamics of strongly interacting matter has become a profound and challenging area of modern physics, both in theory and in experiment. Statistical quantum chromodynamics (QCD), through analytical as well as numerical studies, provides the main theoretical tool; in experiment, high-energy nuclear collisions are the key for extensive laboratory investigations. The field is therefore an area of overlap between statistical, particle, and nuclear physics, conceptually and in the methods of investigation used.

Many young people today are starting their scientific research in this field, and so there is a need for a general introduction, emphasizing in particular the basic concepts and ideas. That is the aim of this book, to explain why we are doing what we are doing. It is not meant to compete with several recent textbooks, which should be consulted to obtain the technical tools of the trade. I want to concentrate here more on the development of the underlying ideas, which I think have given the field a very unique flavor. Moreover, I will concentrate mainly on equilibrium thermodynamics. Nuclear collisions certainly involve non-equilibrium aspects, from thermalization to hydrodynamic flow. But before we can understand these, we will have to understand the striking new features already present in equilibrium.

The general plan of the book is the following. We first introduce the fundamental ideas of strong interaction thermodynamics (Chap. 1) and then summarize the main concepts and methods used in the study of complex systems (Chap. 2). After these general preliminaries, we present some models, i.e., simplified phenomenological pictures, leading to critical behavior in hadronic matter and to hadron-quark phase transitions (Chaps. 3 and 4). Given such a conceptual basis, we can then turn to finite temperature lattice QCD and to the results obtained in computer simulation studies of the lattice formulation (Chap. 5). Complementary to this, we clarify the relation of the resulting critical behavior to symmetry breaking/restoration in QCD (Chap. 6). In Chap. 7, an extension of our considerations to strongly interacting matter at finite baryon density provides the basis for the study of the QCD phase diagram. Following this, we investigate in some detail the main features of a new state of matter, the quark-gluon plasma (Chap. 8). This completes our presentation of bulk equilibrium thermodynamics, although quite a few very interesting aspects have not been dealt with in much detail; in particular, the regime of large baryon density has only been considered quite briefly. But this simply reflects the present state of our knowledge in that area.

Following this more general part of the book, we turn to some specific features which arise when nuclear collisions are considered as a tool for the experimental study of QCD thermodynamics. In Chap. 9, we introduce the conceptual basis for this endeavor; following this, we briefly survey the main probes proposed to investigate the properties of the medium produced in such collisions. Finally, the last two chapters deal with hadron production in high-energy collisions – a topic which, more than 50 years ago, started the intimate relation between strong interaction physics and thermodynamics.

Since our field is so multifaceted, I thought it would be helpful to keep the different chapters self-contained as much as possible, so that a reader could turn to a given topic without having to read in detail all previous chapters. This necessarily implies some repetition, for which I hope for indulgence.

This book is based on three sets of comprehensive lectures, held at the University of Bielefeld in 1995/1996, at the Instituto Superior Técnico (IST) in Lisbon in 2002/2003, and again at Bielefeld in 2010/2011. In addition, it has benefitted much from extensive lectures at a number of schools, organized by CCAST, CERN, Dubna, Frascati, VECC, and others. The course of the years, I think, had a very healthy influence in placing things into some perspective. Let us see what the next years will bring. In the meantime, I hope that what is presented here may help in showing future young researchers that ours is a field with much promise, within and beyond its topical frontiers.

A considerable part of the book was written and edited during two long stays at the IST in Lisbon, made possible by a grant of the Calouste Gulbenkian Foundation. I want to take this opportunity to express my sincere gratitude to Jorge Dias de Deus and João Seixas for their kind hospitality there and to the Gulbenkian Foundation for their generous support.

Next, it is my great pleasure to thank my colleagues for all the help I have received over the years in my attempts to understand this fascinating chapter of physics. My gratitude goes first and foremost to my long-time fellow searchers (whose conclusions may and often do, of course, differ from mine). On the theory side, I thank Rolf Baier, Jürgen Engels, Rajiv Gavai, Sourendu Gupta, Keijo Kajantie, Frithjof Karsch, Dima Kharzeev, Edwin Laermann, Tetsuo Matsui, Larry McLerran, David Miller, Krzysztof Redlich, Vesa Ruuskanen, Esko Suhonen, Bob Thews, Ramona Vogt, and Xin-Nian Wang; more recently, crucial help came from Paolo Castorina and Olaf Kaczmarek. On the experimental side, Francesco Becattini, Peter Braun-Munzinger, Louis Kluberg, Carlos Lourenço, Jürgen Schukraft, Johanna Stachel, and Reinhard Stock have been essential in my search for understanding, and they have moreover made sure that the experimental consequences of whatever theoretical "Hirngespinst" were never ignored or forgotten. This list of names is necessarily incomplete, and I want to express my gratitude also to all the others who have helped me so much.

Furthermore, sincere thanks go to Frithjof Karsch, David Miller, and Krzysztof Redlich, for a careful reading of parts of the draft, and especially to Carlos Lourenço, who read and corrected the entire manuscript with great care. This has certainly helped much in eliminating many wrong and ambiguous formulations; what remains faulty is obviously my responsibility.

Someone once said that being a "normal" person married to a physicist is like being deaf and married to a musician. I dedicate this book to my wife for bearing with me during all those endless hours of unheard music.

Bielefeld, Germany   Helmut Satz
June 2011

Contents

Chapter 1
The Analysis of Dense Matter

> *So there must be an ultimate limit to bodies, beyond perception*
> *by our senses. This limit is without parts, is the smallest possible*
> *thing. It can never exist by itself, but only as primordial part of a*
> *larger body, from which no force can tear it loose.*

> Titus Lucretius Carus: *De rerum natura,*
> liber primus 599, $\sim$ 55 B. C.

Abstract We begin by introducing the search for the ultimate constituents of matter, discussing the possible forms of matter and ist density regimes in the universe. At sufficiently high density, such as prevailed in the early universe, a new state is expected, consisting of deconfined quarks.

What is matter made of? The search for its ultimate constituents has always inspired the imagination. Since antiquity, man has tried to explain the composite macroscopic world in terms of indivisible building blocks on a microscopic scale. Beneath the complexity and irregularity which surround us, we hope to find a hidden world of greater simplicity, in which primordial parts move according to basic laws. This idea turned out to be fruitful beyond all expectations, so that today we find it natural to derive the properties of matter from the dynamics which govern the interaction between some fundamental building blocks.

The analysis of matter thus traditionally begins by asking

- What are the ultimate constituents of matter?
- What are the basic forces between these constituents?

Once we understand the elementary systems and their interactions, we can then ask

- What are the possible states of matter?
- How do transitions between these states occur?

© Springer International Publishing AG 2018

H. Satz, *Extreme States of Matter in Strong Interaction Physics,*

Lecture Notes in Physics 945, https://doi.org/10.1007/978-3-319-71894-1_1

Only in rather recent years has this order of things been put to question. Basic features in the collective behavior of many interacting constituents – of "matter" – may well involve general aspects independent of both the nature of the constituents and of the form of their interactions. Today we are witnessing the emergence of a genuinely new field of physics, the study of complex systems, in which general laws governing collective features dominate over the nature of individual constituents and their interactions.

We want to present here a combination of the more traditional approach, based on the interaction of the fundamental building blocks in strong interaction physics, and the new aspects of collective behavior in complex systems, as encountered in the study of criticality and involving concepts such as spontaneous symmetry breaking, percolation, self-similarity and self-organization.

During the past hundred years, our ideas about the nature of the ultimate constituents of matter have undergone a number of changes. Atoms were found to be divisible into electrons and nuclei. Nuclei in turn consist of protons and neutrons, bound by strong, short-range forces. With the advent of the basic theory of strong interactions, quantum chromodynamics (QCD) [1], has come the conviction that nucleons – and more generally, all strongly interacting particles (hadrons) – are bound states of quarks [2, 3]. An account of all observed hadron states requires the six quark species (or *flavors*) of QCD, denoted as u, d, c, s, b and t for up, down, strange, charm, bottom and top. They are point-like and cannot exist by themselves; they are confined to "their" hadron by a binding potential increasing linearly with quark separation. Hence an infinite amount of energy would be needed to isolate a quark; it is not possible to "split" a hadron into its quark constituents. We are thus closer than ever before to the picture which seemed natural to Lucretius two thousand years ago.

It is of course possible to go on dividing. This has in fact been proposed, with "preons" as subconstituents of quarks and leptons [4, 5], but so far without great resonance; for some recent work, however, see [6]. In any case, in the quark substructure of the strongly interacting particles we encounter for the first time in modern physics the concept of basic constituents without an individual physical existence. So, in a way, we have arrived at the end of the line. The elementary particles in strong interaction physics, the mesons and baryons, remain elementary in the sense that they are the smallest entities which can exist by themselves, as single particle states in the physical vacuum. They have, however, become composite in the sense that they are bound states of quarks, bound so strongly "that no force can tear them loose."

The electromagnetic force holds nuclei and electrons together to form atoms. The nuclei themselves are bound states of nucleons, bound by the much stronger nuclear force. The discovery of radioactivity brought in a much weaker interaction, and gravity is the fourth and weakest force. The formulation of a "theory" for each of these four forces provided the fundamental challenge for more than a century, and the last decades have now seen attempts to unify the results. Electromagnetic and weak interaction theories are united as the electroweak theory [7–9], which together with QCD forms the "standard model" [10]. The different interaction sectors were found to entail quite a number of specific constituents. Besides the quarks forming the bulk of the observable matter of our universe, and the gluons which mediate their interaction, we have, according to our present state of knowledge, as further constituents of the electroweak interaction leptons, photons, heavy vector bosons ($B = Z_0$, $W^\pm$) and the scalar Higgs boson H. A conservative count (no antiparticles etc.) thus leads to the sixteen species of constituents of the standard model shown in Table 1.1; gravitation is not yet in the game, waiting for a "theory of everything".

u	c	t	g
d	s	b	γ
ν_e	ν_μ	ν_τ	B
e	μ	τ	H

Table 1.1 The basic constituents of the standard model

We underline here once more that the quarks, in contrast to all others in this table, cannot exist as isolated physical entities. As color-carrying constituents, they remain "confined" to the proximity of other quarks, close enough so that they can neutralize each other to form colorless spatial regions of hadronic size.

The second major question, addressing the possible states of matter, leads to problems which go beyond the level of elementary interactions. In ancient times they had earth, water, air and fire, as proposed by Empedokles about 450 B.C., and about a century later, Aristotle added the void as a "fifth" state, the "quintessence". The resulting "phase diagram" is illustrated in Fig. 1.1. A very similar scheme had been proposed in Buddhist and Hindu philosophies, where in particular it was noted that the void ("akasha") is what provides space and thus room for all extended substances. It should be noted that the mentioned "states" were then really taken as "forms of matter"; they were not considered as aggregates of many components. This "statistical" part of the story came into play only when the thinking about forms of matter was combined with the ideas of atomism, introduced by Democritos at about the same time.

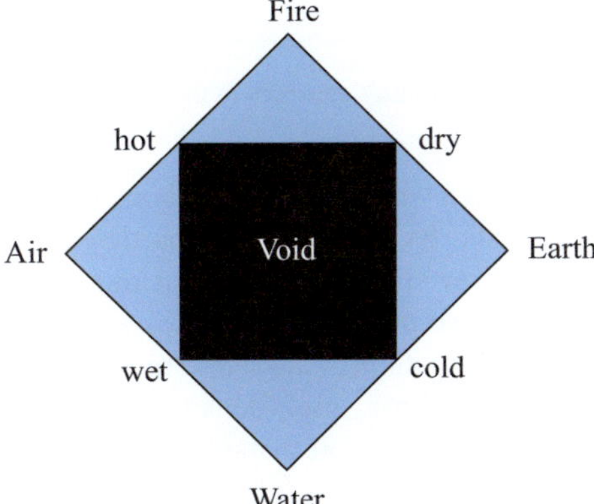

Fig. 1.1 The four states of matter in antiquity, fire, air, water and earth, embedded in the "quintessential" void

The general scheme of antiquity – solid, liquid, gas and plasma, embedded in a void – has survived until today, although now we have in addition also insulators, conductors and superconductors, fluids and superfluids, ferromagnets, spin glasses, gelatines and many more distinct states. In almost all these cases, the knowledge of the elementary interaction neither predicts nor specifies the structure of the possible complex states of many constituents.

We expect things to become even more complicated, when we consider the transitions between the various different states. Recent developments show, however, that this is not necessarily true. On one hand, there are phase transitions of different orders depending on the singular behavior of the partition function, specified in terms of the respective dynamics and the corresponding symmetries, as well as percolation transitions signalling cluster formation and the onset of large-scale connected patterns. On the other hand, self-similarity, scaling and renormalization concepts lead to a universal description of critical phenomena, to a large extent independent of the underlying elementary dynamics. Critical exponents define universality classes which contain quite different interaction forms. Complex systems thus open up new and more general directions in physics.

What happens to matter when we increase its density beyond that encountered in the world around us? Ever since the Big Bang, our universe has been expanding. We thus live in a world of steadily decreasing density. If we could reverse time and let the film run backwards, what kind of matter would we find? The range of densities we encounter in the "observable" world spans over forty orders of magnitude (see Fig. 1.2), from an average taken over the entire present universe, 1 nucleon/m^3, to that inside a heavy nucleus, 10^{44} nucleons/m^3. Within this range, we have some understanding based on experimental information. What happens to matter at still higher densities? That is one of the main topics of this book.

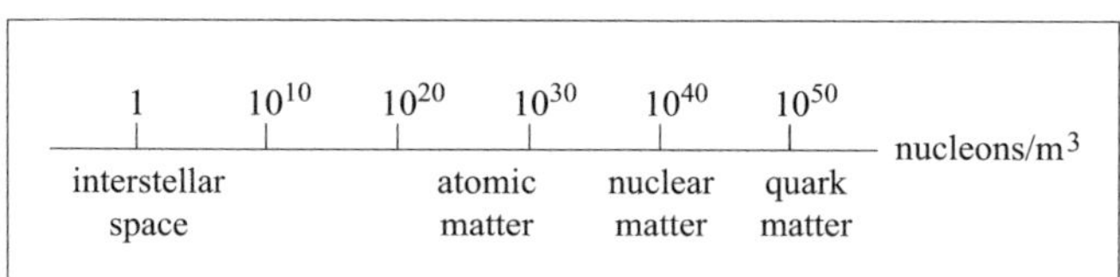

Fig. 1.2 Matter densities in the universe

The high-density limit is, in a way, the many-body counterpart of the high-energy limit of two-body interactions. By colliding two particles at ever higher energies, we study their structure at ever smaller scales. But that is only part of the story: at each scale, we have collective phenomena in many-body systems as well as the dynamic laws of two-body interactions. And while high-energy collisions probe the *dynamics* in the short distance limit, high-density studies are needed to test the *thermodynamics* resulting from this short distance dynamics.

What are the consequences of the quark infrastructure of elementary particles for the behavior of matter at extreme density? That will be one of our basic questions, and we shall try to answer it on various levels of theoretical and experimental understanding. To get a first idea, let us begin with a very simple picture. If nucleons, with their given spatial extension, were both elementary and incompressible, then a state of close packing would constitute the high density limit of matter (Fig. 1.3). If, on the other hand, nucleons are really composite (bound states of point-like quarks), then with increasing density they will start to overlap, until eventually we reach a state in which each quark finds within its immediate vicinity a considerable number of other quarks. It has no way to identify which of these had been its partners in a specific nucleon at some previous state of lower density. Beyond a certain point, the concept of a hadron thus loses its meaning [11], and we are quite naturally led from nuclear matter to a medium whose basic constituents are unbound quarks.

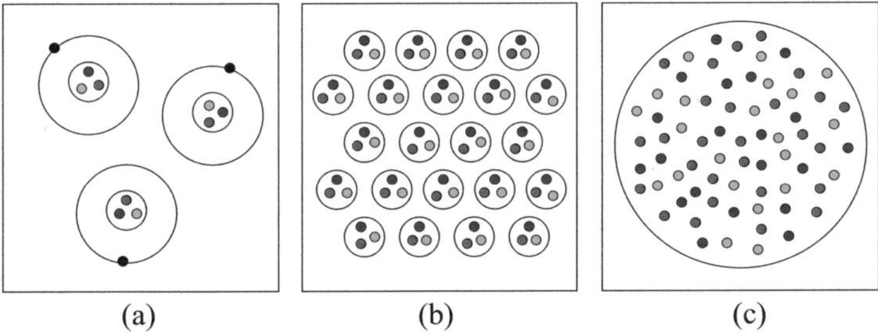

(a)   (b)   (c)

Fig. 1.3 Schematic view of increasing density, from atomic (**a**) to nuclear (**b**) and then to quark matter (**c**)

At first glance, it may seem that in the transition from nuclear matter to quark matter we have ignored the confining forces which bind quarks to form hadrons. However, confinement is a *long-range* feature, which prevents us from isolating a single quark. In the high density situation of Fig. 1.3c, each quark finds very close to it many others and is far from being isolated. The *short-range* aspect of dense matter thus appears to make confinement disappear.

To understand this better, we recall the effect of a dense medium on electric forces. In a vacuum, two electric charges e_0 interact through the Coulomb potential

$$V_e(r) \sim \left\{ \frac{e_0^2}{r} \right\}, \tag{1.1}$$

where r denotes the separation distance. In a dense environment of many other charges, the potential becomes screened,

$$V(r) \sim \left\{ \frac{e_0^2}{r} \right\} \exp\{-r/r_D(n)\}; \qquad (1.2)$$

where $r_D(n)$ is the Debye screening radius; it is a property of the medium and decreases as the charge density n increases. Thus the potential between two test charges a fixed distance apart becomes weaker with increasing density. This occurs because the other charges in the medium cluster around the two test charges and partially neutralize them, thus reducing the effect of the test charge interaction - its range becomes shorter. If instead of the test charges, we place a bound state – e.g., a hydrogen atom – into such a medium, then the screening radius r_D will, with increasing density, eventually fall below the binding radius r_B of the atom. Once $r_D \ll r_B$, the effective force between proton and electron has become so short-ranged that the two can no longer bind. Thus insulating matter, consisting of bound electric charges, will at sufficiently high density become conducting: it will undergo a Mott transition [12], in which charge screening dissolves the binding of the constituents. This leads us to a new state of matter: in addition to solid, liquid, and gas we now have the plasma, a macroscopic system of unbound charged constituents. Again, this is not so different from the picture of the world held in antiquity, with its four "states of matter" earth, water, air and fire.

The interaction of quarks in QCD is also based on an intrinsic charge, the "color"; a quark can exist in one of three different color states ("red, green, blue"). Just as electric charges interact through electrodynamic forces, color charges interact through chromodynamic forces. The main conceptual difference between electrodynamics and chromodynamics lies in the nature of the fields mediating the interaction. Two interacting electrons exchange photons, and these quanta of the electrodynamic field carry no intrinsic charge. Similarly, two interacting quarks exchange gluons, the quanta of the chromodynamic field; but these now carry an intrinsic color charge of their own. A gluon can change a red quark into a blue one, so that it has to be "blue-antired", and so on. Because of their intrinsic charge, the gluons, in contrast to photons, can interact directly with each other as well as with quarks. It is this (non-Abelian) gluon-gluon interaction that makes the difference between electrodynamic and chromodynamic forces. The lines of force radiate spherically from an electric charge, leading to the Coulomb potential as solution of the three-dimensional Laplace equation, while the lines of force radiating from a color charge interact with each other and contract into a string (Fig. 1.4). The resulting one-dimensional Laplace equation has as solution a potential increasing linearly with separation distance r [13],

$$V_q(r) \sim r; \qquad (1.3)$$

providing an effective quark *confinement*.

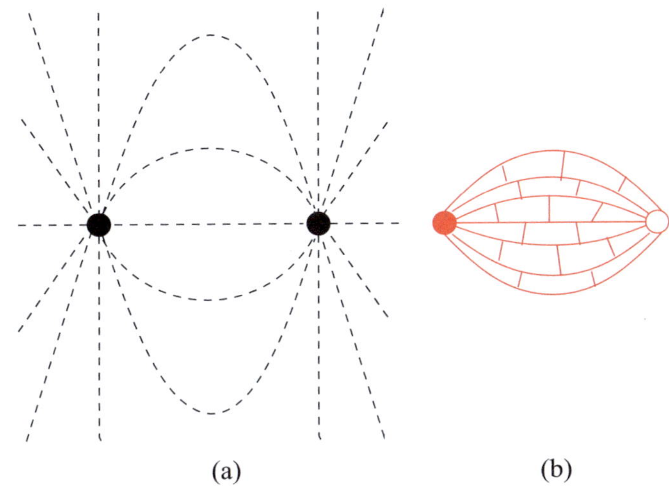

Fig. 1.4 Lines of force between electric charges (**a**) and between color charges (**b**)

Hadrons are color-neutral bound states of colored quarks, and dilute hadronic matter is a color insulator. At a sufficiently high density of color charges, however, we expect color screening to set in, and the resulting exponential damping of the binding force will remove all long-range effects, for the confining potential of the form (1.3) as well as for the Coulomb potential (1.1). We shall return to the relation of quark deconfinement and color screening in much more detail later on; here we only note the general idea [14] that color screening will eventually transform a color insulator into a color conductor, turning hadronic matter into a quark plasma. The transition from insulator to conductor by charge screening in atomic physics is a collective effect, and we thus also expect something like a phase transition at the point of plasma formation in QCD.

Besides deconfinement, i.e., the decoupling of quark binding by color screening, we expect a further transition to occur with increasing density. When atomic matter is transformed from an insulator into a conductor, the effective mass of the conduction electrons undergoes a change. In the insulator, the mass of the electrons in the atoms which make up the material is just the physical electron mass. In the conducting phase, however, the electrons acquire an effective "in-medium" mass different from the actual physical mass. The other conducting electrons, the periodic field of the charged ions and the lattice vibrations combine to produce a mean background field quite different from the vacuum. As a result, the unbound conduction electrons in the metal obtain an effective mass different from the one they have in vacuum. In a similar way, the effective mass of the quarks is expected to change between the confined and deconfined phases. When they are confined in hadrons, the basic quarks acquire through "dressing" with gluons an effective "constituent quark" mass of about 300 MeV: this is about 1/3 of the proton mass or about 1/2 of the ρ meson mass (the pion has a more complex character, to which we shall return later on). On the other hand, the basic quarks in the Lagrangian of QCD are almost massless. This must mean that the mass of the constituent quarks in the confined phase is generated spontaneously, through the confinement interaction itself. Hence it is likely that when deconfinement occurs, this additional mass is "lost" and the quarks revert to the intrinsic mass they have in the Lagrangian.

A shift in the effective quark mass is thus a further transition to look for as the density of strongly interacting matter is increased. Massless fermions – the limiting case of the light quarks in the Lagrangian – possess chiral symmetry: they can be decomposed into independent left- and right-handed massless spin one-half particles, whereas in the case of massive fermions, these two "handed" components are mixed. Therefore, in the limit of vanishing quark mass, the confinement interaction must lead to a spontaneous breaking of this symmetry, which should be restored in the deconfined phase. For this reason, the mass shift transition in QCD is often referred to as chiral symmetry restoration.

With increasing density, we thus expect critical behavior to occur in strongly interacting matter, leading to deconfinement and chiral symmetry restoration. In the study of such phase transitions and of critical phenomena in general, thermodynamics really goes beyond dynamics. In dynamical theories, we deal with isolated systems, and when we combine many such systems to form bulk matter, the correlations between distant constituents generally remain weak. But at critical points the constituents act in a highly collective fashion; the entire macroscopic system becomes correlated, and, as a result, very small parameter changes can produce dramatic changes of state. Water changes little between $50\,°C$ and $0.5\,°C$; but decreasing the temperature half a degree more turns it into ice. It is this new feature – collective action leading to singular behavior for bulk systems – that makes the study of critical phenomena so interesting. Much of the attention in strong interaction thermodynamics is thus centered on the different aspects of the transition from hadronic matter to the quark-gluon plasma.

As we have seen, this transformation is expected to take place once the density of matter has become high enough. Let us have a look at the actual density values involved and see how they compare to those of conventional condensed matter physics. The lowest density we encounter in nature is that of interstellar space, with about one nucleon per cubic meter; this is also the average density of our present universe. In air at sea level, we already have 10^{26} nucleons/m^3, and in dense atomic matter, such as iron, the density increases to about 10^{31} nucleons/m^3. Normal nuclear matter – the material a heavy nucleus is made of – contains 0.17 nucleons/fm^3, i.e., about 10^{44} nucleons/m^3. Statistical QCD suggests, as we shall see later on, that the transition to quark matter should take place for densities of around 2–3 hadrons/fm^3, or some 2–3×10^{45} hadrons/m^3. The regime we want to consider here thus exceeds the density of normal nuclear matter by an order of magnitude or more. Such densities existed in the universe shortly after the Big Bang; hence our subject is the thermodynamics of primordial matter, of the state of our universe before hadron formation occurred.

For the universe, the end of the quark matter era was the birth of the vacuum. In quark matter, quarks are deconfined simply because they can never get far enough away from other quarks and antiquarks to test the long range confining feature of the interquark forces. The space between quarks is therefore not empty; it is more like the interior of a nucleon or meson. One cannot remove quarks and antiquarks and still have the same interquark space. Only when quark matter becomes sufficiently dilute to force quarks and antiquarks to combine to nucleons and mesons, there appears between these hadrons truly empty space, space which can become arbitrarily big without having to change: the physical vacuum.

In non-relativistic physics, a density increase is generally related to compression. In the relativistic regime of high energy strong interactions, however, an energy input will lead to the production of further hadrons and thus also to a particle density increase. So to obtain higher densities, we can either increase the net baryon number density (the number of nucleons minus that of antinucleons, per unit volume), which for cold nuclear matter would just mean compression, or we can "heat" the system, so that collisions between its constituents produce further hadrons. The possible phases of strongly interacting matter thus have to be displayed in a phase diagram as function of its temperature T and its overall baryon density n_B. The latter is generally specified by the corresponding baryochemical potential μ.

With this in mind, let us now return to the fate of the quarks after deconfinement. When the hadrons are dissolved into their quark constituents, the liberated and hence now colored quarks could still interact with gluons to retain a non-vanishing effective mass, i.e., remain massive constituent quarks [15]. This would mean that deconfinement and chiral symmetry restoration do not coincide, since the remaining quark mass implies a continuation of chiral symmetry breaking. We shall see later on that, at least for low baryon density, this does not seem to occur: in that region of QCD, the onset of conductivity coincides with the mass shift of the constituents, just as it does in QED. However, the region of high baryon density and low temperature remains even today the *terra incognita* of QCD, whose possible structure continues to inspire much speculation.

On the one hand, here the liberated quarks might retain their mass for a certain range of baryon density, which would lead to a phase of unbound massive quarks. It is even conceivable that the quark triplets in nucleons, once deconfined, choose to recombine into massive colored quark pairs, locally bound bosonic "diquarks". Hence the massive quark phase could exhibit a structure similar to hadronic matter, with massive quarks as the ground state and diquarks as higher excitations [16].

On the other hand, diquarks are bosons, and, as such, at very low temperature they can form a Bose condensate. In atomic physics this happens when conduction electrons in a low temperature medium couple to form bosonic Cooper pairs; these condense and thus lead to superconductivity. So diquarks could play the role of Cooper pairs in QCD and condense into a color superconductor. The possibility of such color superconductivity has been studied in recent years in considerable detail [17–19].

Putting these concepts together, we obtain a speculative phase diagram of strongly interacting matter which shows a four-phase structure, with one confined and three deconfined states, as illustrated in Fig. 1.5. Along the μ axis, for $T = 0$, we are compressing cold nuclear matter; for $\mu = 0$, along the T axis, we are heating matter of vanishing overall baryon density, containing mesons together with an equal number of baryons and antibaryons. For low T and μ, we have hadronic matter as confined phase. Then deconfinement sets in, resulting at high temperatures in a phase consisting of unbound colored quarks and gluons, the quark-gluon plasma. For increasing baryon density at low temperatures, again we will eventually reach color deconfinement. But now the liberated quarks may retain their mass for a while and form a plasma of massive, unbound quarks. They can moreover also combine to colored bosonic diquarks; at very low temperature, these diquarks could condense to form a color superconductor. Heating such a superconductor would melt the condensate, just as it does for a normal superconductor, and we recover either the massive quark phase or the quark-gluon plasma, both as normal color conductors. The determination of the corresponding "critical" curves, i.e., of the values T_c and μ_c for which the different transitions take place, is one of the main topics of statistical QCD, and we shall return to it in detail in the next chapters. We have written "critical" in quotation marks, since it has turned out that the transitions in question do not necessarily correspond to singular behavior of the partition function, even though thermodynamic observables quite generally show a very rapid change at some "pseudo-critical" point. The actual nature of the transition is thus another aspect to which we have to return in detail.

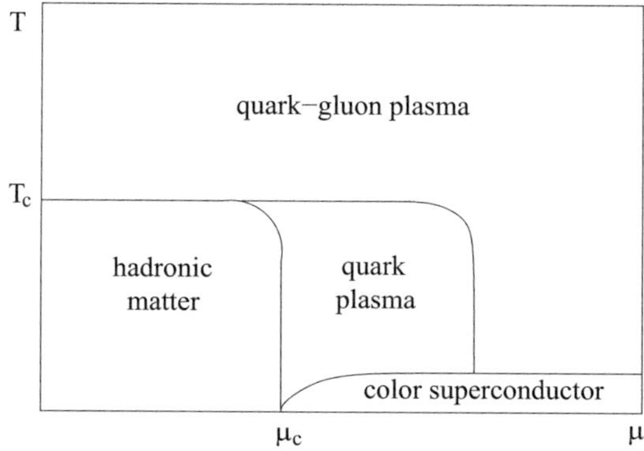

Fig. 1.5 A speculative phase diagram for strongly interacting matter

We have already noted that quark matter made up the early universe. Since dense nuclear matter forms the interior of neutron stars, it is also conceivable that such stars have quark cores. So far, however, the observed features of neutron stars seem to be accountable without invoking quark matter formation. The big bang was long ago, and neutron stars are far away. Is there some way to study strong interaction thermodyamics here and now? Can we create a quark plasma in the laboratory? The rapid growth which the field has experienced in the past two decades was to a very large extent stimulated by the idea that high energy nuclear collisions will produce droplets of strongly interacting matter – droplets large enough and long-lived enough to allow a study of the predictions which QCD makes for macroscopic systems. Moreover, it is expected that the conditions provided in these interactions will suffice for quark plasma formation. Hence the study of matter under extreme conditions has today a strong and multi-faceted experimental side; this, in turn, has stimulated much of the subsequent theoretical developments.

The relevant experiments were initially denoted as ultra-relativistic nucleus-nucleus collisions; they are often also called heavy ion collisions, though not quite correctly: an ion fully stripped of its electrons is a nucleus. The studies began at Brookhaven National Laboratory (BNL) near New York and at the European Organisation for Nuclear Research (CERN) in Geneva around 1986/1987. The first collisions had light nuclei (oxygen, silicon, sulphur) hitting heavy targets (gold, uranium), since light ions could be dealt with using injectors already existing at BNL and CERN. The successful analysis of these experiments provided the basis and motivation for the construction of new injectors of truly heavy nuclei, gold at BNL and lead at CERN; they came into operation in the middle 1990s. These early fixed target experiments were carried out at a center of mass energy of around 5 GeV per nucleon-nucleon collision at the BNL-AGS and around 20 GeV at the CERN-SPS. At the turn of the millennium, the first dedicated nuclear accelerator, the Relativistic Heavy Ion Collider RHIC, started taking data at BNL, with a center of mass energy a factor ten higher, at 200 GeV per nucleon-nucleon collision. And today, the Large Hadron Collider LHC at CERN is providing data with yet another factor of ten higher center-of-mass energy, up to 5500 GeV, or 5.5 TeV, for nuclear collisions. Moreover, now proton-proton collisions at the highest LHC energies of 7 TeV also lead to initial energy densities in the QGP range, so that even there deconfinement may prevail, albeit in smaller volumes.

The work of the different experimental groups working at these facilities has provided an immense wealth of data, and there is little doubt today that in such collisions comparatively large systems are formed, of higher energy density than ever before studied in the laboratory. The detailed analyses of the results have also shown that a number of new features arise which go well beyond standard thermodynamics. Questions of non-equilibrium aspects, of thermalization, evolution and expansion, cooling, flow and many more make a direct application of equilibrium strong interaction thermodynamics anything but straightforward. Any attempt to include these topics here in a comprehensive way seems beyond the scope of the book, and moreover our views here are still so much "under construction" that I simply would not dare to try. My aim is rather to present as clearly and comprehensively as possible the concepts, methods and results of the equilibrium thermodynamics obtained from QCD as the basic theory. These certainly must form the ultimate basis also for the motivation and the understanding of nuclear collision results. In the final four chapters, I then shall address some of the more general problems which arise when one tries to use high energy collisions as a tool to produce a strongly interacting medium. For me, the basic question is whether the systems produced in such collisions are something we can call matter, in the sense of the word used in statistical physics. For further forages into this new terrain, I refer to several excellent textbooks [20–26].

References

1. See e.g., F. Wilczek, in *QCD – 20 Years Later*, ed. by P.M. Zerwas, H.A. Kastrup (World Scientific Publ. Co., Singapore, 1993)
2. M. Gell-Mann, Phys. Lett. **8**, 214 (1964)
3. G. Zweig, Int. J. Mod. Phys. A **25**, 3863 (2010) and earlier references given there
4. J.C. Pati, A. Salam, Phys. Rev. **D10**, 275 (1974)
5. W. Buchmüller, in *Nucleon-Nucleon* and *Nucleon-Antinucleon Interactions*, ed. by H. Mitter, W. Plessas (Springer, Wien/New York, 1985)
6. R.J. Finkelstein, Int. J. Mod. Phys. A **22**, 4467 (2007)
7. See e.g., S. Weinberg, Rev. Mod. Phys. **52**, 515 (1980)
8. See e.g., A. Salam, Rev. Mod. Phys. **52**, 525 (1980)
9. See e.g., S. Glashow, Rev. Mod. Phys. **52**, 529 (1980)
10. See e.g., G. Kane, *Modern Elementary Particle Physics* (Addison-Wesley Publishing Co., Reading, 1993)
11. I. Ya. Pomeranchuk, Doklady Akad. Nauk. SSSR **78**, 889 (1951)
12. N.F. Mott, Proc. Phys. Soc. (Lond.) **A62**, 416 (1949)
13. V.V. Dixit, Mod. Phys. Lett. A **5**, 227 (1990)
14. H. Satz, Nucl. Phys. **A418**, 447c (1984)
15. For recent work on this, see e.g., P. Castorina, R.V. Gavai, H. Satz, Eur. Phys. J. C **69**, 169 (2010)
16. M. Anselmino, S. Ekelin, D.B. Lichtenberg, E. Predazzi, Rev. Mod. Phys. **65**, 1199 (1993)
17. For recent reviews, see e.g., K. Rajagopal, F. Wilczek, *The Condensed Matter Physics of QCD*, hep-ph/001333; published in *At the Frontier of Particle Physics*, ed. by M. Shifman, MIT-CTP-3049, vol. 3
18. For recent reviews, see e.g., M. Alford, Ann. Rev. Nuc. Part. Sci. **51**, 131 (2001)
19. For recent reviews, see e.g., M. Alford et al., Rev. Mod. Phys. **80**, 1455 (2008)
20. C.-Y. Wong, *Introduction to High Energy Heavy Ion Collisions* (World Scientific, Singapore 1994)
21. L.P. Csernai, *Introduction to Relativistic Heavy Ion Collisions* (Wiley, Chichester, 1994)
22. K. Yagi, T. Hatsuda, Y. Miake, *Quark-Gluon Plasma* (Cambridge University Press, Cambridge, 2005)
23. R.Ł. Vogt, *Ultrarelativistic Heavy Ion Collisions* (Elsevier, Amsterdam, 2007)
24. S. Sarkar, H. Satz, B. Sinha (eds.), *The Physics of the Quark-Gluon Plasma*. Lecture Notes in Physics, vol. 785 (Springer, Berlin/Heidelberg, 2010)
25. W. Fiorkowski, *Phenomenology of Ultra-Relativistic Heavy-Ion Collisions* (World Scientific, Singapore, 2011)
26. J. Casalderrey, H. Liu, D. Mateos, K. Rajagopal, U. Wiedemann *Gauge/String Duality, Hot QCD and Heavy Ion Collisions* (Cambridge University Press, Cambridge, 2014)

Chapter 2
The Physics of Complex Systems

> *Great fleas have little fleas*
> *Upon their backs to bite 'em.*
> *And little fleas have lesser fleas,*
> *And so ad infinitum.*
> *And the great fleas themselves, in turn,*
> *Have greater fleas to go on,*
> *While these again have greater still,*
> *And greater still, and so on.*
>
> Augustus de Morgan (1801–1871)
> (paraphrasing Jonathan Swift)

Abstract In the first section of this chapter, we summarize the basic concepts of critical behavior in thermodynamics, using the simplest spin system (Ising model) as illustration. In the second section, we discuss the geometric critical behavior obtained in the formation of clusters of diverging size (percolation).

2.1 Critical Behavior in Thermodynamics

Phase transitions are common everyday occurrences. We are all familiar with the evaporation or the freezing of water, the melting of ice. We know that also metals can melt, that iron can be magnetized, that certain conductors become superconducting at low temperature. Nevertheless, phase transitions have for a long time remained difficult, if not impossible, to treat in physics, because they deal with complex systems which cannot be reduced to a sum of more elementary ones. The 'divide and conquer' method so successful in most other branches of physics breaks down here, so that new forms of analysis were needed. These were developed only in the past fifty years or so, on a conceptual and analytical level (renormalization group theory) as well as in numerical studies (computer simulation), which really became effective only with the advent of high-performance computers. It thus seems worthwhile to

© Springer International Publishing AG 2018
H. Satz, *Extreme States of Matter in Strong Interaction Physics*,
Lecture Notes in Physics 945, https://doi.org/10.1007/978-3-319-71894-1_2

spend a little time on the basics of critical behavior in complex systems. For a more extensive treatment of this topic, several excellent textbooks and surveys are now available; see, e.g., [1–5].

At the transition point, a macroscopic system realizes, so to speak, just how big it is, and hence it refuses to be treated there as just a sum of little systems. For most conditions, a given constituent of a thermodynamic system is correlated only to that of a comparatively small number of other, nearby constituents. At the point of a continuous phase transition, however, it becomes aware of *all* others; the correlation length reaches the size of the system and hence diverges in the thermodynamic limit, for an infinite system. In the case of a first order transition, there are discontinuities at the critical point. The basic feature of all critical phenomena is thus the *singular* or *discontinuous* behavior of observables. When water evaporates, the specific volume of the liquid increases in a discontinuous fashion at the boiling point: the temperature remains constant at $T = T_c$ until all of the medium has turned into steam, thus resulting in a first order (or discontinuous) transition with a mixed phase of liquid and steam bubbles (see Fig. 2.1a). The onset of ferromagnetism provides an example of a second order (or continuous) transition: here the average value of the spin, the magnetization $m(T)$, remains identically zero up to the Curie point $T = T_c$, and for $T > T_c$, it is non-zero, either up or down. Hence $m(T)$ undergoes non-analytic behavior at T_c, and higher derivatives, such as the magnetic susceptibility χ, diverge there (see Fig. 2.1b).

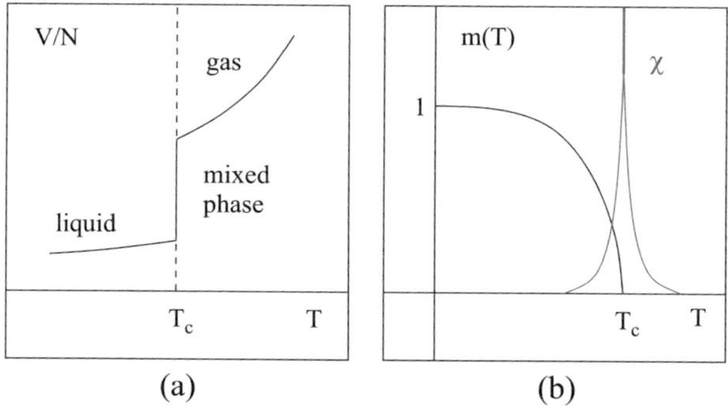

Fig. 2.1 Critical behavior for liquid-gas (**a**) and magnetization (**b**) transitions

Another, complementary approach starts from the observation that a solid has a periodic structure, a certain microscopic order, which is lost when it melts. Similarly, the magnetic state of a ferromagnet is due to its aligned spins; above the Curie point, this alignment disappears. Thus there also seems to be an inherent relation between critical behavior and changes in order or symmetry.

To consider critical behavior in more detail, we turn to what has become the 'falling apple' experience in statistical physics, the *Ising model* [6]. On each of the N^d sites on a d-dimensional lattice, one has spins $s_i = \pm 1$ for all $i = 1, \ldots, N^d$. For

$d = 2$, one specific configuration is illustrated in Fig. 2.2. Between any two next-neighbour spins, there is an interaction specified by $Js_i s_{i+1}$. It is obviously possible to extend this model to include further, more long-range interactions, or have non-uniform couplings $J_{i,j}$, but we shall here restrict ourselves to the original uniform next neighbor version of the model.

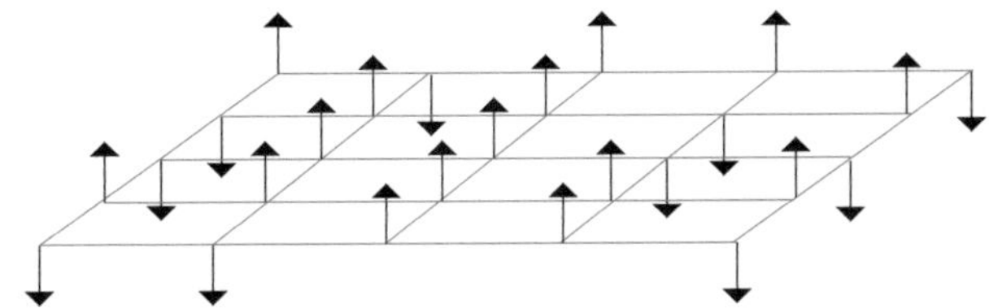

Fig. 2.2   A two-dimensional Ising model configuration on a 5×5 lattice

In the presence of an external field H, the Hamiltonian $\mathcal{H}$ of the Ising model is given by

$$\mathcal{H} = -J \sum_{\{i,j\}} s_i s_j - H \sum_i s_i; \tag{2.1}$$

we recall that the sum in the first term runs over next neighbours only. With $J > 0$, the energy of the system is lower if spins are aligned among themselves ($s_i s_j = 1$); it is also lowest if the spins align themselves with the external field H.

For $H = 0$, the system has a *discrete, global* symmetry: flipping all spins,

$$s_i \to -s_i \; \forall \, i, \tag{2.2}$$

leaves $\mathcal{H}$ invariant. It does not remain invariant, however, under the *local* transformation $s_i \to -s_i$ for some fixed i only. The transformation (2.2) is *discrete*, because we only allow $s_i = \pm 1 \to s_i = \mp 1$, and not for example a *continuous* change in the value of $|s_i|$. The two transformations $s_i \to s_i$ and $s_i \to -s_i$ form the discrete group Z_2, and hence one says that the Ising model is Z_2-symmetric. The introduction of an external field H tends to align the spins in its direction and thereby breaks this symmetry.

The thermal properties of the system are specified by the partition function

$$Z(T,H,N) = \prod_{i=1}^{N^d} \sum_{s_i=\pm 1} \exp\{-\beta \mathcal{H}\} = \prod_{i=1}^{N^d} \sum_{s_i=\pm 1} \exp\{\beta J \sum_{i,j}^{nn} s_i s_j + \beta H \sum_i s_i\}, \tag{2.3}$$

where $\beta^{-1} = T$ denotes the temperature of the system. Since $Z(\beta, H = 0)$ preserves the Z_2-symmetry of the Hamiltonian $\mathcal{H}(\beta, H = 0)$, the same holds for all quantities obtained from it, such as the energy density

$$\epsilon(\beta, H=0) \equiv -\left(\frac{\partial \ln Z(\beta, H)}{\partial \beta}\right)_{H=0}. \tag{2.4}$$

It is thus tempting to ask if also the actual state of the system retains this symmetry at all temperatures. Now the two-dimensional Ising model is exactly solvable, as shown by Onsager in 1944 [7], and we know therefore that this is not the case. At sufficiently high temperatures, the average state of the system shares the symmetry of the partition function: there is *disorder*, with on the average as many up as down spins. However, at

$$T = T_c = \beta_c^{-1} = \frac{2\,J}{\ln[1 + \sqrt{2}]}, \tag{2.5}$$

where J is the coupling strength in Eq. (2.1), *spontaneous symmetry breaking* occurs: below T_c, there is at least partial *order*, on the average there are either more up or more down spins. The system chooses one of these two possibilities, and the Z_2 invariance of the partition function only means that both are equally likely.

Once spontaneous symmetry breaking has taken place, we need an additional parameter to specify the state of the system, to tell us whether it chose up or down. This is provided by the average value of the spin, calculated over the entire lattice; it serves as *order parameter*,

$$m(T, N) = \frac{1}{Z(T, N)} \prod_{i=1}^{N^d} \sum_i \left[\frac{\sum_i s_i}{N^d}\right] \exp\{\beta J \sum_{i,j}^{nn} s_i s_j\} \tag{2.6}$$

which is not invariant under spin flips, since for $s_i \to -s_i \ \forall\ i = 1, \ldots, N^d$, we have

$$m(T, N) \to -m(T, N). \tag{2.7}$$

As a consequence, its average must vanish when the state itself is symmetric,

$$m(T) \begin{cases} = 0 & \text{for the symmetric disordered state,} \\ \neq 0 & \text{for the ordered state with spontaneously broken symmetry.} \end{cases} \tag{2.8}$$

As mentioned, the Z_2-symmetry of $\mathcal{H}(T, H = 0)$ and $Z(T, H = 0)$ makes the up and down directions of the magnetization equally likely; below T_c, the system choses one or the other, making $m(T, H = 0) \neq 0$. In the presence of an external field $H \neq 0$, the magnetization never vanishes, $m(T, H \neq 0) \neq 0$, indicating that the Z_2 symmetry now is always broken. The overall pattern for $m(T, H)$ is illustrated in Fig. 2.3.

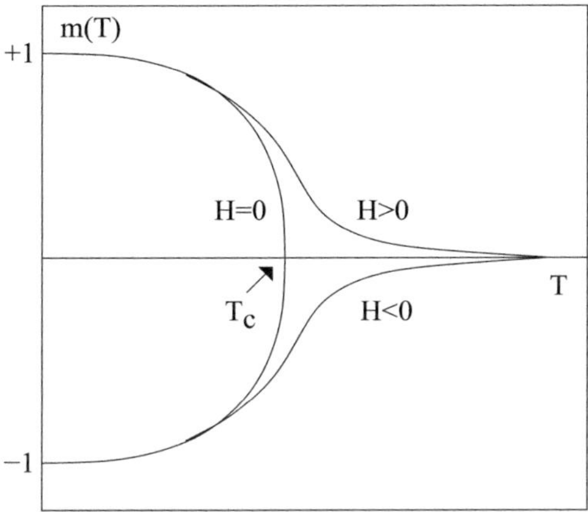

Fig. 2.3 Magnetization in the Ising model without and with external field H

In the thermodynamic limit $N \to \infty$, the magnetization $m(T)$ is therefore not analytic; at $T = T_c$, it is continuous, but it does not vary 'smoothly' with T, it has a 'kink', as seen in Fig. 2.1b:

$$m(T) \begin{cases} \sim (T_c - T)^\beta & \forall\, T < T_c \\[2mm] = 0 & \forall\, T > T_c \end{cases} \tag{2.9}$$

This non-analytic or *singular* behavior is governed by the *critical exponent* β, which specifies how $m(T)$ vanishes for $T \to T_c$.

Other thermal observables of the system also exhibit singular behavior. Let us define the (Gibbs) free energy as

$$F(T, H) = -T \log Z(T, H) \tag{2.10}$$

and as dimensionless measures of the temperature and the external field

$$t = \frac{T - T_c}{T_c} \qquad h = \frac{H}{T}. \tag{2.11}$$

We then obtain as critical behavior for the *specific heat*

$$C_H(t, h = 0) = T^2 \left(\frac{\partial^2 F}{\partial T^2} \right)_{h=0} \sim |t|^{-\alpha}, \tag{2.12}$$

for the spontaneous magnetization already discussed above

$$m(t, h = 0) = \frac{1}{N^d} \left(\frac{\partial F}{\partial h} \right)_{h=0} \sim |t|^{\,\beta} \ \text{ with } t < 0, \tag{2.13}$$

for the susceptibility shown in Fig. 2.1b

$$\chi_T(t, h = 0) = \left(\frac{\partial m}{\partial h}\right)_{H=0} \sim |t|^{-\gamma}, \tag{2.14}$$

and for the magnetization on the critical isotherm

$$m(t = 0, h) = \frac{1}{N^d}\left(\frac{\partial F}{\partial h}\right)_{t=0} \sim h^{1/\delta}, \tag{2.15}$$

defining the set α, β, γ, δ of *global* critical exponents, which specify the critical behavior of the system as a whole. In addition to these global exponents, one also considers two *local* ones, defined in terms of the correlation function $\Gamma(r)$ between two spins separated by a distance r. A rather general form for $\Gamma(r)$ is given by

$$\Gamma(r, t) \sim \frac{e^{-r/\xi}}{r^p}, \tag{2.16}$$

where ξ is the correlation length, with the exponents ν and η defined by

$$\xi \sim |t|^{-\nu} \tag{2.17}$$

for $t \to 0$ and

$$p \equiv d - 2 + \eta, \tag{2.18}$$

at $t = 0$, i.e., when only the power-law part of the correlation function Γ remains. The correlation length ξ specifies how far spins can 'see' each other, and at the critical point, it becomes infinite. In other words, for $t \neq 0$, the correlation length is finite and provides a dimensional scale, beyond which spins do not know of each other; hence the system can be decomposed into a sum of independent subsystems of size $\sim \xi^2$, or more generally ξ^d for systems of d space dimensions. At $t = 0$, ξ diverges, there is no more scale, all spins are correlated, and hence the system cannot be split into independent subsystems. The critical exponent η is related to the fractal behavior at the critical point; it corresponds to the fact that the surface of a correlation domain in a d-dimensional space has a dimension larger than $d - 1$.

The relation between the correlation function and the free energy $F(t, h)$ is best seen in the case of the Ising model. In the exponent of the partition function, we have a term $h \sum_i s_i$; the derivative of $F(t, h)$ with respect to H at $H = 0$ gives us the spontaneous magnetization. If we now replace this by local field interactions, $\sum_i h_i s_i$, then we can write the magnetization as

$$m(t, h = 0) = \sum_i m_i(t, h = 0) = \frac{1}{N^d}\sum_i\left(\frac{\partial F}{\partial h_i}\right)_{h_i=0} \tag{2.19}$$

The second derivative leads to

$$\langle s_i s_j \rangle = \frac{1}{N^d} \left(\frac{\partial^2 F}{\partial h_i \partial h_j} \right)_{h_i = h_j = 0} \tag{2.20}$$

From this, we can define the connected correlation function

$$\Gamma_{i,j}(t, h = 0) = \langle s_i s_j \rangle - \langle s_i \rangle \langle s_j \rangle; \tag{2.21}$$

it provides a measure of the interaction between two spins at sites i and j. Averaging $\Gamma_{i,j}$ over all lattice configurations with fixed separation $i - j = r$ then leads to $\Gamma(r; t)$.

Why does criticality result in *singular* behavior of the thermodynamic observables? The answer for the thermal systems we have studied so far is given by the fact that the phase transition is determined by the onset of spontaneous symmetry breaking, which is an "either-or" phenomenon, not something smooth or gradual. You cannot break symmetry "a little". Equivalently, the correlation length diverges, since now every spin is connected to all others.

There are several reasons why critical exponents are an extremely fruitful way of addressing critical behavior. In condensed matter physics, they describe observables in the vicinity of T_c, where the singular part tends to dominate and hence is most readily measurable. Such measurements resulted in the remarkable observation that the behavior of quite different systems, with quite different critical temperatures, can nevertheless be described in terms of the same critical exponents. That suggested that these exponents are of a more universal nature. This conjecture was supported by a number of relations between the exponents, obtained through rather general thermodynamic considerations. Such "scaling laws" have in the past four decades found a theoretical basis in renormalisation group theory [8, 9].

The basic idea of scaling and renormalisation is that sufficiently near the critical point of a continuous transition, the only relevant scale is the correlation length, which diverges for $t \to 0$. It is therefore the dimensional scale in terms of which we should express all the observables which show singular behavior, and that will then automatically relate critical exponents. To illustrate this, consider the correlation function of two spins on the critical isotherm. On one hand, from Eqs. (2.16), (2.17), and (2.18), it will vanish at large distances as $r^{2-d-\eta}$ for $t \to 0$; on the other hand, it then becomes the product of two spin averages, m^2. We thus have

$$r^{2-d-\eta} \sim m^2 \sim |t|^{2\beta}, \tag{2.22}$$

and if we now set $r \sim \xi \sim |t|^{-\nu}$, we obtain with

$$\beta = \nu(d + \eta - 2)/2 \tag{2.23}$$

one of the mentioned relations between critical exponents. In a similar manner, one finds three further relations,

$$\nu d = (2 - \alpha), \tag{2.24}$$

$$\gamma = \nu(2 - \eta), \tag{2.25}$$

$$\beta\delta = \nu(2 + d - \eta)/2, \tag{2.26}$$

so that, given the space dimension d, appearently only *two* of the six exponents α, β, γ, δ, η and ν remain as independent quantities. Since all the thermodynamic observables also depend on *two* variables, the temperature t and the external field h, this does not seem accidental, and renormalization theory shows that it is not.

Imagine that we change the basic length scale R of our system from R to bR, where b is a constant factor. If we want the physics to remain invariant under such scale changes, then both t and h will have to change accordingly. We thus require that the transformation

$$r \to r' = b\,r \tag{2.27}$$

leads to corresponding transformations

$$t \to t' = b^{y_t}t \ \text{ and } \ h \to h' = b^{y_h}h, \tag{2.28}$$

with some constant exponents y_t and y_h. Consider then a thermodynamic function $\Phi(t, h)$. The function Φ will itself have a certain dimension R^{d_F}, where d_F depends on the quantity considered. Our scale invariance condition thus becomes

$$\Phi(t, h) = b^{-d_F}\Phi(t', h') = b^{-d_F}\Phi(b^{y_t}t, b^{y_h}h). \tag{2.29}$$

For the correlation length ξ at zero field we have $\xi \sim |t|^{-\nu}$, which with

$$\xi(t) = b\,\xi(b^{y_t}t), \tag{2.30}$$

leads to

$$y_t = 1/\nu. \tag{2.31}$$

In a similar way one obtains

$$y_h = d - \beta/\nu. \tag{2.32}$$

With the help of the relations mentioned above, all other exponents can be expressed in terms of y_t and y_h as well; we leave this as an exercise. The critical behavior of the system is therefore completely determined by these two scaling exponents required for temperature and external field.

At this point we should note that the dimensionality dependence of the formulae given above has its limits. It is correct up to $d = 4$; for higher dimensions, the critical behavior becomes that of the so-called mean-field model (see e.g., [1]), which no longer depends on the dimension of the system. The critical exponents for $d \geq 4$ thus remain for all d those obtained above for $d = 4$.

To illustrate what we have just discussed, we return to the two-dimensional Ising model, which, as mentioned, can be solved exactly [7]. The resulting critical exponents are listed in Table 2.1; the specific heat diverges logarithmically at $t = 0$, so that the corresponding critical exponent formally becomes zero. It is readily verified that the exponents satisfy the scaling relations (2.23), (2.24), (2.25), and (2.26) and that they lead to the basic exponents y_t and y_h also listed in the table.

α	β	γ	δ	η	ν		y_t	y_h
0 (log)	1/8	7/4	15	1/4	1		1	15/8

Table 2.1 Critical exponents for the $d = 2$ Ising model

The symmetry properties of the underlying dynamics thus simplify considerably the classification of the critical behavior of different systems. No matter what their nature is (magnetic systems, fluid systems, gauge theories), if they have a continuous phase transition, two critical exponents determine the behavior in the vicinity of the transition point. If these two are the same for two systems, then the critical behavior of these systems is the same. The universality we had encountered above can thus be made quite precise, and one in fact defines as a *universality class* the set of all systems with the same critical exponents. In contrast, the value of the transition temperature is not at all universal – it depends on the details of the dynamics and even on the lattice geometry (e.g., square vs. triangular).

Let us summarize these results. Thermodynamic critical behavior (for simplicity, for continuous transitions) means an onset of spontaneous symmetry breaking, which leads to singular behavior of the partition function and hence of thermodynamic observables. Characteristic critical exponents then define universality classes of different systems sharing the same critical behavior.

2.2 Cluster Formation and Percolation

The critical behavior we have discussed so far was determined by the dynamics and the resulting symmetry of the interacting many-body system. For constituents which have an intrinsic scale, be it as spatial size or through an interaction range, there is a more general *geometric* form of critical behavior: the formation of infinite *connected clusters* or *networks*. The transition of such a system from a disconnected to a connected medium is denoted as *percolation*. For introductions to this topic, see [1–5]; more complete presentations are given in [10–12].

For the simplest example of percolation, we turn once again to the two-dimensional square lattice and treat it as a board for a game like *Go*. We place stones on randomly chosen sites and consider the size of the clusters formed by adjacent occupied sites. Such clusters are said to percolate if they connect two opposite sides. In Fig. 2.4, we show two configurations on an R^2 lattice with $R = 9$. In (a), the occupied sites form only isolated clusters, while in (b) the cluster consisting of filled circles spans the system: there is percolation. The density of occupied sites, i.e., the number of occupied sites N divided by the lattice size, $n \equiv N/R^2$, determines when percolation starts; this critical percolation density n_p is defined in the limit of both N and R becoming infinite. For some simple cases, it can be calculated analytically; otherwise, it has to be determined through numerical studies. Site percolation on a two-dimensional square lattice leads to $n_p = 0.59 \pm 0.01$ [13]. In other words, when somewhat more than half the sites are occupied, there is percolation.

(a)   (b)

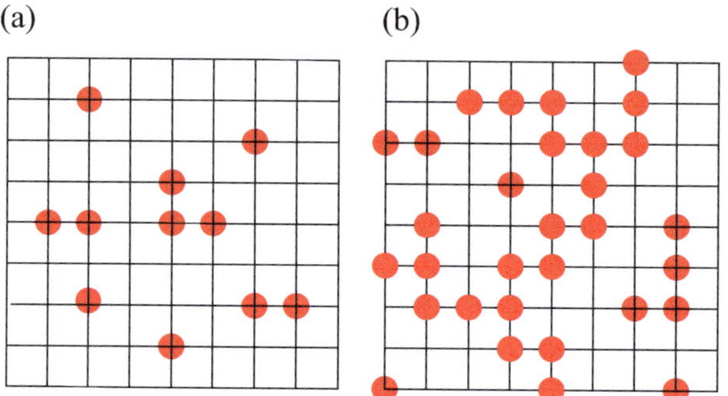

Fig. 2.4 Site cluster configurations on a 9×9 lattice

Another interesting alternative is to establish random connections (*bonds*) between adjacent sites, as shown in Fig. 2.5. If these bonds were metallic and we introduce a voltage difference between opposite sides, then current will begin to flow at the percolation point. In contrast to the site percolation case, this bond percolation problem in two dimensions is exactly solvable, and the critical percolation density is $n_p = 1/2$ [13, 14].

(a)   (b)

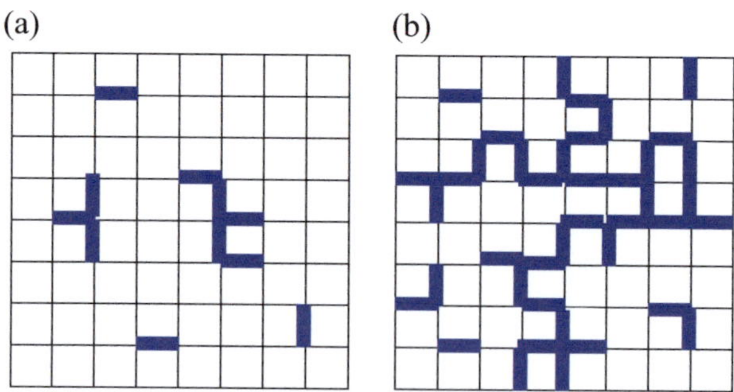

Fig. 2.5 Bond cluster configurations on a 9×9 lattice

To show that percolation does not require a discrete lattice basis, we turn to the continuum percolation of disks in two dimensions, sometimes poetically referred to as "lilies on a pond" (see Fig. 2.6). Here one distributes small disks of area $a = \pi r^2$ randomly on a large surface $A = \pi R^2$, $R \gg r$, with overlap allowed. With an increasing number of disks, clusters begin to form. If the large surface were water and the small disks floating water lilies: how many lilies are needed for a cluster to connect the opposite sides, so that an ant could walk across the pond without getting its feet wet? Given N disks, the disk density is $n = N/A$. Clearly, the average cluster $S(n)$ size will increase with n, and there will be a critical value n_p at which the cluster spans the pond.

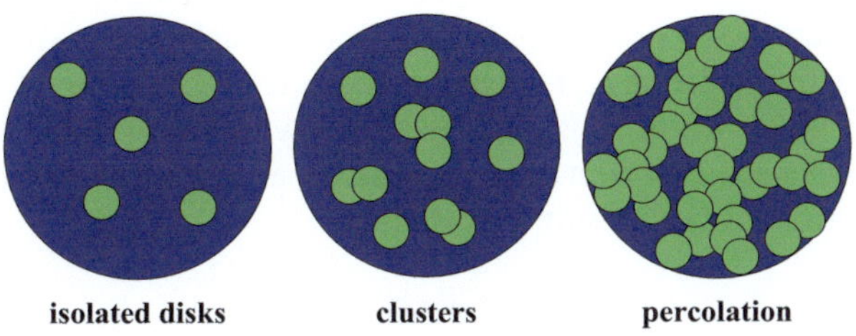

Fig. 2.6 Lilies on a pond

The critical density for the onset of continuum percolation has been determined in numerical studies for a variety of different systems. In two dimensions, disks percolate [13] at

$$n_p = \frac{\eta_p}{\pi r^2} \simeq \frac{1.13}{\pi r^2}, \tag{2.33}$$

i.e., when we have a little more than one disk per unit area. Because of overlap, at this point the pond is not yet covered completely by lilies. In fact, one can show quite generally (for all space dimensions) that the "filling factor" [15]

$$\phi_p = 1 - \exp\{-\eta_p\} \tag{2.34}$$

gives the fraction of the area covered by discs or of the volume covered by spheres; the percolation threshold η_p of course does depend on the space dimension. Hence for our present case, $d = 2$, at percolation only 68% of A is filled, 32% remains empty. We therefore emphasize that n_p is the *average overall* density for the onset of percolation. The density in the largest and hence percolating cluster at this point must evidently be equal to or larger than $n_p/0.68 \simeq 1.66/\pi r^2$; numerical studies show [16]

$$n_p^{\mathrm{cl}} = \frac{\eta_p^{\mathrm{cl}}}{\pi r^2} \simeq \frac{1.72}{\pi r^2}. \tag{2.35}$$

In two dimensions, we moreover have a special situation: when our ant can walk across, a ship can no longer cross the pond, and vice versa. This "fence effect" no longer holds for $d > 2$.

For $d = 3$, the critical density for the percolation spheres is [13]

$$n_p = \frac{\eta_p}{(4\pi/3)r^3} \simeq \frac{0.34}{(4\pi/3)r^3}, \tag{2.36}$$

with r denoting the radius of the little spheres now taking the place of the small disks we had for $d = 2$. At the critical point in three dimensions, Eq. (2.34) now indicates that only 29% of space is covered by overlapping spheres, while 71% remains empty, and here both spheres and empty space form infinite connected networks. The density n_p^{cl} of the largest cluster of spheres at the percolation point must also be much larger than $0.34/V_0$; in fact, we must have $n_p^{\mathrm{cl}} \geq n_p/0.29 \simeq 1.2/V_0$.

For any dimension, we can thus specify the onset point of percolation in two distinct ways: globally, by giving the critical overall density, or locally, by specifying the critical density of the largest cluster at the percolation point [16]. Both $\eta_p(d)$ and $\eta_p^{\mathrm{cl}}(d)$ depend on the dimension of the underlying space. Since the number of nearest neighbors increases with d, η_p decreases, because there are now more ways of making connections; as a result, one obtains

$$\lim_{d \to \infty} \eta_p(d) = 0. \tag{2.37}$$

In other words, the higher the dimension, the more dilute is the system at the percolation point. To estimate the density of the percolating cluster, we use the filling factor, dividing the percolation density by the fraction of occupied space,

$$n_p^{\mathrm{cl}} \sim \frac{n_p}{\phi_p} = \frac{\eta_p}{V_0(d)} \frac{1}{1 - \epsilon^{-\eta_p}}. \tag{2.38}$$

This gives us

$$\lim_{d \to \infty} \eta_p^{\mathrm{cl}}(d) = 1 \tag{2.39}$$

in the limit of large dimensionality. So in this limit, continuum systems reduce to what we had on the lattice: one constituent per intrinsic volume makes a percolating cluster.

For $d = 3$, if we continue to increase the density of spheres, we reach a second critical point at the overall density $\bar{n}_p \simeq 1.24/[(4\pi/3)r^3]$, at which the vacuum stops to form an infinite network: now 71% of space is covered by spheres, and for $n > \bar{n}_c$, only isolated vacuum bubbles remain. A characteristic feature of percolation in $d \geq 3$ dimensions is thus the existence of two percolation thresholds. The first (often called "wet") is the percolation of occupied sites or bonds, or of spheres in the continuum case. The second ("dry") refers to the percolation of empty sites or bonds, or the percolation of empty space in the continuum form. For $n_p < n < \bar{n}_p$, there is percolation for both occupied and empty space. Living creatures benefit much from this feature, since it allows both nerves and blood vessels to span the entire body without crossing each other.

The striking feature in all percolation phenomena is that the cluster size does not grow monotonically with the density of constituents; it grows in a very sudden way, as illustrated in Fig. 2.7 for the two-dimensional disk case. As n approaches some "critical value" n_c, $S(n)$ suddenly becomes large enough to span the pond. In fact, in the limit $N \to \infty$ and $A \to \infty$ at constant n, both $S(n)$ and $dS(n)/dn$ diverge for $n \to n_c$: we have percolation as a geometric form of critical behavior.

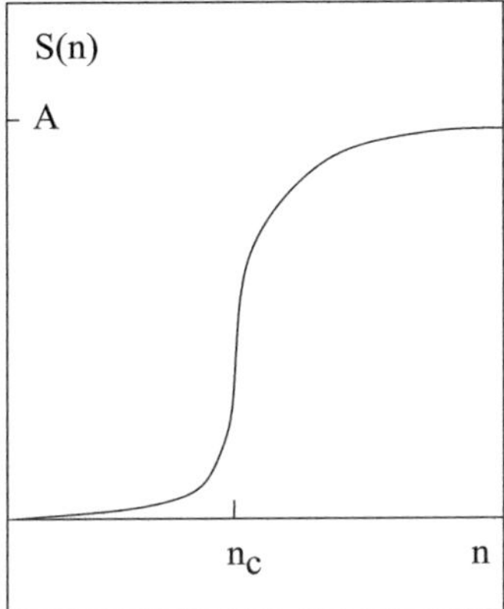

Fig. 2.7 Cluster size vs. density

As in the case of thermal phase transitions, there are many well-known everyday examples of transitions of percolation nature: the *gelatinization* involved in the making of pudding or the boiling of an egg, the onset of conductivity in *random networks* (sometimes compared to an "ant in a labyrinth"), and many others [10]. Instead of spontaneous symmetry breaking, the essential feature now is the spontaneous onset of global connectivity, the transition from disconnected to connected large-scale systems. And again, the singular behavior arises because connection is also an "either-or" phenomenon: you cannot connect things "a little".

To specify the functional form of the critical behavior involved, we consider the probability $P(n)$ that a given constituent (site, bond or disk) is in the infinite cluster. Then

$$P(n) \left\{ \begin{array}{ll} = 0 & \text{for all } n < n_c \\[2ex] \sim (n - n_c)^\beta & \text{for } n \to n_c \text{ from above} \end{array} \right\}, \tag{2.40}$$

constitutes the order parameter for percolation and vanishes for $n \to n_c$ from above as specified by the critical exponent β. The average cluster size (excluding the infinite cluster for $n > n_c$) diverges as

$$\tilde{S}(n) \simeq |n - n_c|^{-\gamma} \tag{2.41}$$

and corresponds to the susceptibility in thermodynamic systems. Other observables as well show singular behavior specified by critical exponents, so that we can define universality classes also for geometric critical behavior [10, 13]. And again, for a given space dimension d, a variety of different systems, including the site, bond and continuum percolation cases mentioned here, lead to the same exponents, connected by scaling relations. For $d = 2$, these are again analytically calculable, and we list them in Table 2.2. For their precise definition, we refer to the listed general references [1, 11, 13]; α, β and γ characterize the behavior of the moments of the cluster size distribution. The basic exponents for percolation are the one governing the cluster size divergence, v, and another, the dimension d_f, determining the "surface" of the diverging cluster,

$$d_f = d - \frac{\beta}{v}. \tag{2.42}$$

The latter gives for $d = 2$ the value $d_f = 91/48 \simeq 1.9$ for the perimeter, instead of the geometric result $d_s = 1$, so that the size of the clusters is fractally increased. This is a reflection of the fact that at the critical point, the "coast line" becomes scale invariant, having the same structure for all resolution scales. It never becomes smooth and so it always remains of a higher dimension.

α	β	γ	δ	v	d_f
$-2/3$	$5/36$	$43/18$	$91/5$	$4/3$	$91/48$

Table 2.2 Critical exponents for $d = 2$ percolation

We have here discussed percolation assuming *randomly* distributed constituents (sites, bonds, disks or spheres); however, the form of the distribution law is not essential for the approach as such – one can as well study percolation using a thermodynamic distribution function, or any other distribution pattern for the constituents.

A perhaps more crucial aspect is the definition of what one calls a cluster. To illustrate the point, let us look in a little more detail at the relation between thermal and geometric critical behavior. This question has been a challenge to statistical physics for a long time; already in 1939 Frenkel [17] suggested that domain fusion and spontaneous symmetry breaking should define the same transition temperature. The point was taken up again later [18], and finally resolved for simple spin models [19–21]. We therefore return briefly to the $d = 2$ Ising model and consider clusters and cluster growth for spins whose distribution is governed by the Ising partition function. In the case of vanishing external field, $H = 0$, it was found that defining a cluster simply as a set of adjacent up or down spins did not lead to the correct thermodynamics. The problem is that thermodynamic clusters consist of correlated constituents, while geometrically, some uncorrelated spins may just happen to point in the same direction. So the "actual" dynamical cluster size has to be smaller than that determined by simple geometric distributions; one has to introduce correlations.

The difficulty was overcome for spin systems by showing that it is possible to define clusters by combination of site occupations with bond couplings, which then allowed a correct account of the Ising dynamics [19, 20]. Once this is done, the percolation of spin clusters and the onset of magnetization are just two different ways of formulating the same transition phenomenon; they lead to identical results for the critical behavior, both for the critical exponents and for the critical temperature.

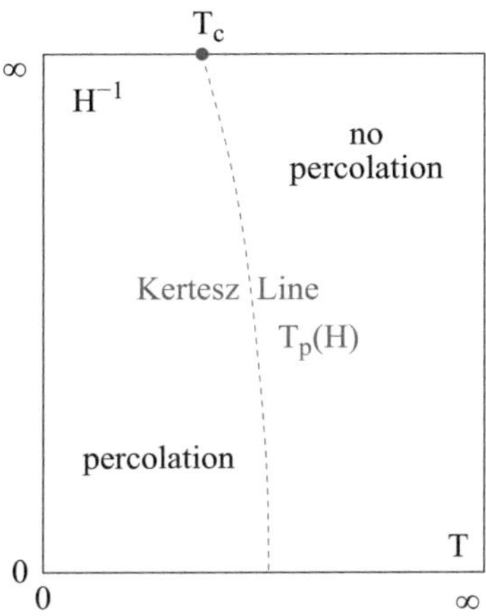

Fig. 2.8 Ising model percolation

For $H \neq 0$, there is no more thermal transition: the magnetization never vanishes (see Fig. 2.3), the symmetry is always broken and the partition function analytic [22]. Nevertheless, the spin cluster percolation transition persists between high and low T for all values of H. The vanishing of the percolation order parameter as function of H, T thus defines a geometric transition line $T_p(H)$, the 'Kertész line' [23] (see Fig. 2.8). Hence percolation and the resulting divergence of cluster variables can occur even when the partition function is non-singular. In other words, the set of all geometric transitions is more general and contains much more than just the conventional thermal phase transitions: there are more critical phenomena in nature than the partition function knows of.

Summarizing percolation, we note that it corresponds to the onset of infinite cluster formation, leading to singular behavior of geometric observables. This behavior is specified through critical exponents defining universality classes.

2.3 Conclusions

We have discussed two types of transitions between different states of many-body systems. Thermodynamic phase transitions are generally related to a change of symmetry (spontaneous symmetry breaking), which in the large volume limit leads to singular behavior of the partition function. Geometric critical behavior (percolation) occurs at the point of formation of infinite connected systems; in the large volume limit, it can, but need not result in a non-analytic partition function. A general scheme of critical behavior, including both these types, seems to be still lacking. It might be that singularities in sub-leading terms of the partition function (surface tension contributions) provide key to the needed generalization [24].

References

1. K. Huang, *Statistical Mechanics*, 2nd edn. (Wiley, New York, 1987)
2. J.J. Binney et al., *The Theory of Critical Phenomena* (Clarendon Press, Oxford, 1992)
3. J.M. Yeomans, *Statistical Mechanics of Phase Transitions* (Clarendon Press, Oxford, 1992)
4. A. Pelissetto, E. Vicari, Critical phenomena and renormalization group theory. Phys. Rept. **368**, 549 (2002)
5. K. Christensen, N.R. Moloney, *Complexity and Criticality* (Imperial College Press, London, 2005)
6. E. Ising, Z. Phys. **31**, 253 (1925)
7. L. Onsager, Phys. Rev. **65**, 117 (1944)
8. L.P. Kadanoff, Physica **2**, 263 (1966)
9. K. Wilson, Phys. Rev. **D10**, 2445 (1974)
10. D. Stauffer, A. Aharony, *Introduction to Percolation Theory* (Taylor and Francis, London, 1994)
11. G. Grimmett, *Percolation* (Springer, Berlin 1999)
12. H. Kesten, *Percolation Theory for Mathematicians* (Birkhauser, Boston 1982)
13. M.B. Isichenko, Rev. Mod. Phys. **64**, 961 (1992)
14. M.P.M. den Nijs, J. Phys. A **12**, 1857 (1979)
15. V.K.S. Shante, S. Kirkpatrick, Adv. Phys. **20**, 325 (1971)
16. S. Digal, S. Fortunato, H. Satz, Eur. Phys. J. C **32**, 547 (2004)
17. J. Frenkel, J. Chem. Phys. **7**, 200 (1939)
18. J.W. Essam, M.E. Fisher, J. Chem. Phys. **38**, 802 (1963)
19. C.M. Fortuin, P.W. Kasteleyn, J. Phys. Soc. Jpn. **26**(Suppl.), 11 (1969)
20. C.M. Fortuin, P.W. Kasteleyn, Physica **57**, 536 (1972)
21. A. Coniglio, W. Klein, J. Phys. A **13**, 2775 (1980)
22. S.N. Isakov, Commun. Math. Phys. **95**, 427 (1984)
23. J. Kertész, Physica A **161**, 58 (1989)
24. J.-S. Wang, Physica A **161**, 249 (1989)

Chapter 3
The Limits of Hadron Physics

> *Wer bist denn du, der du nicht weißt,*
> *daß Phantásien grenzenlos ist?*
>
> Michael Ende, *Die unendliche Geschichte*
>
> *(But who are you, if you do not know*
> *that Phantásia has no limits?*
>
> Michael Ende, *The Never-Ending Story*)

Abstract Here we show that very general aspects of strong interaction physics, such as the intrinsic size of hadrons or the abundant production of resonances of ever increasing masses, can lead to a breakdown of the resulting thermodynamics above a certain temperature. This had inspired Hagedorn to propose an ultimate temperature for hadron physics; today, we interpret it as the first hint for a phase transition to a new state of matter.

3.1 Introduction

In the last chapter, we saw that critical behavior, signalling the transition of a system from one state to another, is generally associated with a discontinuous or singular functional form of some physical observable. Given a complete dynamical theory, it might be possible to derive the complete equation of state, including all phase transitions. But even if we only have an incomplete picture of the fundamental dynamics, we can still look for singular or discontinuous behavior indicating the limit of a certain state of matter. To pass beyond such a singularity, we would then in general need further information about the underlying dynamics. Even without any knowledge of the quark substructure of hadrons, we can thus look for critical behavior of strongly interacting matter, making use only of some phenomenological description of hadronic interactions. Indications for such behavior were in fact found well before QCD had been developed; they have in many ways formed our thinking about dense matter and help us in understanding the results of QCD calculations. We shall therefore start by looking at critical features in hadron physics.

© Springer International Publishing AG 2018
H. Satz, *Extreme States of Matter in Strong Interaction Physics*,
Lecture Notes in Physics 945, https://doi.org/10.1007/978-3-319-71894-1_3

The beginning of statistical hadron physics is marked by three papers. In 1950, Fermi [1] considered an ideal gas of pointlike hadrons. A year later, Pomeranchuk [2] noted the effect of finite hadronic size in limiting the density of such a hadronic medium. Then in 1965, Hagedorn [3, 4] introduced resonance formation as the essential phenomenological input from strong interaction dynamics. Soon afterwards, the dual resonance model [5–7] provided dynamical support for the role of resonances in determining the interaction. The resulting resonance patterns were found to lead to an upper limit for the temperature of hadronic matter [8]; we now consider this to be the transition point to the quark-gluon plasma [9]. By correctly selecting the relevant features of hadron physics, Pomeranchuk and Hagedorn had discovered that the concept of hadronic matter breaks down above a certain temperature. Let us consider these limits of hadron physics.

3.2   The Hadronic Size

In the context of strong interaction physics, Pomeranchuk [2] seems to have been the first to note a rather general limiting feature for media consisting of spatially extensive constituents. For an ideal gas of N identical hadrons, each of which has an intrinsic volume V_0, conceptual difficulties arise once the density $n = N/V$ of the system exceeds $n = 1/V_0$; V denotes the overall system size. In the case of hadrons composed of quarks, we had already noted this problem in Chap. 1, but it persists in general. If the hadrons have a repulsive hard core, they will refuse to be compressed to arbitrarily high densities; a state of dense packing will be the limit. But even before that, one encounters a so-called "jamming" transition, separating a medium of freely roaming constituents from a state where they mutually block each other and thus restrict the mobility. We shall return to this situation in Chap. 7. If the hadrons are fully permeable, then we eventually have arbitrarily many pions in the characteristic volume of a single pion. To avoid this difficulty, Pomeranchuk argued that a hadron gas makes sense only as long as there is at least sufficient volume to give one V_0 to each hadron. We thus obtain

$$n_c \simeq \frac{1}{V_0} = \frac{3}{4\pi R_h^3} \tag{3.1}$$

as the upper limit for the density of hadronic matter.

The partition function for an ideal gas of identical (point-like or fully permeable) hadrons ("pions") of mass m_0, is given by

$$\ln Z_0(T, V) = \left\{ \frac{V}{(2\pi)^3} \int d^3 p \exp\{-\sqrt{p^2 + m_0^2}/T\} \right\} = \frac{VTm_0^2}{2\pi^2} K_2(m_0/T), \tag{3.2}$$

where $K_2(x)$ is the modified Bessel function of the second kind (the Hankel function of pure imaginary argument). From this, the density is obtained through

$$n(T) = \left(\frac{\partial \ln Z_0(T, V)}{\partial V} \right)_T .\tag{3.3}$$

We want to apply the bound (3.1) to determine the limit on the temperature; for this, we have to relate R_h and m_0. Following the idea of Yukawa, the range of the strong force is generally taken to be determined by the lowest mass, so that $R_h \simeq 1/m_0$. Inserting this into Eq. (3.1), the limiting temperature is obtained as the solution of the equation $n(T_c) = n_c$, or

$$\left(\frac{T_c}{m_0} \right) K_2(m_0/T_c) = \frac{3\pi}{2},\tag{3.4}$$

giving $T_c \simeq 1.4\, m_0$. Using the pion mass, this leads to $T_c \simeq 190\,\text{GeV}$, a value quite compatible with what is found today in the thermodynamics based on QCD. In view of the simplistic "derivation" of Eq. (3.1), that is perhaps even surprising. It would seem more reasonable to use percolation arguments to determine the limit, and we shall turn to such an approach in the next chapter.

3.3 The Hadronic Resonance Spectrum

The most important observation in particle physics in the 1960s was that the number of different species of so-called elementary particles, hadrons and hadronic resonances, seemed to grow without limit. Hadron-hadron collisions produced more and more resonant hadronic states of increasing masses. This phenomenon triggered two different theoretical approaches.

The more traditional idea, following the classical atomistic reduction as pursued since antiquity, proposed that there must be a smaller number of more elementary objects, which then bind to form the observed hadrons as composite states. This approach, as we know today, ultimately led to the quark model and to quantum chromodynamics as the fundamental theory of strong interactions; it thus proved once more to be most successful.

A second, truly novel approach asked what such an increase of states would lead to in the thermodynamics of strongly interacting matter [3, 4]. Both the question and its answer, the existence of an ultimate temperature of hadronic matter, are due to Hagedorn [10]. We know today that strong interaction thermodynamics leads to a transition in which hadronic matter turns into a plasma of deconfined quarks and gluons. Statistical QCD describes this quantitatively and provides a detailed description of the limits of confinement. Let us see here how it comes about based on only hadron dynamics.

Most of the stable hadrons observed in high energy collision experiments are the decay products, often through decay chains, of more massive resonant states. The quark infrastructure of hadrons provides an understanding of these resonances as higher excitations of the basic $q\bar{q}$ or qqq bound states, and in fact the wealth of measured states can be accounted for in terms of the predicted excitations.

With higher mass, the observed resonances generally become broader, and so empirically there is a limit to how far up in mass we can determine a resonance structure. Nevertheless, in view of color confinement and the negative outcome of all free quark searches, it seems not unreasonable to assume that the excitation spectrum will continue indefinitely, even though overlap will wash out peaks and make the spectrum look more and more smooth with increasing mass. Hence it is natural to ask how the number $\rho(m)$ of degenerate states for a resonance of given mass m depends on m.

In the low mass regime, we observe that increasing m increases $\rho(m)$; for fixed internal quantum numbers (isospin, baryon number, strangeness, etc.), such an increase arises simply from the $(2J + 1)$ spin degeneracy for states of higher angular momenta. It is in fact observed that resonances of angular momentum J follow a Regge pattern

$$\alpha' m_J^2 - \alpha_0 = J; \ J = 0, 1, 2, 3, \ldots, \tag{3.5}$$

where α_0 specifies the ground state and the universal Regge slope $\alpha' \simeq 1\,\mathrm{GeV}^{-2}$ the growth rate. Moreover, higher mass states can decay into lower mass states, and if we were to count these as degenerate states of the same resonance, this would lead to a further, even stronger increase of $\rho(m)$.

It therefore seems natural to ask for a guiding principle to determine $\rho(m)$ as function of m for all m. There exist two quite distinct approaches for this: the statistical bootstrap model [3, 4] and the dual resonance model [5–7]. Fortunately they provide essentially the same answer – presumably because both are based on the idea of a partition problem of self-similar structure. This means that a given entity is subdivided into smaller entities, and these in turn into still smaller, and so on, with the same composition law at each step.

3.3.1 Partitioning Integers

In our specific case, we thus consider a resonance of large mass m, decaying into n lighter states $m_1, m_2, \ldots, m_n$, each of which again decays into lighter states, and so on (see Fig. 3.1 for the case $n = 3$). Self-similarity means that the decay vertex $V(m; m_1, \ldots, m_n)$, determining the branching pattern, has the same functional form at each step. An example of a pattern *not* of this type would be a decay in which the first step leads to $2^1 = 2$ "children", each of which then produces $2^2 = 4$, those again $2^3 = 8$, and so on. In this case, counting the number of decay products would allow a specification of the decay generation, with 2^n "children" for the nth generation. Self-similar patterns are the same for each generation. We will begin by considering a simple and exactly solvable case of such a branching pattern, which will turn out to be quite relevant for our problem.

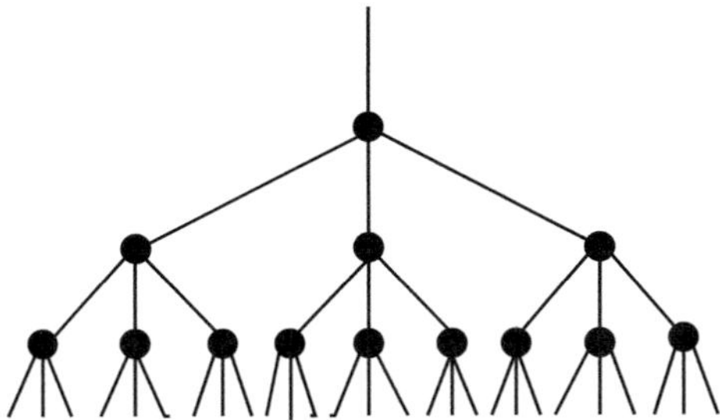

Fig. 3.1 Self-similar decay or composition chain

Consider the number $p(n)$ of ordered partitions of an integer n into integers [11].
To illustrate: for $n = 3$, we have the ordered partitions $3, 2 + 1, 1 + 2, 1 + 1 + 1$,
so that

$$p(n = 3) = 4 = 2^{n-1}. \tag{3.6}$$

Similarly, $n = 4$ gives the partitions $4, 3 + 1, 2 + 2, 2 + 1 + 1, 1 + 3, 1 + 2 + 1, 1 + 1 + 2, 1 + 1 + 1 + 1$, leading to

$$p(n = 4) = 8 = 2^{n-1}. \tag{3.7}$$

This solution can be shown to hold in fact for all n [11], i.e.,

$$p(n) = 2^{n-1} = \frac{1}{2} \, \exp\{n \ln 2\}. \tag{3.8}$$

This is a self-similar form: it specifies the number of ordered ways of partitioning
any integer into integers, independent of its value, and it grows exponentially with n.

We can obtain this result also by solving what is referred to as a "bootstrap
equation" for a function $\rho(n)$ of an integer $n \geq 2$,

$$\rho(n) = \sum_{k=2}^{n} \frac{1}{k!} \prod_{i=1}^{k} \rho(n_i) \, \delta(\Sigma_i n_i - n). \tag{3.9}$$

This is a way to impose self-similarity, by requiring $\rho(n)$ to consist of an arbitrary
number of objects of the same structure, $\rho(n_i)$, with $1 \leq n_i < n$. The name
"bootstrap" refers to the idea that the form of ρ is determined without a further
external input; it pulls itself out of Eq. (3.9) just as the legendary Baron Münch-
hausen extracted himself out of a swamp by pulling on his own bootstraps. The
solution is $\rho(n) = z\, p(n)$, with a normalization z determined by

$$\exp(z/2) - 1 - z = 0, \tag{3.10}$$

giving $z \simeq 2.51$. This illustrates in a particularly simple case the relation between a
bootstrap equation and a partition problem.

The calculation of the number $q(n)$ of *unordered* partitions of an integer n (i.e., not counting permutations, so that for $n = 3$ we have only three instead of four possibilities: 3, 2+1, 1+1+1) is more difficult and can be solved only asymptotically [12]; it leads to

$$q(n) = \frac{1}{4\sqrt{3}\, n} \exp\{\pi \sqrt{2n/3}\}\left[1 + O(\frac{\ln n}{n^{1/4}})\right] \qquad (3.11)$$

and will play a role in our subsequent considerations of the dual resonance model.

3.3.2   The Statistical Bootstrap Model

The first calculation of a self-similar resonance spectrum was given by the statistical bootstrap model of Hagedorn [3, 4], who had assumed that "fireballs consist of fireballs, which consist of fireballs, and so on…". The defining equation for $\rho(m)$ thus is

$$\rho(m, V_0) = \delta(m - m_0) +$$

$$\sum_N \frac{1}{N!}\left[\frac{V_0}{(2\pi)^3}\right]^{N-1}\!\!\int \prod_{i=1}^{N} [dm_i\, \rho(m_i)\, d^3p_i]\, \delta^4(\Sigma_i p_i - p), \qquad (3.12)$$

where m_0 is the lowest hadron mass, presumably the pion, and $V_0 = 4\pi\, r_0^3/3$ specifies the composition volume determined by the range r_0 of the strong interaction. It can be solved analytically [13], giving an exponential growth,

$$\rho(m, V_0) \sim m^{-3} \epsilon^{m/T_H}, \qquad (3.13)$$

with the coefficient T_H determined as solution of the equation

$$(m_0 r_0)^3 \left(\frac{2}{3\pi}\right)\left(\frac{T_H}{m_0}\right) K_2(m_0/T_H) = 2\ln 2 - 1, \qquad (3.14)$$

in terms of two parameters r_0 and m_0. Here again $K_2(x)$ is the Hankel function of pure imaginary argument. Hagedorn had assumed that the interaction range r_0 is specified with the inverse pion mass as scale, $r_0 \simeq 1/m_0$. This leads to

$$T_H \simeq 150\ \text{MeV} \qquad (3.15)$$

for the coefficient governing the exponential increase. It should be emphasized, however, that this is just one possible way to proceed. In the limit $m_0 \to 0$, Eq. (3.14) gives

$$T_H = \left[\frac{3\pi}{4}(2\ln 2 - 1)]\right]^{1/3} \frac{1}{r_0} \tag{3.16}$$

where r_0 as above denotes the range of strong interactions. With $r_0 \simeq 1$ fm, we then have

$$T_H \simeq 195 \,\mathrm{MeV}. \tag{3.17}$$

From this it is evident that the exponential increase persists in the limit of vanishing ground state mass; in fact, its coefficient is only weakly dependent on m_0, provided the strong interaction range r_0 is kept fixed.

3.3.3   The Dual Resonance Model

The most complete description of hadron interactions is provided by the dual resonance model [5–7], which assumes that the analytical structure of the scattering amplitude can in all channels be determined as the sum over all relevant resonances given by the Regge pattern (3.5). For the direct (s) channel of two incident hadrons, this is illustrated in Fig. 3.2, where the intermediate sum refers to all allowed resonance states and the final sum to the different possible decay channels. Extending this description to multiparticle reactions requires that each resonance of mass m, as defined by a Regge trajectory

$$\alpha' m^2 - \alpha_0 = n, \; n = 1, 2, \ldots, \tag{3.18}$$

has a degeneracy determined by the unordered number of partitionings of n. Hence we obtain from Eq. (3.11) that [15]

$$\rho(m) \simeq \left(\frac{\sqrt{2}}{m}\right)\left(\frac{1}{6}\right)^{5/4} \frac{1}{[\alpha' m^2]^{3/4}} \exp\left\{2\pi(2\alpha'/3)^{1/2} m\right\}. \tag{3.19}$$

We again have an exponential increase, but now determined by the resonance growth rate α', instead of the interaction range r_0 of the bootstrap model. Numerically, with $\alpha' = 1\,\mathrm{GeV}^{-2}$, one finds

$$T_V = 1/[2\pi(2\alpha'/3)^{1/2}] \simeq 195 \,\mathrm{MeV} \tag{3.20}$$

for the coefficient governing the exponential increase of the level density $\rho(m)$.

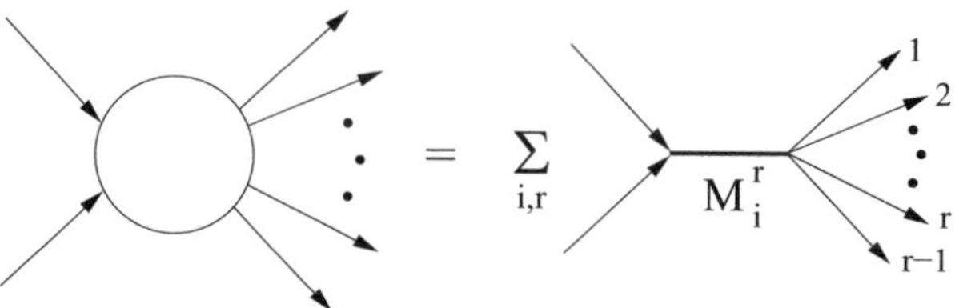

Fig. 3.2 Scattering amplitude as sum over resonances [5–7]

The dual resonance model is closely related to the string model [14], in which the different oscillation modes of strings correspond to the different resonance states, and the string tension σ, with

$$\sigma = \frac{1}{2\pi\alpha'}, \tag{3.21}$$

replaces the Regge slope α'. Both models have been formulated also for dimensions d higher than four, and this is found to lead to a slight decrease of the coefficient T_V, with values as low as 150 MeV for $d = 7$ [15].

3.4 The Ideal Resonance Gas

From the previous considerations, we conclude that a self-similar resonance pattern leads to an exponential growth of the resulting spectrum, with a coefficient $T_H \simeq T_V \simeq 150$–200 MeV. What effect does this have on the thermodynamics of an ideal gas of such resonanes?

We begin by recalling the partition function for a relativistic ideal gas of identical point-like particles of mass m_0, enclosed in a volume V at a temperature T. We shall be concerned with comparatively high temperatures and therefore adopt Boltzmann statistics. The grand canonical partition function is then defined as

$$Z_0(T, V) = \sum_{N=2}^{\infty} \frac{1}{N!} \left\{ \frac{1}{(2\pi)^3} \int_V d^3x \int d^3p \, \exp\{-\sqrt{p^2 + m_0^2}/T\} \right\}^N, \tag{3.22}$$

which becomes

$$Z_0(T, V) \simeq \exp\left\{ \frac{V}{(2\pi)^3} \int d^3p \, \exp\{-\sqrt{p^2 + m_0^2}/T\} \right\} = \exp\{\frac{V}{(2\pi)^3}\phi_0(T)\}. \tag{3.23}$$

Here the ground state momentum factor is

$$\phi_0(T) \equiv \int d^3p \, \exp\{-\sqrt{p^2 + m_0^2}/T\} = 4\pi T m_0^2 K_2(m_0/T). \tag{3.24}$$

For small values of the argument, i.e., for temperatures high enough to neglect the mass, the Hankel function becomes

$$K_2(x) = \frac{2}{x^2}\left[1 + O(x^2)\right].$$

(3.25)

Using this and

$$P(T) = \frac{1}{V} T \ln Z(T, V) \simeq \frac{1}{\pi^2} T^4,$$

(3.26)

we get the familiar Stefan-Boltzmann form for the pressure of an ideal gas. The logarithmic derivative of Eq. (3.22) with respect to T^{-1} yields the energy density of an ideal gas,

$$\epsilon(T) = \frac{-1}{V}\left(\frac{\partial \ln Z(T, V)}{\partial(1/T)}\right)_V = T^2\left(\frac{\partial \ln Z(T, V)}{\partial T}\right)_V \simeq \frac{3}{\pi^2} T^4;$$

(3.27)

the corresponding particle density is

$$n(T) = \left(\frac{\partial \ln Z(T, V)}{\partial V}\right)_T \simeq \frac{1}{\pi^2} T^3$$

(3.28)

and

$$\omega \simeq 3\,T$$

(3.29)

gives the average energy per particle. Hence an increase of the energy density of the system has three consequences: it leads to

- a higher temperature,
- more constituents, and
- more energetic constituents.

We now want to modify this ideal gas form to include the most important feature of strong interaction physics: the abundant formation of resonances in hadron-hadron interactions.

In statistical physics, there is a simple way to do this [16, 17]: one considers an ideal gas whose constituents are not only the basic particles of mass m_0, but also all possible resonances of different masses m they can form. The relative statistical weights $\rho(m)$ of the different constituents thus introduced have to be provided by some compositional or dynamical input. Any interaction between the basic particles is in this way removed at the expense of an increase in the number of different constituents. Crucial for this procedure is the specific form of the relative resonance weights.

The partition function for a relativistic ideal gas of point-like constituents of different masses m, contained in a volume V at a temperature T, is given by

$$Z(T, V) = \exp\left\{ \frac{V}{(2\pi)^3} \int_{m_o}^{\infty} dm\, \rho(m) \int d^3p\, e^{-\sqrt{p^2+m^2}/T} \right\}$$

$$= \exp\left\{ \frac{VT}{2\pi^2} \int_{m_o}^{\infty} dm\, \rho(m)\, m^2 K_2(m/T) \right\} . \tag{3.30}$$

Here m_0 is the stable lowest mass particle ("pion"), and we have again taken Boltzmann statistics. The form (3.30) corresponds to a system of constituents without intrinsic discrete degrees of freedom (such as charge, isospin, or baryon number), which could lead to conserved quantum numbers. Different resonances are thus distinguished only by their masses. The function $\rho(m)$, the resonance level density, determines the relative statistical weights of the different constituents. If we take

$$\rho(m) = \delta(m - m_0), \tag{3.31}$$

we recover the ideal gas form (3.22) for one type of particle only. The presence of further resonance states with masses $m > m_0$ reflects the effect of the interaction, and the level density $\rho(m)$ should in principle be provided by the underlying dynamics.

We have seen above that self-similar models, based on resonance composition [3, 4] or on resonance dynamics [5–7], lead to a resonance spectrum growing exponentially with mass

$$\rho(m) \sim m^{-a} e^{bm}, \tag{3.32}$$

where $b^{-1} \simeq 0.15$–$0.20\,\mathrm{GeV}$ is determined by the basic empirical parameters of hadron physics (hadronic interaction range or Regge slope).

While the exponential increase in m arises quite generally from the underlying self-similar partition problem, the power a of the polynomial factor is found to depend on the details of the approach. In particular, it turns out to be proportional to the space dimension of the partition problem [15, 18]. What then are the thermodynamic consequences of such a level density?

If we have a relativistic gas of only one type of particle as in Eq. (3.22), then an increase of energy density ϵ was seen to produce an increase of both the number of constituents and of their kinetic energy. The increase of the kinetic energy is reflected by a rising temperature; eventually it results in the Stefan-Boltzmann behavior $\epsilon \sim T^4$. Together with the corresponding number density $n \sim T^3$, this implies that the energy per constituent grows linearly with temperature, $\epsilon/n \sim T$. If, on the other hand, we have an exponentially rising level density, then any increase in energy density is strongly favored to go into the formation of heavy resonances, which have the highest statistical weight. We thus expect that the level density Eq. (3.32) will lead to a much slower increase of ϵ with temperature. Let us consider this in more detail.

Inserting the level density (3.32) into the partition function (3.30) yields

$$\ln Z(T, V) \sim VT \int_{m_0}^{\infty} dm\, m^{2-a} e^{bm} K_2(m/T). \tag{3.33}$$

Using the asymptotic form of $K_n(x)$ for large argument,

$$K_n(x) = \left(\frac{\pi}{2x}\right)^{1/2} e^{-x}[1 + O(n^2/x)], \tag{3.34}$$

we obtain

$$\ln Z(T, V) \sim VT^{3/2} \int_{m_0}^{\infty} dm\, m^{(3/2)-a} e^{m(b-(1/T))}. \tag{3.35}$$

For $T > 1/b$, this expression clearly diverges; in other words, $T_H = 1/b$ constitutes an upper bound for the temperature which a resonance gas can attain. The detailed behavior of the system at $T = T_H$ depends on the power a in Eq. (3.32). From Eq. (3.35), the energy density becomes

$$\epsilon(T) \sim VT^{7/2} \int_{m_0}^{\infty} dm\, m^{(5/2)-a} e^{m(b-(1/T))}, \tag{3.36}$$

while

$$C_V(T) = \left(\frac{\partial \epsilon}{\partial T}\right)_V =\sim VT^{3/2} \int_{m_0}^{\infty} dm\, m^{(7/2)-a} e^{m(b-(1/T))} \tag{3.37}$$

gives the specific heat at constant volume.

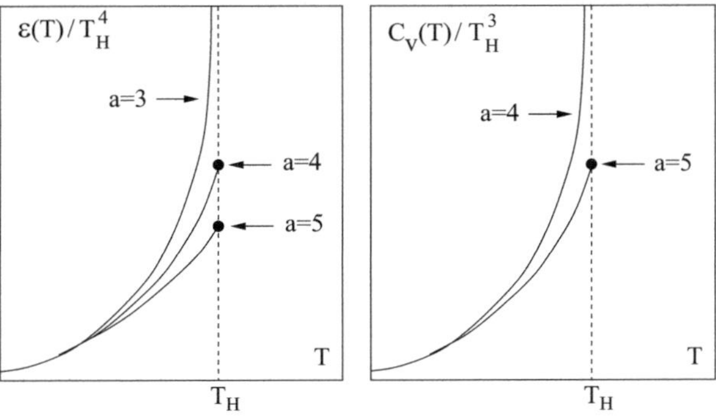

Fig. 3.3 Energy density $\epsilon(T)$ and specific heat $C_v(T)$ of a Hagedorn-type resonance gas

For $a = 3$, the partition function itself exists at T_H, but the energy density $\epsilon(T)$ diverges there, as illustrated in Fig. 3.3. In such a case, only an infinite energy density would bring the system to the temperature T_H. Hagedorn, using the analytical solution [13] with $a = 3$, therefore concluded that $T_H \sim 0.15\,\text{GeV}$ would be the ultimate temperature of all matter [3, 4]. The physical reason for such behavior is evident: the dramatically rising number of high mass resonances effectively absorbs any energy put into the system and thereby prevents a further increase in temperature. The different particle number degrees of freedom simply overwhelm their kinetic counterparts.

Ten years later, however, Cabibbo and Parisi [9] pointed out that the basis for this conclusion was the power a rather than the exponential increase in m. The latter certainly makes T_H a critical temperature in the sense that some derivative of $\ln Z$ with respect to T will diverge as $T \to T_H$. But for $a > 7/2$, also the energy density remains finite at T_H, and hence values higher than $\epsilon_H \equiv \epsilon(T_H)$ could perfectly well exist and lead to temperatures higher than T_H. For example, with $a = 4$, $\epsilon(T_H)$ is finite, but the specific heat $C_V(T)$ diverges at T_H as $(T_H - T)^{-1/2}$, while for $a = 5$, even the specific heat remains finite at T_H. The cases $a = 4$ and 5 are also shown in Fig. 3.3. The singular behavior obtained for different powers a can in fact be studied in terms of the usual critical exponents (see Chap. 2), specifying the behavior of the thermodynamic observables at the critical point $T = T_H$ [18]. The critical structure of the system defined by Eq. (3.35) is discussed in more detail in "Appendix: The Critical Structure of the Hagedorn Gas".

In general, matter can thus exist for $T > T_H$; it is only the specific form of hadronic level density (3.32) which ceases to be meaningful at T_H. The system can undergo a phase transition there into a new state of matter, in which a partition function based on Eq. (3.32) is no longer valid. Quantum chromodynamics today tells us that this new state, beyond the limit of hadron thermodynamics, is a plasma of deconfined quarks, and the critical temperature T_H is the deconfinement point.

We have seen here that an exponentially growing mass spectrum $\rho(m) \sim \exp(bm)$ leads to a singular structure of the partition function and hence to critical thermodynamic behavior at $T_H = 1/b$. It should be emphasized that this was obtained for pointlike or permeable hadrons. If the hadrons have an intrinsic hard-core volume not accessible to other hadrons, then this leads to an excluded volume effect in the partition function. In the particular case that the hard-core size of a resonance grows linearly with its mass, the singularity associated to the exponential resonance growth can no longer be reached [19]. In other words, the box is now "full" before the temperature can attain the value $T_H = 1/b$.

3.5   The Speed of Sound in a Resonance Gas

We conclude this chapter with an interesting consequence of the critical behavior in hadronic matter as $T \to T_H$ [20]. The speed of sound in a gas at constant volume is defined as

$$c_s^2 = \left(\frac{\partial P}{\partial \epsilon}\right)_V = \left(\frac{\partial P}{\partial T}\right)_V \bigg/ \left(\frac{\partial \epsilon}{\partial T}\right)_V = \frac{s(T)}{C_V(T)}, \qquad (3.38)$$

where $s = (\epsilon + P)/T = (\partial P/\partial T)_V$ denotes the entropy density of the system and $C_V(T)$ the specific heat at constant volume. For the ideal gas of pointlike "pions" of mass m_0, this leads to

$$\frac{1}{c_s^2} = 3 + \frac{m_0{}^2 K_2(m_0/T)}{4T^2 K_2(m_0/T) + m_0 T K_1(m_0/T)} \tag{3.39}$$

for the speed of sound. Making use of the small argument limit of the Hankel function,

$$K_n(x) = \frac{2^{n-1}(n-1)!}{x^n}\left[1 - O(x^2)\right], \tag{3.40}$$

shows that for $T \to \infty$, we get $c_s^2 \to 1/3$. The large argument limit (3.34) shows that for $T \to 0$, c_s^2 vanishes linearly with T. The overall behavior for such a pion gas is shown in Fig. 3.4. For a Hagedorn resonance gas with $a = 4$ in the spectrum (3.32), the specific heat diverges at T_H, while the entropy density remains finite. Hence now the speed of sound vanishes at T_H. This is another facet of the reduction of kinetic energy in a gas when more and more resonance states come into play: as we approach the limit of hadronic matter, the medium becomes increasingly inert.

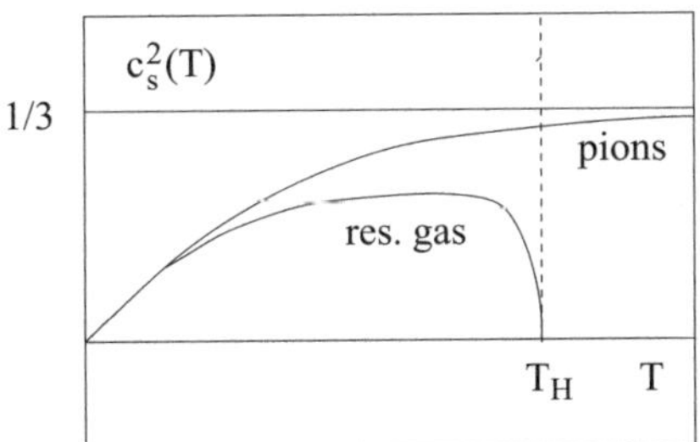

Fig. 3.4 The speed of sound in an ideal pion gas compared to that in an ideal Hagedorn resonance gas $(a = 4)$

To obtain an explicit form of the speed of sound in a Hagedorn gas, we parametrize the resonance spectrum (3.32) with $a = 4$ as

$$\rho(m) = \delta(m - m_0) + A m^{-4} \exp\{m/T_c\}\, \theta(m - 2m_0), \tag{3.41}$$

where the constant A determines the normalization of the resonance contributions relative to the ground state "pion". Using this form, we obtain

$$\frac{1}{c_s^2} = \left(3 + \frac{1}{2\pi^2 T s(T)}\left[3m_0^4 K_2(m_0/T) + A\int_{2m_0}^{\infty} dm\, \exp\{m/T_c\}\, K_2(m/T)\right]\right), \tag{3.42}$$

with

$$s_0(T) = \frac{3m_0^2 T}{2\pi^2}\left[4\,K_2(m_0/T) + \left(\frac{m_0}{T}\right)K_1(m_0/T)\right] \qquad (3.43)$$

and

$$s(T) = s_0(T) + \frac{AT}{2\pi^2}\int_{2m_0}^{\infty} dm\, m^{-2}\,\exp\{m/T_c\}\left[4\,K_2(m/T) + \left(\frac{m}{T}\right)K_1(m/T)\right]$$
$$(3.44)$$

for the entropy density of the pion gas and the Hagedorn gas, respectively. Evaluating this expression leads to the resonance gas form of c_s^2 shown in Fig. 3.4, in comparison with the corresponding pion gas result. The normalization A was here fixed by specifying the energy density at $T = T_H$.

It should be emphasized that the singular behavior we have considered here is a direct consequence of the exponentially rising level density $\rho(m)$. It is of course possible to study an ideal resonance gas for which $\rho(m)$ grows slower, and this is in fact the case in the resonance gas codes based on the states listed in the Particle Data Group compilation (we shall return to such models in Chap. 11). For example, the spin degeneracy alone will lead to $\rho(m) \sim m^2$, and such a system will never show critical behavior. The exponential increase arises only if we count all possible decay channels as degenerate states.

3.6 Conclusions

Hadron thermodynamics, based on phenomenological hadron dynamics, defines its own limits. It does so without any information about the quark infrastructure of hadrons. The intrinsic size of hadrons or self-similar dynamics of hadronic resonance formation result in critical behavior at some temperature T_H.

Since hadron thermodynamics provides its own limits, it cannot tell us what happens beyond these limits. To study the high density state of matter, we have to introduce new dynamics governing that state. In the next chapter, we will therefore consider simple models providing a two-phase structure of strongly interacting matter.

Appendix: The Critical Structure of the Hagedorn Gas

Using the large argument form of the Hankel function, the partition function for an ideal resonance gas with an exponential mass spectrum $\rho(m) \sim m^{-b}\exp\{m/T_H\}$ ("Hagedorn gas") can be written as

$$\ln Z(T) = c_0 V(Tm_0)^{3/2}\int_1^{\infty} dx\, x^{3/2-b}\,\exp\{-x\tau\} = V(Tm_0)^{3/2}F(\tau), \qquad (3.45)$$

where $x = m/m_0$,

$$\tau \equiv m_0 \left(\frac{1}{T} - \frac{1}{T_H} \right) = m_0 \frac{T_H - T}{T T_H}, \tag{3.46}$$

c_0 is a dimensionless constant, and

$$F(\tau) \equiv \int_1^\infty dx\, x^{3/2-b} \exp\{-x\tau\} \tag{3.47}$$

a dimensionless function of τ. As above, m_0 denotes the ground state hadron ("pion"), and to assure a finite pressure and energy density at $T = T_H$, we choose $b > 7/2$. In that case, both $F(0)$ and its first derivative are finite at $\tau = 0$, with

$$F(0) = \frac{1}{b - 5/2}, \quad \left(\frac{dF}{d\tau} \right)_{\tau=0} = -\frac{1}{b - 7/2}. \tag{3.48}$$

The second derivative, however, becomes for $9/2 > b > 7/2$

$$\left(\frac{d^2 F}{d\tau^2} \right) = \int_1^\infty dx\, x^{-(b-7/2)} \exp\{-x\tau\} = \tau^{-(9/2-b)} \Gamma(b - 5/2, \tau), \tag{3.49}$$

where $\Gamma(r, \tau)$ denotes the incomplete Gamma function. At the critical point $\tau = 0$, this gives $\Gamma(b - 5/2, 0) = \Gamma(b - 5/2)$, so that $d^2 F/d\tau^2$ diverges there as $\tau^{-(9/2-b)}$, and all higher derivatives diverge with correspondigly higher powers. Since τ increases when T decreases, all odd derivatives with respect to τ introduce a minus sign compared to derivatives with respect to T. Hence $F(T)$ and all its derivatives with respect to T are positive.

As a result, for $9/2 > b > 7/2$, pressure and energy density are finite and positive at T_H, and approaching T_H, both have positive gradients. The specific heat and all higher derivatives diverge at T_H, the specific heat

$$C_V \sim \tau^{-\alpha} \tag{3.50}$$

with the critical exponent $0 < \alpha = (9/2 - b) < 1$. For a discussion of further critical exponents for the Hagedorn resonance gas, see [21]

References

1. E. Fermi, Progr. Theor. Phys. (Jpn.) **5**, 570 (1950)
2. I.Ya. Pomeranchuk, Doklady Akad. Nauk. SSSR **78**, 889 (1951)
3. R. Hagedorn, Nuovo Cim. Suppl. **3**, 147 (1965)
4. R. Hagedorn, Nuovo Cim. **56A**, 1027 (1968)
5. G. Veneziano, Nuovo Cim. **57A**, 190 (1968)

6. K. Bardakci, S. Mandelstam, Phys. Rev. **184**, 1640 (1969)
7. S. Fubini, G. Veneziano, Nuovo Cim. **64A**, 811 (1969)
8. For a survey, see H. Satz, Fortsch. Physik **33**, 259 (1985)
9. N. Cabibbo, G. Parisi, Phys. Lett. **59B**, 67 (1975)
10. For a general survey of Hagedorn's work, see J. Rafelski, *Melting Hadrons, Boiling Quarks* (Springer Open, 2016)
11. Ph. Blanchard, S. Fortunato, H. Satz, Eur. Phys. J. C **34**, 361 (2004)
12. G.H. Hardy, S. Ramanujan, Proc. Lond. Math. Soc. **17**, 75 (1918)
13. W. Nahm, Nucl. Phys. **45B**, 525 (1972)
14. See e.g., P.H. Frampton, *Dual Resonance Models* (W.A. Benjamin, 1974)
15. K. Huang, S. Weinberg, Phys. Rev. Lett. **25**, 855 (1970)
16. E. Beth, G.E. Uhlenbeck, Physica **4**, 915 (1937)
17. R. Dashen, S.-K. Ma, H.J. Bernstein, Phys. Rev. **187**, 345 (1969)
18. H. Satz, Phys. Rev. **D20**, 582 (1979)
19. V.V. Dixit, F. Karsch, H. Satz, Phys. Lett. **101B**, 412 (1981)
20. P. Castorina, J. Cleymans, D. Miller, H. Satz, Eur. Phys. J. C **66**, 207 (2010)
21. H. Satz, Phys. Rev. **D20**, 582 (1979)

Chapter 4
From Hadrons to Quarks

The time has come, the walrus said,
to talk of many things:
of shoes, and ships, and sealing wax,
of cabbages, and kings,
and why the sea is boiling hot,
and whether pigs have wings.

Lewis Carrol, *Through the Looking Glass*

Abstract In this chapter, we look at some simple models to illustrate how a phase transition from hadronic matter to a quark-gluon plasma can occur. First we determine when hadrons start forming clusters which can be considered as quark matter. After considering the thermodynamics of an ideal hadron gas and of an ideal quark-gluon medium, we introduce bag pressure and baryon repulsion as interaction features to specify under which conditions strongly interacting matter prefers to consist of hadrons and when it wants to turn into a plasma of unbound quarks and gluons. Finally we show that also a simple string model yields localized hadrons at low density, while at high density color charges can move around freely by changing partners. In all cases, very basic physical notions are found to lead to a two-phase structure of matter.

4.1   Cluster Formation in Strongly Interacting Matter

Hadrons are extended objects, but they are composed of quark constituents. We had argued in Chap. 1 (see Fig. 1.2) that hadronic matter should turn into quark matter once the hadronic constituents begin to overlap sufficiently. At what density does that happen? We had seen in Chap. 2 that this is a question of quite general nature, answered by percolation theory.

For our specific problem, hadron clustering, we consider the percolation of spheres of size V_0 in three-dimensional space, allowing overlap of the different spheres [1]; here $V_0 = (4\pi/3)r_0^3$ is the volume of a hadron and $r_0 \simeq 0.8$ fm its

© Springer International Publishing AG 2018
H. Satz, *Extreme States of Matter in Strong Interaction Physics*,
Lecture Notes in Physics 945, https://doi.org/10.1007/978-3-319-71894-1_4

radius. Numerical studies [2], as mentioned, show that hadron percolation occurs at the critical density

$$n_h \simeq \frac{0.34}{V_0} \simeq 0.16 \, \text{fm}^3.$$ (4.1)

Beyond this density, the overlapping hadrons form a connected medium spanning the system. However, because of the overlap, most of space is at this point still empty. To quantify this, we recall Eq. 2.34, which tells us that 71% of space remains empty. Hence the density inside the percolating cluster is much higher and with the result

$$n_c^{\text{cl}} \simeq \frac{1.5}{V_0} \simeq 0.70 \, \text{fm}^{-3},$$ (4.2)

we have on the average three hadrons in the volume of two. This local density is almost three times standard nuclear matter density, and it is therefore indeed tempting to consider such clusters as the expected new form of strongly interacting matter.

Such dense matter begins to appear when the overall density has reached the value (4.1). As the overall density is increased further, the remaining fraction of empty space decreases from its percolation value of 71%, until at the second "dry" percolation point, for

$$n_v \simeq \frac{1.24}{V_0} \simeq 0.58 \, \text{fm}^3,$$ (4.3)

the vacuum is reduced to 29% of space. However, now there exist no longer any connected paths of hadronic size, i.e., a hadron could not freely move from one side of the box to the other. In Fig. 4.1, we illustrate the configurations corresponding to Eqs. (4.1) and (4.3), projecting the three-dimensional case onto two dimensions.

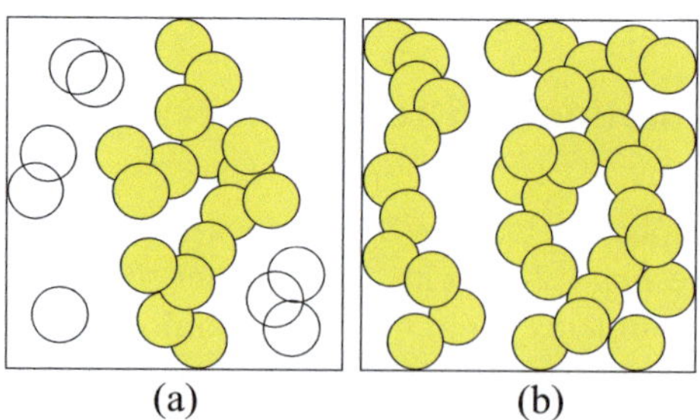

(a)   (b)

Fig. 4.1 Schematic illustration of the onset of hadron percolation (**a**) and the disappearance of vacuum percolation (**b**)

The density region

$$n_h \simeq 0.16 \, \text{fm}^3 \le n \le n_v \simeq 0.58 \, \text{fm}^3$$ (4.4)

is thus something like a geometric "mixed" phase. If we consider the connected (percolating) cluster at n_h as the first occurrence of a deconfined quark-gluon plasma formed through hadron fusion, then with increasing density more and more of space becomes "colored", until at n_v the physical vacuum makes its last appearence as a large-scale feature.

4.2   Ideal Quark-Gluon Plasma and Ideal Hadron Gas

In QCD, the basic units of matter are spin one-half quarks, coming in three intrinsic color degrees of freedom; in other words, quarks can take on three different values of the fundamental charge of the strong interaction. Hadrons are color-neutral bound states of these quarks. The quark baryon number is 1/3, so that mesons are quark-antiquark pairs and baryons three quark bound states. To account for all further observed intrinsic quantum numbers of hadrons – commonly denoted as strangeness, charm, bottom and top – we need in addition six flavor degrees of freedom. Each quark thus has a color label $\alpha = 1, 2, 3$ (the red, green and blue of Chap. 1) and a flavor label $f = 1, 2, \ldots, 6$. The quarks of the different flavors are usually denoted by u (up), d (down), s (strange), c (charm), b (bottom), and t (top). Including antiquarks and the two spin orientations, this gives us a total of $2 \times 2 \times 3 \times 6 = 72$ possible quark degrees of freedom.

To obtain correctly the measured mass values of all hadronic states, it is necessary to associate different rest mass values to the different flavors of quarks. The masses of the usual hadrons arise essentially from the kinetic energy of their almost massless quark constitutents; but for the heavy quark resonances discovered in the last thirty years, this does not work. To give an example: the lowest vector meson state constructed from u and d quarks, the ρ meson, has a rest mass of 0.77 GeV, while the lowest charmonium vector meson J/ψ, made up of a $c\bar{c}$ pair, has a mass of 3.1 GeV. All mass differences can be accommodated if we give the light u and d quarks small masses (10 MeV or less), but much larger masses to the other quark flavors. For the s quark we need a mass around 0.15 GeV, for the c quark 1.5 GeV, and for the b quark 4.5 GeV. Note that these mass values are parameters determined through hadron spectroscopy; they are not measurable quantities. The t quark was first discovered not so long ago and is very much heavier still, around 180 GeV. The occurrence of such non-vanishing quark masses is certainly a flaw in the beauty of QCD, since they have to be put in as outside parameters; in an ideal theory of strong interactions, all hadron masses would arise as "binding energies" of massless quarks, without the need of further dimensional parameters. The conventional hadrons, made up of almost massless u's and d's, come quite close to this. But the fact that quarks actually are not massless and that they are required to exist in six different flavors, of different masses, can certainly not be understood within strong interaction physics. The theory unifying strong and electroweak forces, the "standard model", does determine the specific decay modes of the different hadron species; but even there the origin of the different masses so far remains an open question.

Because of the hierarchy of quark masses, we can for the purpose of thermodynamics very often restrict ourselves to just u and d quarks. The statistical weight determined by the Boltzmann factor, $\exp(-m_q/T)$, gives for the temperature range of greatest interest, up to about $0.2\,\mathrm{GeV}$, a strange quark contribution of still about 50%, while c quarks are already below the 0.1% level. It is of course unproblematic to include more quark species; however, for simplicity we shall here consider for the moment only u and d, which we shall treat as massless. This then leaves us with 12 thermodynamically equivalent quark degrees of freedom, plus 12 more for the antiquarks.

We had already noted that the interaction between quarks is in QCD mediated by gluons – massless vector fields, the counterpart of the photons in quantum electrodynamics (QED). In contrast to the electrically neutral photons, however, the gluons carry an intrinsic color charge; this allows a quark of a given color to be transformed into another color by emission of a single gluon. To achieve this, there are eight different gluonic color states. Since gluons, with the transverse polarization of massless vector fields, have two spin orientations, we get in total $2\times 8 = 16$ gluon degrees of freedom. Together with the 24 "massless" quark and antiquark states (72 if we include all flavors), this completes our set of fundamental QCD constituents for strong interaction thermodynamics.

In the limit of high density, we expect that the screening of the color charge will turn strongly interacting matter into a gas of non-interacting colored quarks and gluons, i.e., into an ideal quark-gluon plasma. The thermodynamics of such a system is quite simple, as we shall now see. The basis for the calculation of all thermodynamic observables is the partition function

$$Z(T,\mu,V) \;\equiv\; \mathrm{Tr}\,\left\{ e^{-(H-\mu N)/T} \right\}. \tag{4.5}$$

Here T, μ and V denote the temperature, the "chemical" potential for the overall baryon number and the volume of the system, respectively; H is the Hamiltonian, N the operator for the net baryon number (number of baryons minus that of antibaryons). The trace is to be carried out over all physical states possible inside the volume V. In terms of the Gibs potential $Z(T,\mu,V)$, the basic thermodynamic observables are defined by

$$P \;=\; \frac{1}{V}(T\ln Z) \tag{4.6}$$

for the pressure,

$$S \;=\; \left(\frac{\partial\,(T\ln Z)}{\partial T} \right)_{\mu,V} \tag{4.7}$$

for the entropy, and

$$N_B \;=\; \left(\frac{\partial\,(T\ln Z)}{\partial \mu} \right)_{T,V} \tag{4.8}$$

for the net quark baryon number. In terms of these quantities,

$$E = -PV + ST + \mu N_B \tag{4.9}$$

gives the overall internal energy of the system.

For an ideal gas of vanishing overall baryon number density, $\mu = 0$, the partition function (4.5) can be calculated in the limit of $m \ll T$ (see Appendix: Bose and Fermi Gas Partition Functions), giving

$$\ln Z_B(T, V) = \frac{\pi^2}{90} VT^3 \left[1 - \frac{15}{4\pi^2} \left(\frac{m}{T} \right)^2 + O((m/T)^4) \right] \tag{4.10}$$

for bosons and

$$\ln Z_F(T, V) = \frac{7}{8} \frac{\pi^2}{90} VT^3 \left[1 - \frac{30}{7\pi^2} \left(\frac{m}{T} \right)^2 + O((m/T)^4) \right]. \tag{4.11}$$

for fermions. Using these results, we can define

$$\ln Z(T, V) = d_b \ln Z_B(T, V) + 2d_f \ln Z_F(T, V) \tag{4.12}$$

as the overall partition function of the ideal quark-gluon plasma at $\mu = 0$; here d_b and d_f denote the internal degrees of freedom for gluons and quarks/antiquarks, respectively; the factor $2d_f$ in Eq. (4.12) takes into account that for $\mu = 0$ quarks and antiquarks contribute the same. With the two transverse spin degrees of freedom for massless gluons, color $SU(3)$ gives $d_b = 2 \times (N_c^2 - 1) = 2 \times 8 = 16$. Retaining only the nearly massless u and d quarks, we have $d_f = 2 \times (N_f = 2) \times (N_c = 3) = 12$ (spin, flavor, color). The high temperature limit $(m/T \rightarrow 0)$ thus gives for the Stefan-Boltzmann form of the QCD partition function

$$\ln Z_q(T, V) = \left(\frac{37\pi^2}{90} \right) VT^3. \tag{4.13}$$

With this we obtain

$$P_q = \frac{37\pi^2}{90} T^4 \simeq 4\, T^4 \tag{4.14}$$

for the pressure,

$$\epsilon_q = \frac{37\pi^2}{30} T^4 \simeq 12\, T^4 \tag{4.15}$$

for the energy density, and hence

$$s_q = \frac{148\pi^2}{90} T^3 \simeq 16\, T^3 \tag{4.16}$$

for the entropy density $s_q = (\epsilon_q + 3P)/T$. If strongly interacting matter at sufficiently high temperatures indeed becomes an ideal gas of quarks and gluons, then these results give the asymptotic values of the corresponding thermodynamic observables.

At this point, let us note two features which hold in general for any ideal gas of massless constituents and without conserved intrinsic quantum numbers. A very useful quantity (and hence one to be encountered again many times) is the interaction measure $\Delta(T) \equiv (\epsilon - 3P)/T^4$. Using Eqs. (4.14) and (4.15), we find

$$\Delta(T) = \frac{\epsilon - 3P}{T^4} = 0 \tag{4.17}$$

as a consequence of the absence of interactions for the massless constituents in the system. From the grand canonical partition function (4.20) we can also obtain the average density n of constituents,

$$n = \left(\frac{\partial \ln Z}{\partial V}\right)_{T,\mu=0} \tag{4.18}$$

which in fact is just equal to (P/T). As a result, the specific entropy, i.e., the entropy per constituent, becomes

$$\frac{S}{N} = \frac{s}{n} = 4 \tag{4.19}$$

for an ideal gas of massless constituents.

For $\mu \neq 0$, but given massless, non-interacting bosons and fermions, Eq. (4.5) can again be evaluated in closed form (see e.g., [3]), leading to

$$T \ln Z(T, \mu_q, V) = (N_c^2 - 1)\left(\frac{\pi^2 V T^4}{45}\right)$$

$$+ N_f N_c \left(\frac{V}{6}\right)\left[\left(\frac{7\pi^2 T^4}{30}\right) + \mu_q^2 T^2 + \left(\frac{\mu_q^4}{2\pi^2}\right)\right], \tag{4.20}$$

with N_c and N_f for the number of color and flavor degrees of freedom, as above. Note that μ_q now denotes the baryochemical potential for quarks, i.e., of constituents of baryon number 1/3. The density of strongly interacting matter, given by $n(T, \mu_q) = \ln Z(T, \mu_q, V)/V$, can thus be increased either through heating (increasing T) or through compression (increasing μ_q).

After having considered the case $\mu_q = 0$, we now turn to the other extreme, baryonic matter at $T = 0$. The system then contains no gluons and (choosing $\mu_q > 0$) no antiquarks. Hence we get

$$P_q = \frac{1}{2\pi^2}\mu_q^4 \tag{4.21}$$

for the pressure,

$$\epsilon_q = \frac{3}{2\pi^2}\mu_q^4 \tag{4.22}$$

for the energy density, and

$$n_q = \frac{2}{\pi^2}\mu_q^3 \tag{4.23}$$

for the quark density; since quarks have a baryon number of 1/3, this gives

$$n_B^q = \frac{2}{3\pi^2}\mu_q^3 \tag{4.24}$$

for the net baryon number density $n_B^q = n_q/3$ of cold quark matter. The entropy vanishes for $T = 0$. Equations (4.21), (4.22), (4.23), and (4.24) provide the basic thermodynamic observables for large μ_q, if cold nuclear matter at sufficiently high density becomes quark matter.

So far, we have treated strongly interacting matter at very high density, where color screening was expected to produce a system of non-interacting quarks and gluons. With decreasing density, i.e., for lower T and/or μ_q, the interaction between the basic constituents of QCD becomes stronger, eventually hadron formation sets in, and at sufficiently low densities, we expect to find a rather dilute and hence weakly interacting gas of hadrons. For the low density limit, we thus need the thermodynamics of an ideal hadron gas. Let us therefore now also consider the two extreme cases, $\mu = 0$ (meson gas) and $T = 0$ (cold nuclear gas), for this state of matter.

The partition function for an ideal gas of mesons is obtained immediately from Eq. (4.20), if we consider only pions and assume that the temperature is high enough to neglect the pion mass. We then have $d_f = 0$ and $d_b = 3$, corresponding to the three pionic charge states, and obtain

$$P_\pi = \frac{\pi^2}{30}T^4 \simeq \frac{1}{3}T^4, \tag{4.25}$$

$$s_\pi = \frac{4\pi^2}{30}T^3 \simeq \frac{4}{3}T^3, \tag{4.26}$$

$$\epsilon_\pi = \frac{\pi^2}{10}T^4 \simeq T^4 \tag{4.27}$$

for the pressure, the entropy density and the energy density of a pion gas, respectively. Equations (4.25), (4.26), and (4.27), together with (4.14) and (4.15), give us the expected low and high temperature limits of strongly interacting matter at $\mu = 0$; corrections for finite constituent masses are in both cases obtainable through Eqs. (4.10) and (4.11). The main observation to be made here is that if with increasing temperature the system passes from a dilute pion gas to a dense ideal quark-gluon plasma, the quantities $P/T^4, s/T^3$ and ϵ/T^4 grow by approximately a factor twelve. This just reflects the difference in the number of intrinsic degrees of freedom of the respective constituents in the two limiting situations, with three different intrinsic pion states and 37 effective degrees of freedom in a quark-gluon plasma containing massless quarks and antiquarks of two different flavors.

For a system of nucleons at $T = 0$, but of a density high enough to neglect the mass of the nucleons, we can again use Eq. (4.20) to obtain the relevant thermodynamical quantities. With protons and neutrons of two spin orientations each, we get

$$P_n = \frac{1}{6\pi^2}\mu^4 \tag{4.28}$$

for the pressure, and

$$\epsilon_n = \frac{1}{2\pi^2}\mu^4 \tag{4.29}$$

for the energy density; the entropy again vanishes, while

$$n_B = \frac{2}{3\pi^2}\mu^3 \tag{4.30}$$

gives the net baryon number density. In Eqs. (4.28), (4.29), and (4.30), the baryochemical potential μ refers to the nucleons: it corresponds to the energy necessary to bring one additional nucleon into the system. Since a nucleon consists of three quarks, we have $\mu = 3\mu_q$ in terms of the quark baryochemical potential μ_q.

The corresponding expressions for a non-vanishing nucleon mass M can also be given in closed form (see [3]); in particular, we obtain

$$n_B = \left(\frac{2}{3\pi^2}\right)\left(\mu^2 - M^2\right)^{3/2} \tag{4.31}$$

for the baryon density of a cold nucleon gas, and

$$P_n = \left(\frac{1}{6\pi^2}\right)\left\{\left[1 - \left(\frac{M^2}{\mu^2}\right)\right]^{1/2}\left[1 - \left(\frac{5}{2}\right)\left(\frac{M^2}{\mu^2}\right)\right]\right.$$
$$\left. + \left(\frac{3}{2}\right)\left(\frac{M}{\mu}\right)^4 \ln\left[\left(\frac{\mu}{M}\right)\left(1 + \left[1 - \frac{M^2}{\mu^2}\right]^{1/2}\right)\right]\right\}\mu^4 \tag{4.32}$$

for the corresponding pressure.

4.3   Confinement and Bag Pressure

So far we have considered the thermodynamics of an ideal quark-gluon plasma and of an ideal hadron gas, i.e., neglecting in both cases any possible interaction. Now we know from general thermodynamic stability conditions that – given a choice – a system will always be in that state for which the thermodynamic potential $\Omega \equiv -T \ln Z$ has a minimum; equivalently, it chooses the state of highest pressure. How do the two states which we had looked at compare in this respect? From Eqs. (3.17) and (3.27) we find at all temperatures

$$\frac{P_q}{P_\pi} \simeq 12 \qquad . \tag{4.33}$$

for the ratio of pressures of the plasma with respect to the pion gas. This would seem to indicate an intrinsic instability of conventional hadronic matter, in favor of a quark-gluon plasma. Apparently there must be some essential interaction feature which assures the existence of our hadronic world at conventional densities.[1]

It is the difference between the physical vacuum and the ground state in the deconfined phase which constitutes this crucial missing feature. We had already noted in Chap. 1 that the lowest possible state in deconfined QCD matter is in fact not the physical vacuum, in which only color-neutral objects can exist [5]. The bag model of hadrons [6] provides a particularly good illustration of this feature and introduces in a qualitative way some aspects we will later encounter again in statistical QCD. Let us therefore have a look at this model. It considers hadrons as small bubbles – *bags* – within which an otherwise non-interacting quark-antiquark pair or three quark system is confined. The confinement force is represented as a pressure B, which the physical vacuum exerts on the region containing the quarks. For a static hadron, this pressure must be just compensated by that of the quarks on the bag surface. The rest mass M_n of a hadron therefore consists of two parts:

$$M_h = BV_h + np_0 = \left(\frac{4\pi R_h^3}{3}\right) B + \frac{2n}{R_h}. \tag{4.34}$$

Here the first term is the energy of the empty bag, which is just the difference in energy density between physical vacuum and lowest state of QCD; $V = 4\pi R_n^3/3$ is the bag volume. The second term gives the kinetic energy np_o of n massless quarks and/or antiquarks, with $p_o^2 - \mathbf{p}^2 = 0$. By the uncertainty relation, $|\mathbf{p}| \sim 1/R_h$; the

---

[1]It is a priori not impossible that quark matter might indeed be the thermodynamically stable form, making our present universe just a metastable bubble in the true vacuum. The production of even a drop of quark matter could then trigger the transition to the stable ground state. This would make the attempt to create quark matter by nuclear collisions a most risky enterprise: we would not be around to see the successful outcome. Fortunately, rate estimates of nucleus-nucleus collisions based on cosmic ray studies indicate [4] that such collisions must have taken place sporadically in the history of the universe, apparently without having disturbed the vacuum.

factor 2 in Eq. (4.34) comes from the wave function of a quark inside a spherical bag. Requiring the rest mass of the hadron to have a minimum (which is equivalent to making bag pressure and kinetic energy of the quarks compensate each other), we set

$$\frac{dM_h}{dR_h} = 0 \tag{4.35}$$

and thus get

$$R_h = \left(\frac{n}{2\pi B}\right)^{1/4}. \tag{4.36}$$

For a nucleon $n = 3$ and for a meson $n = 2$, so that we immediately obtain from this the ratio of nucleon to meson radii,

$$\left(\frac{R_n}{R_m}\right) = \left(\frac{3}{2}\right)^{1/4} \simeq 1.11. \tag{4.37}$$

Inserting Eq. (4.36) into (4.34), we find

$$M_h = \left(\frac{4\pi R_h^3}{3}\right) B + \left(4\pi R_h^3\right) B = 4\left(\frac{4\pi R_n^3}{3}\right) B = \frac{8}{3}(2\pi n^3 B)^{1/4}, \tag{4.38}$$

so that the mass of a hadron consists of three parts quark kinetic energy and one part bag energy. As counterpart to the universal bag pressure B, Eq. (4.38) gives us

$$\epsilon_h \equiv \left(\frac{M_h}{V_h}\right) = 4B. \tag{4.39}$$

as the universal energy density of hadrons. From Eq. (4.38) we further obtain

$$\left(\frac{M_n}{M_m}\right) = \left(\frac{3}{2}\right)^{3/4} \simeq 1.36 \tag{4.40}$$

for the ratio of nucleon to meson mass. The bigger mass of the nucleon is accompanied by a bigger radius (see Eq. (4.37)), so that the energy density of nucleons and mesons is the same, as given by Eq. (4.39).

We note here that the ratios of nucleon to meson mass, Eq. (4.40), and of the corresponding radii, Eq. (4.37), are given as absolute numbers, independent of the value of B. Since the bag pressure is not really an observable quantity, this is a very reasonable feature of the bag model – a feature which, as we shall see later on, is a reflection of the fact that QCD does not contain any dimensional parameters and hence predicts only ratios of observables.

Concerning the predicted numerical values of masses and radii, we have to note that the bag model presented here is the very simplest form – it provides just one type of baryon (a generic nucleon) and one type of meson (a generic ρ meson). From Eq. (4.38) we get 1.36 for the ratio of nucleon to rho mass; experimentally it is only 1.23. An even greater discrepancy arises for the (dimensionless) product of nucleon mass and radius. From Eqs. (4.36) and (4.38), it is 8.0; the experimental value (averaging over nucleon and $\Delta(1236)$ mass) is around 5. Equation (4.36) can be used to estimate the bag pressure; with 0.85 fm for the nucleon radius, we obtain $B^{1/4} \simeq 0.2\,\text{GeV}$. The resulting "nucleon" mass is 1.9 GeV, while the experimental average over nucleon and delta states is 1.2 GeV. The inclusion of angular momentum effects, perturbative quark-quark interactions, etc., improves the model and adds the fine structure necessary to get the level splitting for the observed nucleon and meson states (see e.g., [7]). However, it can never provide a value of the pion mass as low as the actual one; the reason for this is that the pion is only in part a quark-antiquark bound state. Its "other" origin, as a Goldstone boson associated to the approximate chiral symmetry of QCD, tends to make it massless, and the observed value is a combination of both origins. We shall return to this topic in Chap. 6.

The bag pressure thus plays an important role in obtaining a reasonable phenomenological description of much of the hadron spectrum. If we really want to compare the thermodynamics of non-interacting quarks and gluons with that of an ideal meson gas, we also have to include the effect of the physical vacuum on the plasma. Using the bag model, this means that the partition function of the quark-gluon plasma system, Eq. (4.5), has to be replaced by

$$T \ln Z_B(T, \mu, V) \equiv T \ln Z_q(T, \mu, V) - BV, \tag{4.41}$$

taking into account the bag pressure B. This gives us

$$P_B = P_q - B = \frac{37\pi^2}{90} T^4 - B \simeq 4T^4 - B \tag{4.42}$$

instead of Eq. (4.14), thus reducing the "outward" kinetic quark-gluon pressure by an "inward" pressure of the physical vacuum on the quark-gluon medium. The energy density now becomes

$$\epsilon_B \simeq \frac{37\pi^2}{30} T^4 + B \simeq 12T^4 + B, \tag{4.43}$$

instead of Eq. (4.16); the entropy density remains unchanged by such vacuum structure corrections. We also note that, with a finite difference between the physical vacuum and the ground state of the deconfined matter phase, the high temperature limits stay the same.

The comparison of pressures for quark-gluon plasma and pion gas at finite temperatures looks quite different, once the bag pressure is included. In Fig. 4.2, we show P_B, from Eq. (4.42), and P_π, from Eq. (4.25), as a function of T^4. At low temperatures, the pion gas now gives the higher pressure; then there is a cross-over when the two pressures become equal, at

$$T_c = \left(\frac{45}{17\pi^2}\right)^{1/4} B^{1/4} \simeq 0.72\, B^{1/4},\qquad(4.44)$$

and after that, the quark-gluon plasma dominates. In terms of a bag model of hadrons, we have for $T < T_c$ isolated bags, i.e., hadrons, in which quarks and gluons are confined to a restricted volume. For $T > T_c$, these bags fuse together into one big bag in which quarks and gluons can move around freely [8].

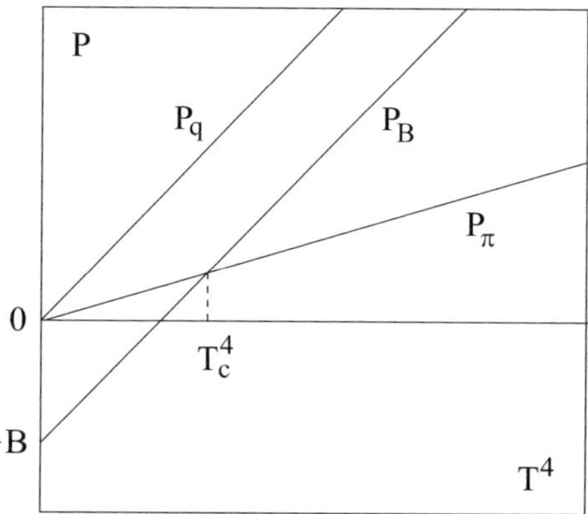

Fig. 4.2 Pressure of an ideal gas of massless pions (P_π) compared to that of an ideal quark-gluon plasma without (P_q) and with (P_B) bag pressure

Since a physical system will aways choose the state of highest pressure, we arrive at a two-state picture of mesonic matter. It arises from the competition of two effects: the confining pressure of the physical vacuum reduces the pressure of the quark-gluon state relative to that of a pion system, the higher number of intrinsic degrees of freedom enhances it. At low temperatures, the first effect wins, at high temperatures the second, and in between, we have – by construction – a first order phase transition. To obtain a value for T_c in physical units, we have to relate B to the hadron masses or radii. From Eq. (4.44) we get

$$T_c \simeq 0.14\,\text{GeV},\qquad(4.45)$$

using $B^{1/4} \simeq 0.2\,\text{GeV}$ obtained from a nucleon radius of 0.85 fm. This value of the critical temperature is not at all unreasonable; it agrees quite well with the Hagedorn temperature found in Chap. 3, and we shall later on get very similar results from statistical QCD. It is in fact surprising that the simplest form of the bag model does so well on this.

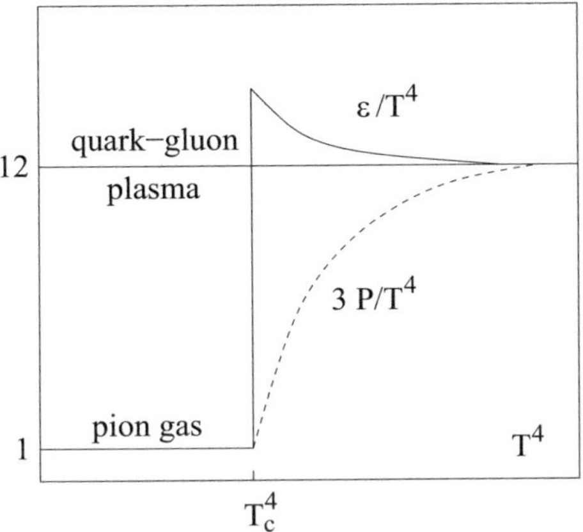

Fig. 4.3 Energy density and pressure in an ideal gas model with bag pressure

The resulting behavior of the energy density is shown in Fig. 4.3. By construction, the transition is here of first order: at T_c, the energy density increases abruptly by the latent heat of deconfinement, $\Delta\epsilon$. Using Eq. (4.44), its value is found to be

$$\Delta\epsilon = \epsilon_{qg}(T_c) - \epsilon_\pi(T_c) = 4B, \tag{4.46}$$

so that it is determined completely by the bag pressure measuring the level difference between the physical and the colored quark-gluon vacuum.

The quantity $\Delta = (\epsilon - 3P)/T^4$ was already introduced above as interaction measure; it is the trace of the energy-momentum tensor divided by T^4, and for an ideal gas of massless constituents, it generally vanishes, as seen in Eqs. (4.14) and (4.15). However, introducing the bag pressure as interaction effect into our model of the ideal plasma of massless quarks and gluons, we find for $T \geq T_c$

$$\Delta = \frac{\epsilon - 3P}{T^4} = \frac{4B}{T^4}, \tag{4.47}$$

again specified by the bag pressure and not zero. This is related to the so-called trace anomaly and indicates the dynamical generation of a dimensional scale; we shall return to it later on.

4.4 Nucleon Repulsion and Excluded Volume

What happens when the density of cold nuclear matter is increased, i.e., what is the phase structure as a function of μ at $T = 0$? Let us first compare the pressures (4.21) and (4.28). Making use of $\mu_n = 3\mu_q$, we find

$$\frac{P_q}{P_n} = \frac{1}{27}. \tag{4.48}$$

Since the pressure of the confined state of nucleons far exceeds that of cold quark matter, we should now expect nuclear matter to be the limiting high density state. By using Eq. (4.28), we have neglected the nucleon mass, and that is for low values of μ_n certainly not a good approximation. By using Eq. (4.21), we have also not taken the bag pressure into account, again something not reasonable for low μ_q. But inclusion of the nucleon mass or of the bag pressure does not change the asymptotic result: for $\mu \to \infty$, the nuclear matter phase always wins. The reason for this is quite clear: at $T = 0$, the pressure is just the Fermi pressure for the ground state of spin one-half particles, both for quarks and for nucleons. In the case of a quark system, we can divide this pressure among 2×3 Fermi seas, corresponding to the three quark color states of u and d quarks, and this leads at fixed $\mu_q = \mu_n/3$ to a lower value than found for the nucleon system with the two Fermi seas of protons and neutrons. In this case, the smaller number of degrees of freedom of the hadronic state thus turns out to be an advantage.

Since we do expect that the system will be in a quark phase at sufficiently high density [9, 10], we must again be ignoring some important interaction effect, and such an effect indeed exists: in contrast to mesons, which attract each other to form resonances, nucleons strongly repel each other at short distances, and this dynamical repulsion must be included in our description [11]. In Eq. (4.28) we had assumed that nucleons are arbitrarily compressible; Eq. (4.30) shows that the number of nucleons per unit volume diverges when $\mu_n \to \infty$. Let us see what happens if we give an intrinsic hard core volume V_{hc} to a nucleon. We first want to consider the qualitative features of introducing nuclear repulsion and therefore again neglect the mass of the nucleon. For N point-like nucleons, the hard core amounts to reducing the available volume from V to $(V - NV_E)$, where V_E is the effective volume per nucleon in a state of random close packing. For spherical nucleons, this is found to be $V_E \simeq 2\,V_{\mathrm{hc}}$ [12]. With such a restriction, the nucleon density is modified

$$\frac{N_b}{V} \to \frac{N_B}{V - N_B V_E} = \frac{n_B}{1 - n_B V_E} = n_b \qquad (4.49)$$

and hence we obtain

$$n_B(\mu) = \frac{n_b(\mu)}{1 + n_b(\mu) V_E} \qquad (4.50)$$

for the density $n_B(\mu)$ of hard-core nucleons in terms of the corresponding expression $n_b(\mu)$ for point-like constituents. From Eq. (4.50) we see that now $n_B \to 1/V_E$ when the density of point-like nucleons diverges, so that we never get more than one nucleon per effective excluded volume. The corresponding expression for the pressure becomes

$$P_B = \frac{P_b}{1 + n_b V_E}. \qquad (4.51)$$

Equations (4.49) and (4.51) imply that the pressure of hard-core nucleons grows less strongly with μ than that of point-like nucleons, due to the vanishing of the exclusion factor $1/(1 + n_b V_E)$. Using the ideal gas forms at $T = 0$ (see Eqs. (4.28) and (4.30)) we obtain from Eqs. (4.50) and (4.51) for large μ

$$P_B \simeq \frac{\mu}{4V_E} \tag{4.52}$$

instead of Eq. (4.28). The pressure of a system of hard-core nucleons now falls for large μ below that of the quark plasma. We thus get again a change of regimes: at low density, the pressure is greater for the hadronic state and at high density for the quark state. Hence the system will undergo a transition from nuclear matter to quark plasma at the value of μ for which

$$P_B(\mu) = P_q(\mu/3), \tag{4.53}$$

and the presence of nuclear repulsion prevents a return to a hadronic state at high pressure.

We note at this point that the derivation of the effect of a hard-core in a grand-canonical treatment was in fact heuristic and only approximative. A correct treatment [13] leads in addition to a shift of the baryochemical potential,

$$\mu \to \mu_B = \mu - P_B(\mu)V_E, \tag{4.54}$$

which becomes approximately equal to μ only in the limit of small excluded volume $V_E \to 0$. In general, all nucleon observables become functions of the shifted variable (4.54), such as $P_b(\mu_B)$ or $n_b(\mu_B)$.

Moreover, for a detailed view of the transition region, our considerations up to now are insufficient because we have neglected both the nucleon mass on the hadronic side and the bag pressure in the quark state. Including the nucleon mass M, we get on the hadronic side the baryon density

$$n_B(\mu) = \left[\left(\frac{3\pi^2}{2(\mu_B^2 - M^2)^{3/2}} \right) + V_E \right]^{-1}. \tag{4.55}$$

and a corresponding pressure $P_B(\mu)$ specified by Eqs. (4.32) and (4.51). On the quark matter side, the pressure becomes

$$P_Q = \frac{1}{2\pi^2}\mu_q^4 - B = \frac{1}{2\pi^2}\left(\frac{\mu}{3}\right)^4 - B, \tag{4.56}$$

adding the effect of the bag pressure to Eq. (4.21). To determine the transition point, we now have to equate the two pressures,

$$P_B(\mu) = \frac{P_b(\mu_B)}{1 + n_b(\mu_B)V_E} = \frac{1}{2\pi^2}\left(\frac{\mu}{3}\right)^4 - B = P_Q(\mu), \tag{4.57}$$

with μ_B given as a function of μ by Eq. (4.54). The solution of this equation gives us the critical baryochemical potential $\mu_c(V_E, B, M)$ for the transition from baryonic to quark matter at $T = 0$. The resulting functional behavior is schematically illustrated in Fig. 4.4.

The pressure of nuclear matter vanishes for $\mu \leq M$; hence the quark matter pressure must satisfy $P_q \leq 0$ in that region. This requires $B^{1/4} \geq M/[3(2\pi^2)^{1/4}] \simeq 0.15$ GeV, which is in accord with spectroscopic estimates of the bag pressure. The actual transition point depends quite sensitively on the values of the bag pressure and the hard core. To illustrate, we choose $R_b = 0.4$ fm and hence get an excluded volume $V_E = 0.54$ fm^3; random dense packing thus implies a nucleon density $n_R^{dp} \simeq 1.9$ fm^{-3} or more than ten times standard nuclear density. In our two phase picture, however, this limit is never reached, since at the value of μ_n determined by the equality of baryonic and quark pressures, Eqs. (4.57) and (4.56), the system turns into quark matter. We shall return in Chap. 7 to a further discussion of the detailed deconfinement pattern for baryonic matter.

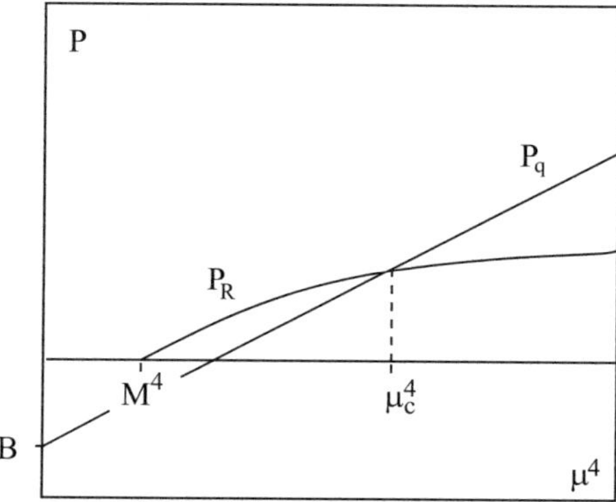

Fig. 4.4 Transition between nuclear matter with hard core repulsion and quark matter with bag pressure, at $T = 0$

4.5   Strings and Flip-Flop

In the bag model, the confining force binding quarks to hadrons was represented as an effect of the physical vacuum on the quark system. The string model provides an alternative picture of confinement, in terms of a potential increasing linearly with interquark separation, as in Eq. (1.3). As simplest case we consider a confined quark-antiquark system, i.e., a generic meson; its energy is given by [14]

$$M_h = 2\sigma R_h + \left(\frac{3}{R_h}\right). \tag{4.58}$$

Here R_h is the radius of the meson, so that $d = 2R_h$ is the distance between quark and antiquark; σ is the string tension, i.e., the confining force, which we assume to be constant. The second term corresponds, as in the bag model counterpart (4.34), to the kinetic energy $2p_0$ of quark and antiquark, and is as above obtained by the uncertainty relation from the energy $p_0 \sim 1/R$ of a massless quark. The constant in the relation between p and R is again model-dependent; we have here used the form $p = 3/2R$ for a three-dimensional quantum oscillator. Minimizing the energy as in Sect. 3.4, we get

$$R_h = \left(\frac{3}{2\sigma}\right)^{1/2} \tag{4.59}$$

for the meson radius and

$$M_h = 2\sqrt{6\sigma} \tag{4.60}$$

for its mass. Once more we thus have a model building extended hadrons from pointlike massless quarks; these hadrons again have an intrinsic size and mass determined by a force exerted on the quarks. Here it is the string tension σ, in Sect. 3.4 it was the bag pressure B; either determines radius and mass of the hadron.

Let us now look at the thermodynamics of such string-like hadrons [15, 16]. In comparing a quark-gluon system confined by bag pressure to an ideal hadron gas, we had from the outset two phases; the requirement of minimal free energy (maximal pressure) allowed nature to choose which it preferred in what region. Here we have a system with one given kind of string dynamics, and we want to ask whether it behaves differently at different densities. Why should it?

The total energy E of a system of N quarks and antiquarks, coupled by a string potential, is

$$E = \sum_{i=1}^{2N} p_i + \sigma \sum_{i=1}^{N} d_i, \tag{4.61}$$

where the first term gives the kinetic and the second the potential energy; d_i denotes the distance between quark and antiquark in the ith pair, and p_i is the absolute value of the ith constituent, quark or antiquark. We want to consider the grand canonical average of this energy, $\langle E \rangle$, for a system contained in a volume V at temperature T. To be in thermal equilibrium, the system arranges itself in such a way that $\langle E \rangle$ is minimal, which makes the nearest quarks and antiquarks combine into pairs in order to reduce the potential energy. In determining this minimum, however, we have to take the uncertainty principle into account, which forbids small pairs of low relative quark-antiquark momenta. Hence to reduce the potential energy through almost pointlike pairs, we must have very energetic constituents, and this in turn in a grand canonical system means high temperatures and densities. So in the high temperature limit, we expect an ideal gas of essentially free constituents, since the high density makes the potential energy negligible compared to the kinetic. At low temperatures, on the other hand, both the number of constituents and their momenta decrease, and so the crucial aspect is to minimize the potential energy. As a result, the favored configurations consist of isolated quark-antiquark pairs, whose minimum size is given by Eq. (4.59) – i.e., of the "hadrons" of the string model.

Two qualitatively different states of matter thus seem to appear quite naturally from string dynamics. We have suppressed, however, an interesting question concerning the high temperature state: since the potential between paired quarks and antiquarks never vanishes (for a constant string tension σ), how can these constituents move freely through the system and in particular get away from their respective bound-state partners? The answer is known as "flip-flop" mechanism and brings us back to the picture discussed in Chap. 1. At high density, each quark will find in its immediate vicinity a number of antiquarks, all about equally far away and each coupled to some quark of its choice. Since all separation distances are about equal, a rearrangement of bindings (see Fig. 4.5) does not change the energy of the system and hence is equally probable. By a succession of flips from one partner to the next, with associated flops on the part of the partners, any given constituent can thus move freely throughout the medium.

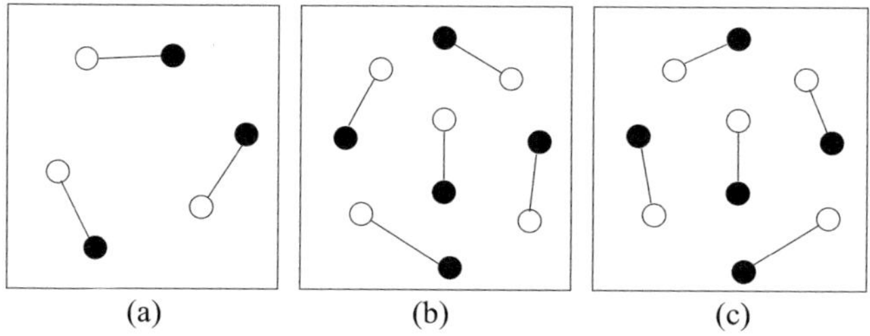

(a)   (b)   (c)

Fig. 4.5 String gas at low density (**a**) and for two different high density configurations (**b**) and (**c**) obtained through flip-flop

While a system of quark-antiquark pairs thus does change its qualitative features between low and high temperature, it is not obvious that such a change corresponds to critical behavior. The analogy to percolation theory leads one to expect this, and numerical studies indeed always show at least a very rapid cross-over, for specific systems even a discontinuity [16].

4.6   Conclusions

We saw that very simple models based on extended hadrons made up of quark constituents can lead to a phase transition from hadronic matter to a quark-gluon plasma. The underlying reasons were:

- In percolation theory, the onset of hadron connectivity leads to the formation of clusters of dense strongly interacting matter, which eventually cover all space, leading to the disappearence of the physical vacuum.

- In the bag model, bag pressure makes the pion gas the dominant phase at low temperature, while at high temperature the quark-gluon phase with its higher number of intrinsic degrees of freedom is favored.
- At zero temperature, cold nuclear matter has a higher Fermi pressure than cold quark matter, but the baryon repulsion between nucleons makes the quark state take over at high density.
- In the string model, there is pairing into isolated hadronic bound states at low density, while at high density the change of partners allows free quark flow even for non-zero string tension.

Let us now see to what extent these phenomenological features can be derived from what we believe is the real basis of strong interaction thermodynamics: statistical QCD.

Appendix: Bose and Fermi Gas Partition Functions

The partition function for an ideal gas of particles of mass m at temperature T in a volume V is given by [17]

$$\ln Z_s(T, V) = -s \frac{V}{2\pi^2} \int_0^\infty dp\, p^2\, \ln\left[1 - s\exp(-\sqrt{p^2 + m^2}/T)\right], \qquad (4.62)$$

with $s = +1$ for bosons, $s = -1$ for fermions. For simplicity we assume in both cases no internal degrees of freedom. Expanding the logarithm as a power series, we obtain

$$\ln Z_s(T, V) = s \frac{V}{2\pi^2} \sum_{n=1}^\infty \frac{1}{n} \int_0^\infty dp\, p^2 \left(s\, \exp(-\sqrt{p^2 + m^2}/T)\right)^n; \qquad (4.63)$$

the first term in this sum constitutes the Boltzmann limit. The momentum integration can be carried out to give

$$\ln Z_s(T, V) = s \frac{m^2 V T}{2\pi^2} \sum_{n=1}^\infty \frac{s^n}{n^2} K_2(nm/T), \qquad (4.64)$$

where $K_2(x)$ is the modified Hankel function of second order. In the limit $x \to 0$, it becomes

$$K_2(x) = \frac{2}{x^2}\left[1 + \frac{x^2}{4} + O(x^4)\right]. \qquad (4.65)$$

Inserting this into Eq. (4.64), we get

$$\ln Z_s(T, V) = s\frac{VT^3}{\pi^2}\left[\sum_{n=1}^{\infty}\frac{s^n}{n^4} + \frac{1}{4}\left(\frac{m}{T}\right)^2\sum_{n=1}^{\infty}\frac{s^n}{n^2} + O((m/T)^4)\right]. \tag{4.66}$$

The sums in Eq. (4.66) are Riemann zeta functions [18]:

$$\zeta(r) = \sum_{n=1}^{\infty}\frac{1}{n^r} \tag{4.67}$$

and

$$(1 - 2^{(1-r)})\zeta(r) = \sum_{n=1}^{\infty}\frac{(-1)^{n+1}}{n^r} \tag{4.68}$$

In particular, we have

$$\sum_{n=1}^{\infty}\frac{1}{n^4} = \zeta(4) = \frac{\pi^4}{90}, \tag{4.69}$$

$$\sum_{n=1}^{\infty}\frac{(-1)^{n+1}}{n^4} = (1 - 2^{-3})\zeta(4) = \frac{7}{8}\frac{\pi^4}{90}, \tag{4.70}$$

$$\sum_{n=1}^{\infty}\frac{1}{n^2} = \zeta(2) = \frac{\pi^2}{6}, \tag{4.71}$$

$$\sum_{n=1}^{\infty}\frac{(-1)^{n+1}}{n^2} = (1 - 2^{-1})\zeta(2) = \frac{1}{2}\frac{\pi^2}{6}. \tag{4.72}$$

Inserting this into Eq. (4.66) gives as leading terms

$$\ln Z_B(T, V) = \frac{\pi^2}{90}VT^3\left[1 - \frac{15}{4\pi^2}\left(\frac{m}{T}\right)^2 + O((m/T)^4)\right] \tag{4.73}$$

for the partition function for bosons and

$$\ln Z_F(T, V) = \frac{7}{8}\frac{\pi^2}{90}VT^3\left[1 - \frac{30}{7\pi^2}\left(\frac{m}{T}\right)^2 + O((m/T)^4)\right]. \tag{4.74}$$

for fermions. In both cases, the first term is the Stefan-Boltzmann limit. Comparing Eqs. (4.66) with (4.73), we see that the numerical coefficient of the bosonic partition function in the limit $m \to 0$ is $\pi^2/90 \simeq 0.1097$ for correct Bose-Einstein statistics, while it becomes $1/\pi^2 \simeq 0.1013$ in the Stefan-Boltzmann limit. This provides the justification for using the latter form whenever $m \ll T$.

References

1. T. Celik, F. Karsch, H. Satz, Phys. Lett. B **97**, 128 (1980)
2. See e.g., M.S. Isichenko, Rev. Mod. Phys. **64**, 961 (1992)
3. J. Cleymans, R.V. Gavai, E. Suhonen, Phys. Rep. **130**, 217 (1986)
4. P. Hut, Nucl. Phys. A **418**, 301c (1984)
5. T.D. Lee, in *Statistical Mechanics of Quarks and Hadrons*, ed. by H. Satz (North-Holland, Amsterdam, 1981)
6. A. Chodos, R.L. Jaffe, K. Johnson, C.B. Thorn, V.F. Weisskopf, Phys. Rev. D **9**, 3471 (1974)
7. K. Johnson, in *Statistical Mechanics of Quarks and Hadrons*, ed. by H. Satz (North-Holland, Amsterdam, 1981)
8. J. Baacke, Acta Phys. Polonica B **8**, 625 (1977)
9. J.C. Collins, M.J. Perry, Phys. Rev. Lett. **34**, 1353 (1975)
10. B.A. Freedman, L.D. McLerran, Phys. Rev. **16**, 1169 (1977)
11. J. Cleymans, K. Redlich, H. Satz, E. Suhonen, Z. Phys. C **33**, 151 (1986)
12. G.Y. Onoda, E.G. Liniger, Phys. Rev. Lett. **22**, 2727 (1990)
13. D. Rischke et al., Z. Phys. C **51**, 485 (1991)
14. E. Eichten, K. Gottfried, T. Kinoshita, J. Kogut, K.D. Lane, T.-M. Yan, Phys. Rev. Lett. **34**, 369 (1975)
15. H. Miyazawa, Phys. Rev. D **20**, 2953 (1979)
16. V. Goloviznin, H. Satz, Yad. Fiz. **60**, 523 (1997) [Phys. Atom. Nucl. **60**, 449 (1997)]
17. See, e.g., K. Huang, *Statistical Mechanics*, 2nd edn. (Wiley, New York, 1987)
18. See, e.g., I.S. Gradshteyn, I.M. Ryzhik, *Tables of Integrals, Series, and Products* (Academic, New York/London, 1965)

Chapter 5
Statistical QCD

*El universo (que otros llaman la biblioteca) – es ilimitado y
periódico. Si un eterno viajero le atravesara en qualquier
dirección, comprobaria al cabo de los siglos que los mismos
volúmenes se repiten en el mismo desorden que, repetido, seria
un orden: el Orden.*

Jorge Luis Borges, *La Biblioteca de Babel*

*(The universe (which others call the library) – is unlimited and
periodic. If an eternal voyager were to traverse it in any
direction, he would find, after many centuries, that the same
volumes are repeated in the same disorder which, since
repeated, would be an order: order itself.)*

(Jorge Luis Borges, *The Library of Babel*)

Abstract We now come to the basic theoretical problem: given QCD as strong interaction *dynamics*, derive strong interaction *thermodynamics*. After looking at some essential features of chromodynamics, we develop statistical QCD at finite temperature in the lattice formulation and discuss how it can be evaluated by computer simulation. We then use this method to study the thermodynamics of strongly interacting matter at vanishing baryon number density.

5.1 The Gauge Field Theory of Strong Interactions

Quantum chromodynamics describes the interaction of quarks and gluons. It does so in the form of a gauge field theory very similar to quantum electrodynamics (QED), which deals with the interaction of electrons and photons. In both cases we have spin one-half matter fields (electrons or quarks) interacting by exchange of massless vector gauge fields (photons or gluons). In QCD, however, the intrinsic color charge is associated with the non-Abelian gauge group SU(3), in place of the Abelian group U(1) for the electric charge of QED. As a consequence, there are quarks of three different color charge states (e.g. red, blue and green), and the corresponding

© Springer International Publishing AG 2018

H. Satz, *Extreme States of Matter in Strong Interaction Physics*,
Lecture Notes in Physics 945, https://doi.org/10.1007/978-3-319-71894-1_5

antiquarks; the gluons, which transform according to the adjoint representation of
SU(3), carry eight charges (3 × 3 combinations of the type red/anti-red, red/anti-
blue, etc., with the unitarity restriction $UU^+ = 1$ removing one combination). In
contrast, QED contains only one charged fermion and its antiparticle, together with
uncharged photons. The intrinsic charge of the gauge field is the crucial difference
between QCD and QED, where photons carry no charge. It allows direct interactions
between gluons, whereas photons cannot interact without intermediate electrons.
The direct gluon interaction contracts the lines of force between two color sources
into a "flux tube" or "string" (see Fig. 1.4). As a consequence, the three-dimensional
Poisson equation, which in non-relativistic QED leads to the Coulomb potential
$V \sim 1/r$, now effectively becomes one-dimensional, with the confining form $V \sim r$
as solution.

The Lagrangian density defining QCD is

$$\mathcal{L} = -\frac{1}{4} F^a_{\mu\nu} F^{\mu\nu}_a - \sum_f \bar{\psi}^\alpha_f \left[i\delta_{\alpha\beta}\gamma^\mu \partial_\mu - \frac{g}{2}(\lambda_a)_{\alpha\beta}\gamma^\mu A^a_\mu \right] \psi^\beta_f , \qquad (5.1)$$

with

$$F^a_{\mu\nu} = (\partial_\mu A^a_\nu - \partial_\nu A^a_\mu - g f^a_{bc} A^b_\mu A^c_\nu) . \qquad (5.2)$$

Here A^a_μ denotes the gluon vector field of color a ($a = 1,2,\ldots,8$) and ψ^α_f the quark
spinor field of color α ($\alpha = 1,2,3$) and flavor f. All indices are summed over the
relevant ranges. The structure functions f^a_{bc} are fixed by the color gauge group, whose
generators we denote by λ_a; they satisfy the commutation relation

$$[\lambda_a, \lambda_b] = i f^c_{ab} \lambda_c. \qquad (5.3)$$

The generators λ_a are eight 3×3 matrices in color space, generally denoted as Gell-
Mann matrices; they are the $SU(3)$ generalization of the three 2×2 Pauli matrices
for $SU(2)$. If we would set all $f^a_{bc} = 0$, the Lagrangian density (5.1) would reduce to
that of QED; it is just the last term in Eq. (5.2) which provides the self-interaction
among the gluons. In Fig. 5.1, we compare some Feynman graphs for the interaction
of the charged constituents in QED and QCD to illustrate this effect. We note that
the gluons are "flavor-blind"; their coupling to quarks does not depend on the quark
flavor.

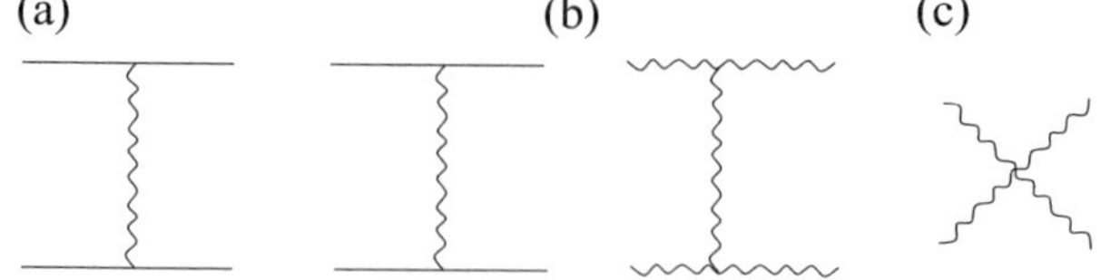

Fig. 5.1 Lowest order Feynman diagrams in QED (**a**) and QCD (**b, c**)

Here and in the following we restrict ourselves to the effectively massless u and d quarks, which give us all mesons and baryons without further intrinsic quantum numbers (such as strangeness, charm, etc.). Hence the Lagrangian (5.1) contains no mass term. Strange quarks and quarks of still "higher" flavors not only have a non-negligible mass, but they also have to be formed in quark-antiquark pairs, in order to conserve the further intrinsic quantum numbers. Because of this, they are thermodynamically somewhat (strange) or considerably (charm and beauty) suppressed at temperatures up to and including the transition region.

The presence of N_f quark species with non-vanishing masses m_f would add a term

$$\mathcal{L}_m = \sum_{f=1}^{N_f} m_f \bar{\psi}_f^{\alpha} \psi_{\alpha,f} \tag{5.4}$$

to Eq. (5.1); note that since quark masses are connected to physical observables (hadron masses), they are "color-blind". Since $\bar{\psi}\psi$ is related to the quark mass, it can be used to check if the state in which the system finds itself modifies the quark mass. Even for $m_f = 0$ in the Lagrangian, it is still possible that the system is in a specific state which makes $\langle \bar{\psi}\psi \rangle \neq 0$. In this case, we speak of the spontaneous generation of an effective quark mass: massless gluons "dress" massless quarks such as to give them a non-vanishing effective mass. We will return to this phenomenon later on, when we consider the spontaneous breaking and restoration of chiral symmetry. We should note here, however, that with the Lagrangian (5.1) containing only massless quarks we are considering an idealised world, in which e.g. pions are also massless. To obtain the real physical world with experimentally observed hadron masses, a term of the form (5.4) must be added to (5.1), with $m_f \neq 0$ for *all* quark flavors.

The Lagrangian (5.1) and (5.2) determines all of strong interaction dynamics in terms of one dimensionless coupling constant, g. It therefore cannot provide us with a dimensional scale, and so QCD predicts only the ratios of physical quantities – it does not give the absolute value of any observable in terms of physical units. How can we then obtain an experimentally useful prediction for the deconfinement temperature or for the energy density necessary for the formation of a quark-gluon plasma?

In QCD, all hadrons are color-neutral bound states of three quarks or of a quark-antiquark pair, and as the fundamental theory of strong interactions, QCD must predict the ratios of all hadron masses as well as the ratios of these masses to other observables, such as the deconfinement temperature. In addition to T_c, we therefore have to calculate as well some hadron mass whose value is known in physical units. From the calculated ratio of temperature and hadron mass, we then get T_c in MeV. Crucial for this prediction is thus the precision of the hadron mass calculations involved, and these can of course be tested through comparison with the known hadron mass ratios. At present, tuning of the quark mass in extensive hadron mass studies, including virtual quark loops (see e.g. [1, 2]) lead to masses for both mesons and baryons which converge to the known experimental values.

In addition to hadron spectroscopy, QCD must also describe hadronic scattering processes. The successful application of *perturbative* QCD to hard scattering processes at large momentum transfer was in fact decisive in establishing it as the basic theory for strong interactions, and tests of the theory in high energy experiments have confirmed this with remarkable precision (see e.g. [3]). On the other hand, it must be noted that little is known today about any direct quantitative applications of QCD to *non-perturbative* hadronic scattering processes.

Given the Lagrangian density (5.1), the formulation of statistical QCD becomes, at least in principle, a well-defined problem. We have to calculate the partition function

$$Z(\beta, V) = \mathrm{Tr}\exp\{-\beta E\}, \tag{5.5}$$

with the Hamiltonian E determined from $\mathcal{L}$; we have here assumed a vanishing baryon number density, i.e., the same number of baryons as of antibaryons. In the trace, we have to sum over all physical states accessible to a system within a spatial volume V; $\beta^{-1} = T$ denotes the physical temperature. Once $Z(\beta, V)$ is obtained, we can calculate all thermodynamical observables in the usual fashion. Thus

$$P = \frac{1}{V}T\ln Z \tag{5.6}$$

leads to the pressure, while differentiating with respect to the inverse temperature

$$\epsilon = (-1/V)\left(\frac{\partial \ln Z}{\partial \beta}\right)_V \tag{5.7}$$

gives the energy density.

But how can these expressions actually be evaluated? After all, we are faced with a relativistic, interacting quantum field theory. In QED, we encounter divergences both for small (infrared) and for large (ultraviolet) momenta, and as a result, we have to renormalize in order to get finite results. We therefore expect that renormalisation will be necessary for QCD as well. But there is a further, more serious, problem. The standard evaluation method for QED – perturbation theory – is not applicable to the study of critical behavior, because long range correlations and multi-particle interactions are of crucial importance here, and we thus cannot assume interaction terms to be small. We therefore need a new approach to the solution of a relativistic quantum field theory – a non-perturbative regularisation scheme. So far, there is only one method available which fulfills these requirements: the lattice formulation introduced by K. Wilson [4]. It gives us the thermodynamic observables, such as the energy density (5.7) or the pressure (5.6), in a form that can be evaluated numerically. Let us see how this form is attained and how it can be evaluated by computer simulation [5, 6]; this is of course a topic on its own, so we refer to [7–9] for more detailed presentations.

5.2   Lattice QCD at Finite Temperature

The lattice formulation of statistical QCD is obtained in four steps. First we replace the usual Hamiltonian form (5.5) of the partition function by an equivalent form in terms of a Euclidean path integral. This means that instead of summing over all the possible many-particle states (in the trace of Eq. (5.5)), we sum over all possible field configurations of the system. Any quantum mechanical partition function can be expressed in this way as a functional integral over field configurations [10]. For a field theory such as QED or QCD, the partition function then becomes [11]

$$Z_E(\beta, V) = \int \mathcal{D}A \, \mathcal{D}\psi \, \mathcal{D}\bar{\psi} \, \exp\left\{-\int_V d^3x \int_0^\beta d\tau \, \mathcal{L}_E(A, \psi, \bar{\psi})\right\}. \qquad (5.8)$$

Here the vector fields for the gluons are defined as

$$A_\mu = (1/2) \sum_{a=1}^{8} \lambda_a A_\mu^a \qquad (5.9)$$

and thus are also 3×3 matrices in color space. This form involves directly the (Euclidean) Lagrangian density defining the theory; we do not have to calculate first the Hamiltonian H needed in the form (5.5), and by integrating over field configurations, we do not have to project onto the allowed physical states in the trace (5.5). The spatial integration in the exponent of Eq. (5.8) is performed over the entire volume V of the system; in the thermodynamic limit it becomes infinite. The time component x_0 is "rotated" to become pure imaginary, $\tau = ix_0$, thus turning the Minkowski manifold, on which the fields A and ψ are originally defined, into a Euclidean space. The integration over τ in Eq. (5.8) runs over a finite slice whose thickness is determined by the temperature $T = \beta^{-1}$ of the system. The finite temperature behavior of the partition function in the Euclidean form thus becomes a finite size effect in the imaginary time direction. In the thermodynamic limit (infinite volume), the integration over space and temperature extends to infinity only for $T \to 0$. Equation (5.8) is derived from the trace form (5.5). As a consequence, vector fields have to be periodic and spinor fields antiperiodic at the boundaries of the imaginary time integration.

The idea of temperature as imaginary time might seem curious at first sight. Feynman's basic idea was that the time evolution of, e.g., a single quantum mechanical spin variable $s = \pm 1$,

$$\ldots, +1, +1, -1, +1, -1, -1, -1, +1, +1, -1, \ldots$$

has the same structure as a one-dimensional linear chain of spins at some temperature β^{-1}. The time evolution of the single spin is governed by the factor $\exp\{-iHt\}$, where H denotes the relevant Hamiltonian, while the temperature dependence is given by $\exp\{-\beta H\}$. This suggests the relation $\beta = it$ between inverse temperature and imaginary time.

Next, the Euclidean $(\mathbf{x}, \tau)$ manifold is replaced by a discrete lattice, with N_σ points and spacing a_σ in each space direction, and with N_τ points and spacing a_τ for the τ axis. The overall space volume thus becomes $V = (N_\sigma a_\sigma)^3$ and the inverse temperature $\beta = N_\tau a_\tau$. The spinor quark fields ψ and $\bar\psi$ are defined on each of the $N_\sigma^3 N_\tau$ lattice sites. To assure the gauge invariance of the formulation, the gauge fields A must, however, be defined on the links connecting each pair of adjacent lattice sites; we shall return to this point in just a moment. Clearly the lattice thus introduced can only be an intermediate step; the results we really want are those of continuum QCD. In the end we will thus have to extrapolate all lattice results to the *continuum limit* defined by $N_\sigma \to \infty$, $a_\sigma \to 0$ and $N_\tau \to \infty$, $a_\tau \to 0$, with V and β held constant.

In the third step, the gluon fields are replaced, through a "change of variables", by the corresponding gauge group variables,

$$U_{ij} = \exp -[ig(x_i - x_j)^\mu A_\mu (\frac{x_i + x_j}{2})], \tag{5.10}$$

with x_i and x_j denoting two adjacent lattice sites; thus U_{ij} is an SU(3) matrix assigned to the link between these two sites. The integration over the fields A now becomes one over U, i.e., over the eight real "Euler" angles needed to specify an SU(3) matrix.

After these three steps, the QCD partition function on the lattice is given by

$$Z(N_\sigma, N_\tau; g^2) = \int \prod_{\text{sites}} d\psi \, d\bar\psi \prod_{\text{links}} dU_{ij} \, \exp\{-S(\psi, \bar\psi, U)\}. \tag{5.11}$$

The action S in Eq. (5.11) has the form

$$S = S_G + S_Q, \tag{5.12}$$

where S_G describes the purely gluonic part (the integral over the first term of Eq. (5.1)), while S_Q comes from the interaction of quarks and gluons (the integral over the second term of Eq. (5.1)). Let us look at each in some detail.

The gluon action S_G corresponds to the pure gauge field term $F^a_{\mu\nu} F^{\mu\nu}_a$ in the Lagrangian density (5.1). Wilson's form for S_G is

$$S_G = \frac{6}{g_\sigma^2} \frac{a_\tau}{a_\sigma} \sum_{P_\sigma} (1 - \frac{1}{3} \, Re \, Tr \, U_{ij} U_{jk} U_{kl}^+ U_{li}^+) \tag{5.13}$$

$$+ \frac{6}{g_\tau^2} \frac{a_\sigma}{a_\tau} \sum_{P_\tau} (1 - \frac{1}{3} \, Re \, Tr \, U_{ij} U_{jk} U_{kl}^+ U_{li}^+). \tag{5.14}$$

It contains two distinct coupling parameters g_τ and g_σ; these are necessary in order to treat the spatial and temporal lattice spacings a_τ and a_σ as independent variables – which we have to do to differentiate with respect to temperature and volume, as in Eqs. (5.7) and (5.6). Once this differentiation is carried out, we can set $a_\tau = a_\sigma \equiv a$;

we then also recover one "isotropic" coupling $g_\tau = g_\sigma \equiv g$. As already mentioned, the color $SU(3)$ matrices U_{ij} "live" on links connecting the lattice sites i and j. The products occurring in Eq. (5.14) correspond to those of matrices on the smallest closed path on the lattice.

The form (5.14) of the gluonic action actually corresponds to that of a gauge invariant Ising model [12]. In the usual Ising model, the Hamiltonian is a sum over the interactions of spins on nearest-neighbor sites (i, j),

$$\mathcal{H}_I = -J \sum_{(i,j)} s_i s_j, \tag{5.15}$$

where $s_i = \pm 1$ and J measures the strength of the interaction. H_I is invariant under the *global* transformation $s_i \to -s_i$ for all i; it does not remain so under the gauge-like *local* transformation $s_i \to -s_i$ for only one fixed i, leaving all other s_i unchanged. However, if the spins are associated to the links between sites, rather than to sites, then the product of the four spins around a closed path containing site i does remain invariant if we flip the spins on all links touching site i, since two spins in the product change signs (Fig. 5.2). In this way, Wegner formulated a gauge invariant form for the Ising model [12]: spins live on links instead of sites, and next-neighbor products $s_i s_j$ are replaced by the products $s_i s_j s_k s_l$ of the four spins on the links around the smallest closed path on the lattice,

$$\mathcal{H}_W = -J \sum_{(i,j,k,l)} s_i s_j s_k s_l. \tag{5.16}$$

The smallest loops on the lattice are generally called "plaquettes". In Eq. (5.14), the spin variables s_i are generalized to the SU(3) color group matrices U, and hence a plaquette now is expressed in terms of four such matrices associated to adjacent links on the lattice. In addition, the lattice has now become four-dimensional, and so the two terms in Eq. (5.14) correspond to summations over plaquettes in space-space and space-time planes, respectively. Just as the action (5.16) is invariant under local spin flips, the form (5.14) is invariant under gauge transformations $U_{ij} \to V_i U_{ij} V_j^+$, with $V \in SU(3)$.

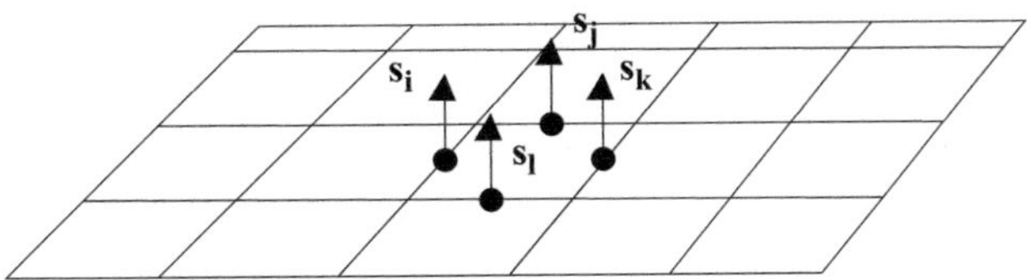

Fig. 5.2 Spins on links forming a plaquette

The quark action S_Q has the schematic form

$$S_Q = \sum_f \bar{\psi}_f Q_f(U) \psi_f, \tag{5.17}$$

where Q is a matrix connecting the quark fields $\psi(x)$ and $\bar{\psi}(x)$ over all lattice points x; it therefore is of size $N_c \times N_f[N_\tau N_\sigma \times N_\tau N_\sigma]$. The matrix Q has to be specified in terms of the $SU(3)$ color group elements $U_{i,j}$ living on the lattice links connecting the sites on which the quark fields are defined. One possible form for Q was proposed by Wilson; it is given by

$$S_Q = \sum_f \bar{\psi}_f (1 - \kappa M) \psi_f, \tag{5.18}$$

where the interaction matrix M coupling quarks through gluons depends on the direction of the link on which the gluon "lives",

$$M_{\mu,nm} = (1 - \gamma_\mu)U_{nm}\delta_{n,m-\hat{\mu}} + (1 + \gamma_\mu)U^+_{mn}\delta_{n,m+\hat{\mu}} . \tag{5.19}$$

Here $\hat{\mu}$ is a unit vector along the lattice link in the μ-direction. At finite temperature, the quark-gluon coupling strength κ, the so-called "hopping parameter", also depends on the link direction, just as the gluon coupling g in Eq. (5.14). The expression κM is a short-hand notation for

$$\kappa M \equiv \kappa_\tau M_0 + \kappa_\sigma \sum_{\mu=1}^{3} M_\mu . \tag{5.20}$$

As before, the hopping parameter also reduces to one variable for $a_\tau = a_\sigma = a$: $\kappa_\tau = \kappa_\sigma \equiv \kappa$. The basic Lagrangian in Eq. (5.1) contains only one coupling g; the apparence of a separate coupling κ in Eq. (4.15) is a consequence of the quark mass. Thus for $m_q \to 0$, κ must be expressible in terms of g. For massless quarks we have in fact [13]

$$\kappa(g) = \frac{1}{8}[1 + 0.11g^2 + O(g^4)] \tag{5.21}$$

in the limit of small g. At the finite values of g in actual calculations, the relation may be more complex, however.

The form (5.17), inserted into Eq. (5.11), gives us something like an exponential quadratic form in the fields ψ and $\bar{\psi}$; the simplicity is disturbed "only" by the fact that these fields are non-commuting Grassmann variables. Nevertheless, as fourth and final step in obtaining the lattice formulation of QCD, the integration can be carried out [14]; for N_f quarks of equal mass this results in the partition function

$$Z(N_\sigma, N_\tau; g^2) = \int \prod_{\text{links}} dU[\det Q(U)]^{N_f} \exp\{-S_G(U)\}. \tag{5.22}$$

We recall that $Q(U)$ connects the fields ψ and $\bar{\psi}$ over the whole lattice, so that it is of dimension $N_c[(N_\sigma^3 N_\tau) \times (N_\sigma^3 N_\tau)]$. The evaluation of the determinant of such a large matrix turns out to be one of the main technical problems in the numerical evaluation of lattice QCD. Exponentiating the determinant, we obtain

$$Z(N_\sigma, N_\tau; g^2) = \int \prod_{\text{links}} dU \exp\{-[S_G(U) + \tilde{S}_Q(U)]\}, \tag{5.23}$$

where $\tilde{S}_Q = -N_f\{\ln(\det Q)\}$ corresponds to an effective quark action defined in terms of the fermion matrix Q (see Eq. (5.22)); this presupposes that $\det Q$ is real and positive, which is generally the case for systems containing quarks and antiquarks in equal numbers, i.e., for vanishing baryon number density.

With Eqs. (5.14), (5.18) and (5.22), (5.23), we have a completely defined lattice formulation of the QCD partition function. It gives us $Z(N_\tau, N_\sigma, a_\tau; a_\sigma, g_\tau, g_\sigma; \kappa_\tau, \kappa_\sigma)$, but not all variables are independent. To obtain the desired physical partition function $Z(\beta, V)$ and from this the resulting thermodynamic observables, we first carry out the differentiations such as needed in Eqs. (5.7) and (5.6), and then set $a_\tau = a_\sigma = a$, $g_\tau = g_\sigma = g$, $\kappa_\tau = \kappa_\sigma = \kappa$. We must now relate the "isotropic" coupling g to the "isotropic" lattice spacing a; then $V = (N_\sigma a)^3$ and $\beta^{-1} = N_\tau a$ give us $Z(\beta, V)$ from $Z(N_\tau, N_\sigma, g, \kappa)$. The necessary relation between coupling and lattice spacing is obtained through the following reasoning. We want our lattice formulation to provide results which are independent of the specific lattice used in the evaluation; in particular, they should not depend on lattice size and lattice spacing. To achieve this, we must change the coupling strength g when we change the lattice spacing a. In general, renormalisation group theory assures us that we have lattice independence around $g = 0$, provided g and a are related through the Callan-Szymanzik equation

$$a\frac{dg(a)}{da} = \beta(g), \tag{5.24}$$

where $\beta(g)$ is a function of g only. In the limit $g \to 0$, $\beta(g)$ can be determined perturbatively, and the Callan-Szymanzik equation can then be solved. The result is $a\Lambda_L = R_0(g)$, with

$$R_0(g) \equiv \exp\left\{-\frac{4\pi^2}{(33 - 2N_f)}\left(\frac{6}{g^2}\right) + \frac{(459 - 57N_f)}{(33 - 2N_f)^2}\ln\left[\frac{8\pi^2}{(33 - 2N_f)}\left(\frac{6}{g^2}\right)\right]\right\}; \tag{5.25}$$

here Λ_L is an integration constant of dimension $(\text{length})^{-1}$. In quantitative studies, we must make sure that at the coupling values used, this solution (for $g \to 0$) is already justified; otherwise we have to resort to non-perturbative methods to determine $\beta(g)$ [15]. Once the relation between a and g is established, we have the physical partition function $Z(\beta, V)$ in terms of temperature $T = \beta^{-1} = 1/(N_\tau a)$ and volume $V = (N_\sigma a)^3$ from the lattice form $Z(N_\tau, N_\sigma, g)$ for given N_τ, N_σ, and g, with $\kappa(g)$ determined by Eq. (5.21).

All physical quantities calculated in this way are given in units of the lattice scale Λ_L. As already mentioned, the starting Lagrangian (5.1) contains no dimensional parameter, and hence Λ_L is arbitrary. We can thus either calculate dimensionless ratios of observables, so that the Λ_L's cancel in numerator and denominator, or calculate a measured quantity, such as the proton mass or the critical temperature for deconfinement, and then use this to obtain a scale Λ_L in physical units.

To illustrate this procedure, we note that the inversion of Eq. (5.25) gives in leading order the customary form

$$\alpha_s(T) \equiv \frac{g^2}{4\pi} \simeq \frac{6\pi}{(33 - 2N_f) \ln(1/a\Lambda_L)} = \frac{6\pi}{(33 - 2N_f)[\ln(T/T_c) + \ln(T_c/\Lambda_T)]} \tag{5.26}$$

of the "running" strong interaction coupling $\alpha_s(T)$ as function of temperature. Here we fix the scale factor $\Lambda_T = \Lambda_L/N_\tau$ by determining in a given lattice calculation the critical temperature $T_c = 1/N_\tau a$ in lattice units and then insert its value in physical units.

To what extent is the lattice formulation just described equivalent to the continuum form of statistical QCD? By letting $a = (x_i - x_j)$ go to zero everywhere, we do indeed recover the continuum formulation. The converse is not true, however; neither the gluon action S_G in Eq. (5.14) nor the quark action S_Q in Eq. (4.15) are unique. Various other forms have been considered, and they all give the same continuum limit. In fact, we can use this freedom and try to find forms for which finite cut-off effects are particularly small ("improved actions" [16]). All physical results should of course be independent of the specific choice of action, and finite temperature thermodynamics provides a particularly sensitive test of this "universality". So far, it appears to be well satisfied.

The quark action leads to some additional problems. If we simply put the quarks on the lattice by associating a spinor field ψ with each lattice site, then the derivative in the Lagrangian (4.1) leads to the appearance of 16 degenerate fermions per flavor [17, 18]. To avoid this "species doubling", the quark action (5.19) is chosen such as to give 15 of these quarks a mass m which becomes infinite in the continuum limit $g \to 0$, so that they then no longer contribute. Such a procedure, however, has more than just esthetic difficulties. The continuum Lagrangian (4.1) with massless quarks is invariant under chiral transformations: the four-spinors describing massless fermions can be decomposed into two independent left-handed and right-handed two-spinors. This chiral symmetry is explicitly broken by Wilson's lattice formulation (5.18), to be recovered only in the continuum limit. It can moreover be shown [19] that any lattice formulation of a theory with quarks leads either to species doubling or to non-local derivatives, if it is to preserve chiral symmetry. To avoid species doubling in a local theory, one thus has to break chiral symmetry at finite lattice spacing. Because of these difficulties, the choice of action is to some extent governed by the problem under consideration and by technical features of the evaluation. An often employed alternative to the Wilson form uses "staggered fermions" [18]. Here the four components of the spinor are distributed over different adjacent sites of a hypercube on the lattice, which reduces the number of additional species. Moreover, it retains at least part of the chiral symmetry of the continuum Lagrangian (4.1). After integrating out the spinor fields, the Kogut-Susskind formulation also leads to the partition function (5.22), but with a different form for the fermion determinant. More recently, still other forms of the quark action have been considered [20–22]. This of course makes it all the more important to check that different formulations lead to the same physical predictions.

5.3   Lattice QCD at Finite Baryon Number Density

In Eq. (5.8), we had expressed the QCD partition function as a functional integral

$$Z_E(\beta, V) = \int \mathcal{D}A \, \mathcal{D}\psi \, \mathcal{D}\bar{\psi} \, \exp\{-S(A, \psi, \bar{\psi}, \beta)\} \qquad (5.27)$$

with a weight determined by the QCD action

$$S(A, \psi, \bar{\psi}, \beta) = -\int_V d^3x \int_0^\beta d\tau \, \mathcal{L}(A, \psi, \bar{\psi}). \qquad (5.28)$$

The (inverse) temperature β thus enters as the limit for the integration over the imaginary time. This formulation corresponds to a grand canonical system of vanishing overall charge and baryon number density, i.e., both electric and baryonic charges of the quarks add up to zero. If we want to consider a system of non-zero baryon number density, in which the total number of quarks minus that of antiquarks does not vanish, the Lagrangian $\mathcal{L}$ in Eq. (5.28) has to be modified correspondingly. In statistical mechanics, the Boltzmann factor $\exp -\beta \mathcal{H}$ becomes

$$\exp\{-\beta \, (\mathcal{H} - \mu \mathcal{N})\}, \qquad (5.29)$$

where $\mathcal{H}$ is the Hamiltonian and $\mathcal{N}$ the net conserved charge number. For our case, this means that the action (5.28) now becomes

$$S(A, \psi, \bar{\psi}, \beta) = -\int_V d^3x \int_0^\beta d\tau \, \{\mathcal{L}(A, \psi, \bar{\psi}) - \mu \mathcal{N}_b(\psi, \bar{\psi})\} \,. \qquad (5.30)$$

with

$$\mathcal{N}_b = \bar{\psi} \gamma_0 \psi \qquad (5.31)$$

denoting the overall baryon number density (quarks minus antiquarks) and μ the associated baryochemical potential. In Hamiltonian statistical mechanics, temperature and chemical potential enter in a very similar fashion, essentially as Lagrangian multipliers assuring that on the average energy and charge quantum number are conserved. In contrast, in the functional integral formulation this "symmetry" is removed, with the temperature becoming an integration limit. We will see below that this feature leads to serious difficulties for the numerical evaluation of all observables at non-zero μ.

5.4   The Computer Simulation of Gauge Field Thermodynamics

With Eq. (5.23), we have *formulated* the QCD partition function on the lattice. How can we now actually *evaluate* this function and calculate the physical observables of interest? To introduce the basic idea of the evaluation procedure, computer simulation [5, 6], let us first look at a simplified case, the thermodynamics of pure gauge theory. We thus consider a system of gluons only, without any real or virtual quarks. Since gluons can interact directly, they can form bound states ("glueballs" or gluonium states) and undergo a transition from a gluonium gas to one of deconfined gluons. This simplified world is therefore far from trivial – in contrast to the corresponding system in QED, a gas of photons, which in the absence of electrons cannot interact. As we shall see later on, pure gauge field thermodynamics in fact provides a particularly transparent illustration of deconfinement physics.

Without quarks, the partition function (5.11) on the lattice becomes

$$Z(N_\sigma, N_\tau; a_\tau/a_\sigma; g_\sigma^2, g_\tau^2) \;=\; \int \prod_{\text{links}} dU_{ij} \, \exp\{-S_G(U)\}, \tag{5.32}$$

with $S_G(U)$ given by Eq. (5.14). To calculate from this the energy density (5.7) on a lattice of fixed size, we have to differentiate $\ln Z$ with respect to $\beta \sim a_\tau$. This leads to

$$\epsilon/T^4 \;=\; 18 N_\tau^4 \, [g^{-2}(\bar{P}_\sigma - \bar{P}_\tau) + c'_\sigma(\bar{P} - \bar{P}_\sigma) + c'_\tau(\bar{P} - \bar{P}_\tau)]. \tag{5.33}$$

The corresponding form for the pressure is obtained by differentiation with respect to $V \sim a_\sigma^3$ and yields

$$P/T^4 = \epsilon/3T^4 - 6 N_\tau^4 a \left(\frac{dg^{-2}}{da} \right) [\bar{P}_\tau + \bar{P}_\sigma - 2\bar{P}]. \tag{5.34}$$

Here $\bar{P}_\sigma$ and $\bar{P}_\tau$ denote the lattice averages of space-space and space-time plaquettes, i.e.,

$$\bar{P}_\sigma \;=\; (3 N_\sigma^3 N_\tau Z)^{-1} \int \prod_{\text{links}} dU_{ij} \exp[-S_G(U)] \left[\sum_{P_\sigma} (1 - \frac{1}{N_c} \, Re \, Tr \, UUU^+U^+) \right],$$
$$\tag{5.35}$$

and similarly for $\bar{P}_\tau$. The anisotropy constants c'_σ and c'_τ in Eq. (5.33) arise from the differentiation of the couplings g_σ and g_τ with respect to a_τ. In the perturbative limit, they can be calculated explicitly [23]. The energy density (5.33) is normalized to zero at $T = 0$ by subtracting from the lattice derivative of $\ln Z(N_\sigma, N_\tau)$ the zero temperature form obtained from $\ln Z(N_\sigma, N_\tau = N_\sigma)$. This leads in Eqs. (5.33) and (5.34) to the "symmetric" plaquette averages $\bar{P}$, calculated on a N_σ^4 lattice, which for large N_σ is a good approximation of the zero temperature case. Since ϵ/T^4 is entirely given in terms of *differences* of plaquette averages, it is clear that it vanishes as $T \to 0$, where $N_\sigma = N_\tau$ makes all plaquette averages equal.

After performing the differentiations necessary to obtain ϵ, we set $a_\sigma = a_\tau$, which in turn gives $g_\sigma = g_\tau = g$, and with this the forms shown in Eqs. (5.33) and (5.34). As noted before, the energy density is thus entirely determined in terms of plaquette averages, i.e., of lattice averages of four adjacent gauge group elements U around the smallest closed loop in the lattice. This is the gauge theory equivalent of the conventional Ising model, where the energy density involves the lattice average of the product of two adjacent spins. In both cases, the energy density is thus determined in terms of *local* quantities – lattice averages of products of two or four *adjacent* lattice variables. Such short-distance observables are therefore much easier to calculate than potentials or correlation functions, which involve the behavior of $r \to \infty$.

To evaluate the energy density (and other such thermodynamic observables) in gauge field thermodynamics, we now have to find a way to calculate the plaquette averages. This problem is of course structurally very similar to the evaluation of corresponding quantities in the study of spin systems; let us therefore return once more to the Ising model to have a first look at the idea of computer simulation, and then consider how it can be applied to gauge field theory.

In the Ising model, the energy of the system for a given configuration is given by the Hamiltonian (5.15). We want to calculate lattice averages of thermodynamic observables of the type

$$\langle O \rangle = \frac{\sum_{\{s_i = \pm 1\}} \exp[-\beta H_I]\, O(\{s_i\})}{\sum_{\{s_i = \pm 1\}} \exp[-\beta H_I]}, \tag{5.36}$$

with $\beta = 1/T$. The summations here run over all configurations, i.e., over all sets of N spin values, such as $s_1 = +1, s_2 = +1, s_3 = -1, \ldots, s_N = -1$. Defining the configuration probability $P(\{s_i\})$,

$$P(\{s_i\}) \equiv \frac{\exp[-\beta H_I]}{\sum_{\{s_i\}} \exp[-\beta H_I]}, \tag{5.37}$$

we can rewrite this as

$$\langle O \rangle = \sum_{\{s_i\}} P(\{s_i\})\, O(\{s_i\}) \tag{5.38}$$

and thus consider it as an average of $O(\{s_i\})$ over configurations weighted with the probabilities $0 < P(\{s_i\}) < 1$. The actual evaluation by Monte Carlo computer simulation is now carried out with the help of the so-called Metropolis algorithm [24]. We create on the computer a specific configuration $\{s_i\}$ on a lattice of N^3 sites, e.g., $s_i = 1\ \forall\ i$. Starting from this configuration, we go through the entire lattice site by site. Since the spin at a given site can take on two equally possible values, we generate at each i a random number $0 < n \leq 1$, and if $n \leq 0.5$, we retain the spin assignment s_i at that site; otherwise, we flip $s_i \rightarrow -s_i$. If this procedure leads to a new configuration $\{s_i'\}$, we ask whether the ratio of configuration probabilities $P(\{s_i'\})/P(\{s_i\})$ is greater or smaller than a randomly chosen number $0 \leq r \leq 1$. If it is greater, we discard the old configuration in favour of the new one; if it is smaller, we retain the old one. Applying this procedure once to all lattice sites, i.e., taking one "trip" through the whole lattice, is generally called a "sweep" or an "iteration". After sufficiently many iterations, we attain "equilibrium": further iterations do not lead to any systematic changes, neither in the average of $P(\{s_i\})$ over the lattice, nor in that of any observable we wish to "measure". At this point, we generate as large as possible a number of "equilibrium" configurations, and these (not the "transient" configurations leading to equilibrium) we use to calculate, or rather measure, the observables of interest. Starting from the given microscopic dynamics, Monte Carlo simulation thus consists of three steps:

- from some given initial configuration, we randomly generate new configurations until $P(\{s_i\})$ is maximized, i.e., equilibrium is attained;
- we now generate a large number of different equilibrium configurations of the system;
- we then calculate the value of any observable of interest on each of these configurations and average over all configurations.

Note that the third step in general means first calculating the lattice average (e.g., of a plaquette) on a given configuration and then averaging this over all available equilibrium configurations. In other words, we create a stationary world according to the given dynamics, and on the different possible configurations of this world we "measure" our observables, just as an experimentalist measures them in the "real" world. If the underlying dynamics is correct, the two worlds, and thus also the two sets of results for the observables, must agree.

It is quite straightforward now to extend this method from the Ising model to gauge field theory. The variables which determine a given configuration now are the $SU(3)$ gauge group matrices $U_{i,j}$, instead of the spins in the Ising model, and they "live" on links between two adjacent sites i, j, and not on the sites themselves. In the Ising model, to average over all configurations meant summing over $s_i = \pm 1$ for all s_i; in $SU(3)$ gauge theory, we have to "sum" over all the possible matrices $U_{i,j}$ – which here means integrating over the eight Euler angles necessary to characterize such a matrix. Thus Eq. (5.37) is now replaced by

$$\langle O \rangle = \int \prod_{\text{links}} dU_{i,j}\, P(\{U_{i,j}\})\, O(\{U_{i,j}\}), \qquad (5.39)$$

where

$$P(\{U_{i,j}\}) \equiv \frac{\exp[-S_G(\{U_{i,j}\})]}{\int \prod_{\text{links}} dU_{i,j} \, \exp[-S_G(\{U_{i,j}\})]} \tag{5.40}$$

and $dU_{i,j}$ denotes the mentioned integration (including the correct weight measure) over the eight angles specifying the matrix $U_{i,j}$.

To carry out such a Monte Carlo procedure, we again start from a given configuration – for example, $U_{i,j} = \mathbf{1} \; \forall i,j$, and then proceed with our trip through the lattice: at each link, we pick at random new values for the eight angles and retain them if $P(\{U'\})/P(\{U\})$ is larger than some random number between 0 and 1; otherwise we discard them. After attaining equilibrium, we again generate a large number of configurations to be used for the actual "measurement" of the observables of interest.

The procedure we have just sketched remains the same also for full QCD, including quarks. In this case, however, we have to evaluate for each configuration at each lattice point the determinant of the fermion matrix, whose size is proportional to that of the entire lattice. This of course complicates the calculations considerably, and so many studies began with simplified cases, such as pure gauge theory thermodynamics (no quarks) or QCD in the "quenched" (or "valence quark") approximation. In the latter, one neglects the fermion determinant, setting $det \, Q = 1$, but retains any quark variables in the observable operator O. It can be shown that this corresponds to all virtual quark contributions. For many aspects, such simpler versions already give the correct qualitative picture; but to reach this conclusion, we need the results of full QCD with dynamical quarks. Moreover, there are features which do depend on the presence of virtual quarks, in particular the value of the threshold temperature for deconfinement. So the ultimate goal will always be the large-scale study of thermodynamics based on full QCD.

Before turning to the results so far obtained in the computer simulation of QCD thermodynamics, we have to point out one rather serious technical limitation of the method. We had seen that a crucial element in getting to equilibrium lattice configurations is the existence of a configuration probability $0 < P(U) < 1$ as a weight to determine if the new configuration is closer to equilibrium than the old. In pure gauge theory, $P(U) \sim \exp\{-S_G(U)\}$ was determined by the gluonic action $S_G(U)$. In full QCD, we have instead $P(U) \sim \exp\{-[S_G(U) + S_Q(U)]\}$, where $S_Q = -N_f \ln(det \, Q)\}$ is the exponentiated fermion determinant in Eq. (5.22). This exponentiation makes sense only if the determinant is positive, and unfortunately this is true only in the case of vanishing overall baryon number density, i.e., for systems containing an equal number of quarks and antiquarks. For non-zero baryon number density, or equivalently, for non-vanishing baryonic chemical potential μ, the determinant becomes complex [25, 26], and although the integral with the imaginary part as weight vanishes, the remaining real part is no longer positive-definite. The standard Metropolis procedure for getting to equilibrium from some given starting configuration therefore breaks down: we do not have a weight any more to tell us which way to evolve. Clearly this is only a technical difficulty –

it is not the simulation method as such which breaks down, but only the specific procedure for producing equilibrium configurations. Hence there were numerous attempts to overcome this difficulty, but only very recently first results on QCD thermodynamics at finite baryon density have been put forward [27–30], and it is not really clear yet how to provide an error estimate of the method nor specify out to what baryon density the results can be trusted. We will later summarize the present state of these studies, but it should be kept in mind that all results so far remain preliminary.

5.5 The Deconfinement of Quarks and Gluons

In this section, we summarize the present status of deconfinement in QCD thermodynamics at vanishing baryon number density. As in the previous section, we shall start with pure gauge theory and then extend our considerations to full QCD.

5.5.1 SU(3) Gauge Theory

The first lattice results on gauge field thermodynamics appeared over 30 years ago; they actually considered a model simplified still further, SU(2) gauge theory. In this case, one has only three Euler angles to specify an SU(2) matrix, in contrast to the eight Euler angles of SU(3), and this reduced the computational effort sufficiently to allow calculations on the computers available at that time. Even then, the results were obtained on quite small lattices [31, 32]. Today, detailed studies provide us with the continuum limit for the equation of state for SU(3) gauge theory, obtained by extrapolating the results of calculations on different lattice sizes [33, 34]. In contrast to previous simulations, they also do not rely on the perturbative limit for the relation between lattice spacing a and coupling g. In place of the perturbative solution (5.25) of the Callan-Symanzik equation (5.24), one determines a more general $R(g)$ by requiring some calculated physical observable to remain invariant under changes of g in the actual range of couplings used. Together with the requirement $R(g) \rightarrow R_0(g)$ for $g \rightarrow 0$, this allows a numerical determination of $\beta(g)$ in the non-perturbative region. In Fig. 5.3, we show the behavior obtained for the energy density ϵ as function of the temperature. The calculations [33, 34] were performed on lattices of size $16^3 \times 4$ and $32^3 \times N_\tau$, with $N_\tau = 6$ and 8, and then extrapolated to the continuum limit $N_\sigma \rightarrow \infty$, $N_\tau \rightarrow \infty$. Looking at this energy density and remembering the patterns found in Chap. 3, we see quite clearly a sudden transition from a low temperature regime with a small number of degrees of freedom to a high temperature phase with many more degrees of freedom. The low-temperature system should here be a gas of quite massive glueballs, the high-temperature state one of deconfined gluons. Detailed studies show a two state signal at the transition point [35, 36], so that we have a first order transition. We define as critical temperature T_c the value of T at the discontinuity and use this value as scale in Fig. 5.3. In the continuum limit, the latent heat of the transition is $\Delta\epsilon \simeq 2\,T_c^4$, so that in spite of the rapid increase of ϵ, the system does not jump directly all the way up to the Stefan-Boltzmann limit,

$$\epsilon = \frac{8\pi^2}{15}T^4 \simeq 5.3\,T^4, \tag{5.41}$$

obtained from Eq. (4.13). At $T/T_c \simeq 1.5\text{--}2$, it has come to within some 15%–20% of this limit; but a further temperature increase leads only to a very slow increase of ϵ. Even at $T/T_c \simeq 5$, a discrepancy of nearly 10% remains. Recent studies [37] find that above this region, the thermodynamics approaches a limit which can be described through modified perturbative calculations; we will come back to this in Chap. 8.

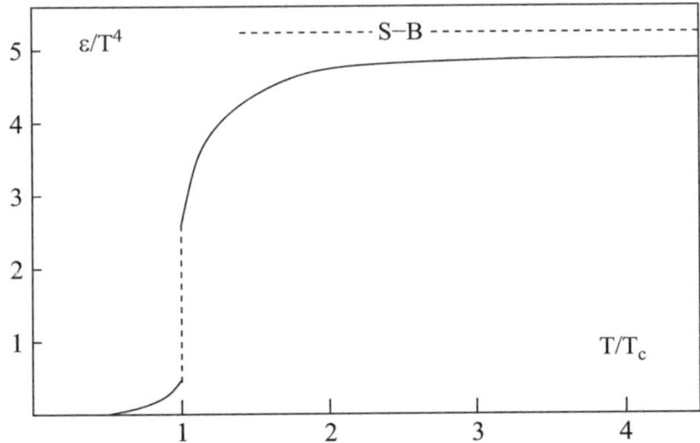

Fig. 5.3 Energy density and pressure in SU(3) gauge theory [33, 34]

We now return to the issue of color deconfinement. So far, we have identified the point of discontinuity in ϵ as the transition point for the corresponding phase change. The justification for this is that at T_c, the energy density quickly increases from a low hadronic value toward the number of degrees of freedom of an ideal gas of gluons of three colors and two spin orientations. The confinement status can also be tested in another way: by considering the dependence of the binding of a static quark-antiquark pair $Q\bar{Q}$ on the distance of separation [38–40]; by static we here mean that both quark and antiquark are very heavy. We imagine such a pair put into our SU(3) gluon gas and consider the correlation function

$$\Gamma(r,T) \equiv \exp\{-F_Q(r,T)/T\}, \tag{5.42}$$

where $F_Q(r, T)$ is the free energy of the pair at a separation r. At $T = 0$, this free energy is usually parametrized in the form [41]

$$F_Q(r, T = 0) = \sigma\, r - \frac{\alpha}{r},$$ (5.43)

where σ is the string tension and the second term takes into account short distance Coulomb-like interactions and large distance transverse string vibrations; for the latter, one has the universal form $\alpha = \pi/12$ [41, 42]. For $0 < T < T_c$, this form is expected to remain valid, but both the string tension σ and the Coulombic coupling α will now become temperature dependent. In particular, it is generally argued that $\sigma(T_c) = 0$ signals the onset of deconfinement. However, for $T > T_c$ it is in fact not quite correct to assume that $F(r, T)$ vanishes at large distances [43]. The free energy $F_Q(r, T)$ specifies the contribution of the two static charges introduced into the medium, i.e., it is the difference in free energy between a system with a static $Q\bar{Q}$ pair and one without it. But even above deconfinement, at some $T > T_c$, two static color charges separated arbitrarily far will still polarize the gluonic medium in their immediate surrounding, and this will make the system different from one without such charges. The result is an effective dressing $Q(r, T)$, whose value depends on the range of the polarization effect. Hence we extend Eq. (5.43) to finite temperatures as

$$F_Q(r, T) \simeq \sigma(T)r - \frac{\alpha(T)}{r} + 2Q(r, T).$$ (5.44)

The polarization term $Q(r, T)$ vanishes for small r, since the two static quarks then neutralize each other in color charge and are not seen by the medium. With increasing r, $Q(r, T)$ increases, until in the large distance limit it becomes the polarization cloud of a single quark. Given the form (5.44), it is now possible to consider $\sigma(T \geq T_c) = 0$. The resulting free energy for different temperatures is illustrated in Fig. 5.4. From Eqs. (5.42) and (5.44) we can conclude that

$$\Gamma(T) \equiv \lim_{r \to \infty} \Gamma(r, T)$$ (5.45)

will signal the onset of deconfinement in the form

$$\Gamma(T) = \begin{cases} 0 & \text{for } T \leq T_c, \\ \exp\{-2Q(T)/T\} > 0 & \text{for } T > T_c. \end{cases},$$ (5.46)

with $Q(T)$ defined as the large distance limit of $Q(r, T)$. We thus find that $\Gamma(T)$ constitutes a deconfinement order parameter in the sense of statistical mechanics, much like the magnetization $m(T)$ in the Ising model.

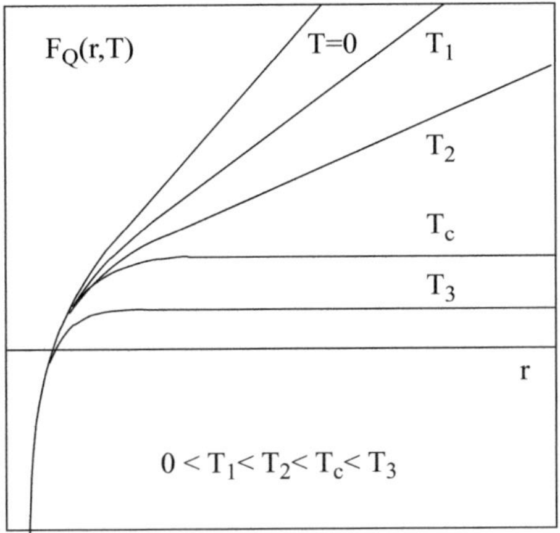

Fig. 5.4 Free energy of a static $Q\bar{Q}$ pair in SU(3) gauge theory

On the lattice, the correlation function $\Gamma(r, T)$ can be expressed [38–40] in terms of the so-called Polyakov loop,

$$L(r) \equiv \frac{1}{3} \operatorname{Tr} \prod_{\tau=1}^{N_\tau} U_{r,\tau}, \tag{5.47}$$

where r denotes some spatial lattice site on a time-plane labelled by τ. The product in Eq. (5.47) runs over N_τ matrices U on links along a time axis passing through the site r. In terms of the notation $U_{i,j}$ used above, $U_{r,\tau}$ thus is the matrix associated to the link from $x_i = (\vec{r}, \tau)$ to $x_j = (\vec{r}, \tau + 1)$. Because of the periodicity, the gluon fields at $\tau = 0$ and $\tau = \beta$ are equal, making $L(x)$ a closed loop and hence gauge invariant. It can be shown to correspond to a static quark at site x, so that

$$\Gamma(r, T) \sim |\langle L(0)L^+(r)\rangle| \tag{5.48}$$

leads to the desired correlation function. More precisely, $|\langle L(0)L^+(r)\rangle|$ specifies the free energy $F_Q(r, T)$ of a static quark-antiquark pair at temperature T and separation r, up to a lattice-size dependent renormalization $c(g, N_\tau)$,

$$-\ln|\langle L(0)L^+(r)\rangle| = c + \frac{F_Q(r, T)}{T}. \tag{5.49}$$

To determine $c(T)$, we note that at very short distances, for $rT \ll 1$, there will not be any temperature-dependent modifications of the interquark potential. In this region of r, the Coulomb term $\alpha(T = 0)/r$ will thus provide the dominant contribution to $V(r, T)$ in Eq. (5.44) and can be used to define a correctly normalized Polyakov loop [43].

We now want to study on the lattice the behavior of $|\langle L(0)L^+(r)\rangle|$ in the limit of large quark-antiquark separation, and hence define

$$\lim_{r\to\infty} |\langle L(0)L^+(r)\rangle| \equiv L(T)^2 \exp\{c(g,N_\tau)\} \tag{5.50}$$

to determine the onset of deconfinement. The renormalization by $\exp\{c(g,N_\tau)\}$ removes the lattice size dependence of the Polyakov loop expectation value. In Fig. 5.5 we show the behavior of the renormalized $L(T)$ in pure SU(3) gauge theory. It vanishes for $T \leq T_c$ and then becomes finite in a discontinuous fashion at the same temperature T_c at which $\epsilon(T)$ showed a jump, and it thus confirms that $T = T_c$ indeed signals the deconfinement point.

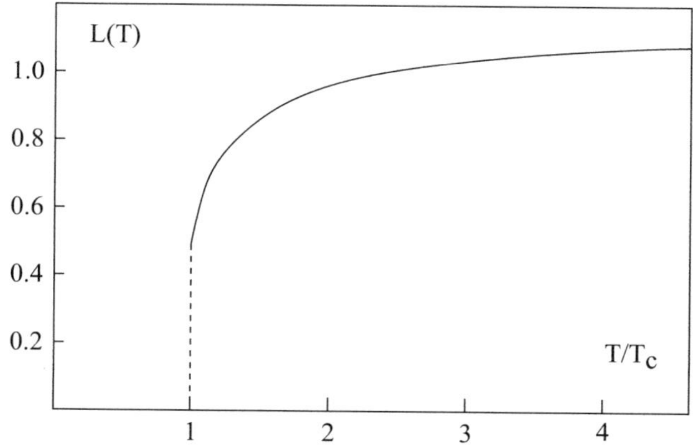

Fig. 5.5 Renormalized Polyakov loop in SU(3) gauge theory

We will return to the deconfinement transition and its relation to spontaneous symmetry breaking in the next chapter. But we do want to note the actual value of the transition temperature. As already mentioned, to this end we have to fix a lattice scale in physical units, by comparing some calculated observable to its measured value. Now pure $SU(3)$ gauge theory of course does not correspond to any real physical world; one therefore generally uses the string tension as scale, retaining its "physical" value. The deconfinement temperature in pure $SU(3)$ gauge theory is then found to be [44]

$$T_c = (0.640 \pm 0.015)\,\sqrt{\sigma}; \tag{5.51}$$

it is obtained by a detailed analysis using different forms of the lattice action and different lattice sizes to reach the continuum limit. If we use the value $\sqrt{\sigma} \simeq 470 \pm 10\,\text{MeV}$, as presently obtained in $(2+1)$ flavor QCD [48–51], this leads to a deconfinement temperature of about 300 MeV.

5.5.2   *Full QCD*

Let us now turn to the thermodynamics of full QCD, beginning with several caveats which are fortunately being removed with the improved performance of computing facilities. As already indicated, we have to choose a specific lattice formulation for the quarks, and each form has its problems [19]. Today, one has quite extensive calculations using different lattice actions, and the agreement of the results provides a good estimate of their precision. A further problem of the computer simulation is that it requires non-vanishing quark masses, and the smaller these masses are, the longer the simulation takes. Early calculations had therefore used quite large u and d quark masses. More recently, calculations with physical quark masses were carried out [45], using one heavier (s) and two light (u, d) quarks, for a variety of lattice sizes, so that the continuum limit could be obtained. In Fig. 5.6 we show the results for the energy density, the entropy density and the pressure. We also indicate the present value for the transition temperature T_c; we shall return shortly to the determination of T_c used here. Figure 5.6 shows that again in a very narrow temperature interval there is a rapid increase from the low values of hadronic matter to much higher values approaching the ideal quark-gluon plasma limit. Deconfinement thus remains clearly evident also in full QCD.

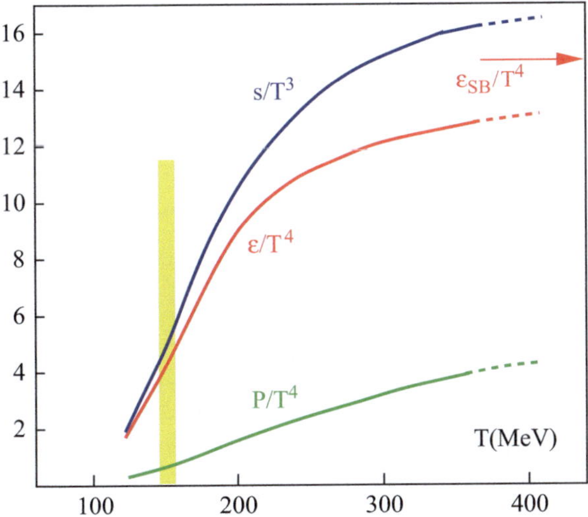

Fig. 5.6 Energy density, entropy density and pressure in full QCD with $(2+1)$ physical quarks [45]. The ideal gas energy density is indicated by ϵ/T^4

As already mentioned, in initial studies the masses of the light quarks were not as low as required for a correct pion mass. At present, there are several evaluations of full QCD based on one heavy and two light quark species, and in these studies, the light quark mass is tuned such as to approach the correct pion mass. Such studies can then begin to address the two essential questions: how should the critical point for deconfinement be defined, and what is its actual value in physical units?

In the case of pure gauge theory, we had established deconfinement as a genuine phase transition, with $L(T)$ as order parameter vanishing for $T \leq T_c$ and non-zero above T_c. For this, it was crucial that below T_c the potential between two static quarks diverged in the large distance limit, causing $L(T)$ to vanish. In full QCD, in the presence of light quarks, this is no longer the case. Once the free energy $F_Q(r)$ become sufficient to excite a quark-antiquark pair from the vacuum, it becomes energetically favorable to produce such a pair and have the constituent quark combine with the static antiquark, the constituent antiquark with the static quark. The result is that we now have two "light-heavy" mesons, something like a $D\bar{D}$ or $B\bar{B}$ system, and these mesons can be separated arbitrarily far without any energy expenditure. Taking the light quarks to be massless, the required energy is that needed to "dress" both the light and the heavey quarks, i.e., about $300\,\mathrm{MeV}$ per quark. In other words, for $F_Q(r) \simeq 4M_q \simeq 1.2\,\mathrm{GeV}$, the string connecting the static quark to the static antiquark "breaks". This occurs already for a static pair in vacuum, and hence becomes even easier at finite temperature. As a result, the free energy $F(r, T)$ now has the form shown in Fig. 5.7.

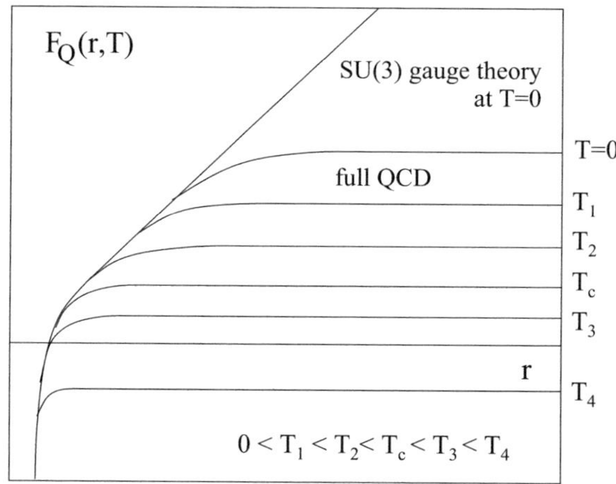

Fig. 5.7 Free energy of a static $Q\bar{Q}$ pair in full QCD

Since in full QCD, string breaking prevents $F_Q(r, T)$ from diverging for $r \to \infty$ at any temperature, $L(T)$ does not vanish below T_c. It continues, however, to vary sharply (even discontinuously for $N_f \geq 3$ and vanishing quark masses) at the inflection point of the energy density, even though it now remains finite for $T < T_c$. A rough estimate of the behavior in the confinement regime can be obtained by setting $F_Q(r, T) \simeq 4M_q$ in the Polyakov loop,

$$L^2(T) \simeq \exp -\{4M_q/T\}, \quad T \leq T_c, \tag{5.52}$$

instead of having it vanish there, as in pure gauge theory. In the deconfinement region, there remain only the polarization effects around the two static quarks, causing a sharp increase of $L(T)$ as we approach the transition point. We thus get the schematic pattern shown in Fig. 5.8. The resulting susceptibility, i.e., the temperature derivative of $L(T)$, shows a pronounced peak at a "critical" temperature T_c. Taking into account the "inversion" of the temperature variable between spin and gauge theories, the behavior of $L(T)$ here is quite similar to that found in the Ising model in the presence of a small magnetic field.

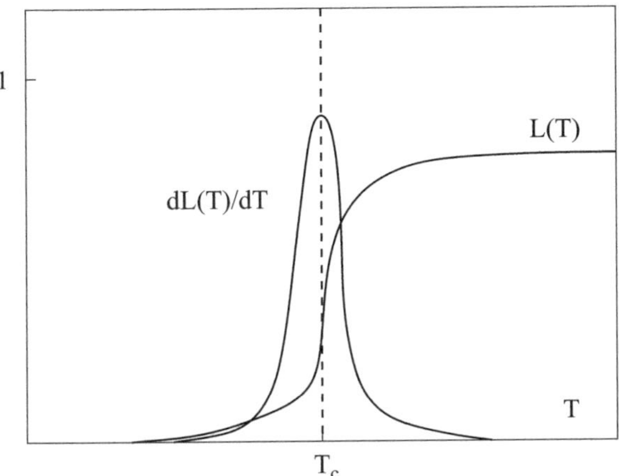

Fig. 5.8 Schematic view of the Polyakov loop and its temperature derivative in full QCD

For quite some time, deconfinement and the corresponding critical or pseudo-critical temperature were in fact defined in terms of the position of the peak in the temperature derivative of $L(T)$; the value of T_c used in Fig. 5.6 was determined this way. In some cases, such a definition is evidently correct: thus, for a first order transition at $T = T_c$, all observables show some form of discontinuous behavior at that temperature. However, we shall see in the next chapter that deconfinement is more naturally specified through the temperature variation of the energy density, i.e., the specific heat $C_V(T) = (\partial\epsilon/\partial T)_V$. It can be written as

$$\frac{C_v(T)}{T^3} = 4\frac{\epsilon(T)}{T^4} + T\left(\frac{\partial(\epsilon/T^4)}{\partial T}\right)_v ; \tag{5.53}$$

in the high temperature limit, we thus get $C_v(T)/T^3 \to 4\epsilon(T)/T^4$. In Fig. 5.9, we show a schematic view of recent results for the energy density in the (2+1) flavor case, based on almost physical quark masses ($m_q/m_s = 0.05$ for the ratio of light to strange quark) [46, 47]. As already indicated, finite temperature lattice studies require a further calculation to fix the result in physical units. One can thus calculate a specific hadron mass in lattice units and fix the latter through the experimental mass value. At present, one generally uses the mass splitting in heavy quark bound states (charmonia or bottomonia) for this purpose, since it is very precisely known [48–51]. The scale in Fig. 5.9 is obtained in this fashion. For the moment, we define the peak position of the specific heat shown in this figure to be the deconfinement point, and thus obtain

$$T_c \simeq 160 \pm 15 \text{ MeV}. \tag{5.54}$$

It turns out that also this definition has its problems, and we will have to modify it somewhat. Nevertheless, a striking feature of statistical QCD is that the modification of definitions has remarkably little effect on the actual value of the "transition" point.

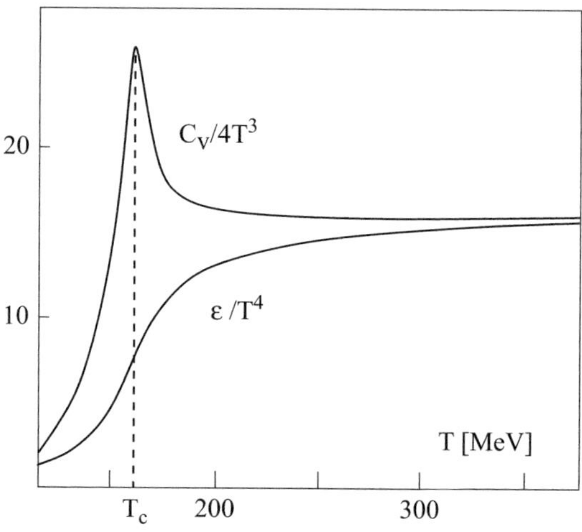

Fig. 5.9 Energy density and specific heat in full QCD [46, 47]

In the following chapter we shall address the symmetry aspects of the hadron-quark transition, in pure gauge theory as well as in full QCD. We will show that in the limit of massless quarks, chiral symmetry restoration provides us with a well-defined transition and a corresponding order parameter.

5.6 Conclusions

We saw how statistical QCD can be formulated on the lattice, leading to a partition function quite similar to that of a generalized spin system. The resulting thermodynamics can therefore be obtained numerically through computer simulation. This evaluation gives in particular

- a sharp transition in energy density, from a confined hadronic phase with few degrees of freedom to a deconfined plasma with many;
- for pure SU(3) gauge theory, well-defined critical behavior (first order phase transition), for full QCD a very rapid change in behavior allowing an approximate determination of the transition point;

- a transition temperature decreasing from 300 MeV for SU(3) gauge theory to 160 MeV for the physically most interesting case of QCD with two light (or two light plus one heavy) quark flavors.

We will now turn to the connection of critical behavior in statistical QCD to underlying symmetries, and we shall also try to understand the connection between deconfinement and chiral symmetry restoration.

References

1. C.T.H. Davies et al., Phys. Rev. Lett. **92**, 022001 (2004)
2. S. Dürr et al., Science **322**, 1224 (2008)
3. G. Altarelli, Ann. Rev. Nucl. Part. Sci. **39**, 357 (1989)
4. K. Wilson, Phys. Rev. D **10**, 2445 (1974)
5. M. Creutz, Phys. Rev. Lett. **43**, 553 (1979)
6. M. Creutz, Phys. Rev. **D21**, 2308 (1980)
7. H.J. Rothe, *Lattice Gauge Theories*. World Scientific Lecture Notes in Physics, vol. 74, 3rd edn. (World Scientific, Singapore, 2005)
8. I. Montvay, G. Münster, *Quantum Fields on a Lattice* (Cambridge University Press, Cambridge, 1994)
9. C. Gattringer, C.B. Lang, *Quantum Chromodynamics on the Lattice* (Springer, Berlin/Heidelberg, 2010)
10. R. Feynman, Phys. Rev. **80**, 440 (1950)
11. C. Bernard, Phys. Rev. D **9**, 3312 (1974)
12. F.J. Wegner, J. Math. Phys. **10**, 2259 (1971)
13. N. Kawamoto, Nucl. Phys. B **190**[FS3], 617 (1981)
14. P.T. Matthews, A. Salam, Nuovo Cim. **12**, 563 (1954)
15. J. Engels et al., Phys. Lett. B **252**, 625 (1990)
16. K. Symanzik, Nucl. Phys. B **226**, 187 and 205 (1983)
17. K. Wilson, in *New Phenomena in Subatomic Physics*, ed. by A. Zichichi (Plenum Press, New York, 1977)
18. J. Kogut, L. Susskind, Phys. Rev. **D11**, 395 (1975) and **D16**, 3031 (1977)
19. H.B. Nielsen, M. Ninomiya, Nucl. Phys. B **185**, 20 (1981)
20. D.B. Kaplan, Phys. Lett. B **268** 342 (1992)
21. H. Neuberger, Phys. Lett. B **417**, 141 (1998)
22. H. Neuberger, Phys. Rev. D **87**, 5417 (1998)
23. F. Karsch, Nucl. Phys. B **205**[FS5], 285 (1982)
24. N. Metropolis et al., J. Chem. Phys. **21**, 1087 (1953)
25. P. Hasenfratz, F. Karsch, Phys. Lett. B **125**, 308 (1983)
26. R.V. Gavai, Phys. Rev. D **32**, 519 (1985)
27. Z. Fodor, S.D. Katz, JHEP **0203**, 014 (2002)
28. Ph. de Forcrand, O. Philipsen, Nucl. Phys. B **642**, 290 (2002)
29. M.-P. Lombardo, Phys. Rev. D **67**, 014505 (2003)
30. C.R. Allton et al., Phys. Rev. D **68**, 014507 (2003)
31. J. Engels, F. Karsch, I. Montvay, H. Satz, Phys. Lett. B **101**, 89 (1981)
32. J. Engels, F. Karsch, I. Montvay, H. Satz, Nucl. Phys. B **205**, 545 (1982)
33. G. Boyd et al., Phys. Rev. Lett. **75**, 4169 (1995)
34. G. Boyd et al., Nucl. Phys. B **469**, 419 (1996)
35. T. Çelik et al., Phys. Lett. B **125**, 411 (1983)
36. J. Kogut et al., Phys. Rev. Lett. **51**, 869 (1983)

37. Sz. Borsanyi et al., arXiv:1104.0013 [hep-ph]
38. L.D. McLerran, B. Svetitsky, Phys. Lett. B **98**, 195 (1981)
39. L.D. McLerran, B. Svetitsky, Phys. Rev. D **24**, 450 (1981)
40. J. Kuti, J. Polónyi, K. Szlachányi, Phys. Lett. B **98**, 199 (1981)
41. E. Eichten et al., Phys. Rev. D **17**, 3090 (1978) and **21**, 203 (1980)
42. M. Lüscher, G. Münster, P. Weisz, Nucl. Phys. B **180**, 1 (1981)
43. O. Kaczmarek et al., Phys. Lett. B **543**, 41 (2002)
44. M. Teper, hep-th/9812187
45. A. Bazavov et al. (HotQCD), Phys. Rev. D **90**, 094503 (2014)
46. A. Bazavov et al., Phys. Rev. D **80**, 014504 (2009)
47. M. Cheng et al., Phys. Rev. D **81**, 054504 (2010)
48. M. Wingate et al., Phys. Rev. Lett. **92**, 162001 (2004)
49. A. Gray et al., Phys. Rev. D **72**, 094507 (2005)
50. C. Aubin et al., Phys. Rev. D **70**, 094505 (2004)
51. M. Cheng et al., Phys. Rev. D **77**, 014511 (2008)

Chapter 6
Broken Symmetries

Seht ihr den Mond dort stehen?
Er ist nur halb zu sehen,
und ist doch rund und schön!
So sind wohl manche Sachen,
die wir getrost belachen,
weil unsre Augen sie nicht sehn.

Matthias Claudius, *Sämmtliche Werke des*
Wandsbecker Bothen (1774)

(Do you see the moon up there?
Only half of it is visible,
and yet it is round and beautiful.
Quite a few things are like that,
and we joke about them,
because our eyes cannot see them.)

(Matthias Claudius, *Complete Works of the*
Messenger from Wandsbeck (1774))

Abstract Recalling the relation between critical behavior and spontaneous symmetry breaking, we show that deconfinement in SU(N) gauge theory is based on the breaking of a global ZN symmetry. Next we turn to the chiral symmetry of the QCD Lagrangian in case of massless quarks and discuss the spontaneous breaking and eventual restoration of this symmetry. We then consider the relation of the two kinds of symmetry breaking in determining the transition behavior as well as the general transition pattern as a function of the input quark masses.

6.1 Symmetry Breaking and Critical Behavior

In Chap. 2 we had seen that critical behavior in thermodynamics is quite generally related to the onset of spontaneous breaking of a global symmetry of the system. In the Ising model, the self-organised alignment of spins is an example of this: the system finds itself in a stable thermodynamic state which breaks a global symmetry

H. Satz, *Extreme States of Matter in Strong Interaction Physics*,
Lecture Notes in Physics 945, https://doi.org/10.1007/978-3-319-71894-1_6

of its own Hamiltonian. The quantity we had used to establish this symmetry breaking, the *order parameter* $m(T)$, vanishes when the state of the system shares the symmetry of the Hamiltonian, and it becomes non-zero when the symmetry is broken. In the latter case, it is evident that an additional parameter (besides T) is necessary to specify the state of the system, since for a given $T < T_c$ there exist two different but equally probable possibilities. It is the order parameter which plays this role; it defines the phase in which the system finds itself, disordered or ordered, and if ordered, up or down. In the vicinity of phase transition, the order parameter is therefore a particularly important thermodynamic quantity.

Concerning "order": in the case of the Ising model, disordered spin configurations lead to $m = 0$, while any configuration with some degree of order leads to a non-vanishing m. Hence here $m \neq 0$ does in fact mean that the corresponding phase is at least partially ordered. But the term *order parameter* is used quite generally for a function which defines the phase structure of a system by vanishing in a certain temperature range and remaining non-zero in another – even when there is no immediate relation to order or disorder. A familiar example is the liquid-gas transition, where the order parameter is defined as the difference of the densities in the two states, $\Delta\rho \equiv \rho_L - \rho_G$. For $T < T_c$, the densities of the two states differ, so that $\Delta\rho \neq 0$; above T_c, there is no longer any difference and hence $\Delta\rho = 0$.

Returning to the Ising model, we note another relation between order and spontaneous symmetry breaking. If the external magnetic field H is not zero, the spins will always be partially aligned, so that $m(T, H)$ never vanishes for $H \neq 0$ (see Fig. 2.2). The Z_2 symmetry of the Hamiltonian is now broken explicitly by the presence of $H \neq 0$. If we gradually reduce the strength of the external field, what does $m(T, H)$ do in the limit $H \to 0$? It becomes evident that one can understand spontaneous symmetry breaking as the limiting case of explicit symmetry breaking, as seen in the relation

$$m(T, H\!=\!0) = \lim_{H \to 0} \left\{ \lim_{N \to \infty} \frac{T}{N^2} \left(\frac{\partial \ln Z}{\partial H} \right) \right\}. \tag{6.1}$$

Spontaneous symmetry breaking can thus also be interpreted as an order enforced by a self-organized field, created by the system without any "outside help".

In the case of continuous phase transitions, we had seen that critical exponents determine the behavior in the vicinity of the transition point and thereby define as *universality class* the set of all systems with the same critical exponents. We shall see shortly that this allows us to predict in statistical QCD the behavior at color deconfinement or at chiral symmetry restoration in terms of the magnetisation pattern in spin models.

It would of course be nice to have a similar classification scheme for discontinuous (first order) transitions; unfortunately this is at present still lacking. So far, one can only say that all first order transitions in systems of a given spatial dimension approach the thermodynamic limit in the same way, i.e., show the same finite-size scaling behavior.

Before we begin to apply these general considerations to the specific critical behavior encountered in QCD, let us see what happens when we extend them to the case of *continuous* global transformations. We had seen that a *discrete* global symmetry, specifically the up-down invariance under the Z_2 group transformations, led to the possibility of spontaneous symmetry breaking when the system had to choose between two equivalent lowest energy configurations which individually break the symmetry. In case of invariance under Z_3 transformations, there are three such equivalent states, corresponding to the spin orientations $s_i = \exp\{n\,i(2\pi/3)\}$, with $n = 0, 1, 2$. For general Z_N, there are N degenerate configurations, and as $N \to \infty$, the number of degenerate states of different symmetry becomes infinite. For continuous symmetry groups, we thus get a continuum of equivalent configurations, and at this point, the mode of symmetry breaking changes. The system, in addition to choosing one out of a continuum of equivalent states rotated relative to each other, in order to indicate its symmetry breaking, develops a wave consisting of a superposition of these states, travelling through the medium. In spin systems with a continuous global symmetry, it is thus the presence of such spin waves which signal spontaneously broken symmetry, not a specific symmetry-breaking ("up" or "down") state. The development of a massless excitation as consequence of a spontaneously broken continuous symmetry is not restricted to spin systems, however; another familiar example are the sound waves (phonons) in crystals as a result of the breaking of the continuous translation invariance of a fluid. In general, the excited states formed through the spontaneous breaking of a continuous symmetry are referred to as Goldstone excitations [21]. We shall see further down that in an ideal world of strong interactions only, pions are the Goldstone excitations caused by the breaking of the continuous chiral symmetry of the QCD Lagrangian with massless quarks.

6.2 The Deconfinement Transition

In Chap. 5, we had considered the correlation function for a static quark-antiquark pair in a purely gluonic medium of temperature T,

$$\Gamma(r, T) \sim |\langle L(0)L^{+}(r)\rangle| \sim \exp\{-F_{Q\bar{Q}}(r, T)/T\}, \tag{6.2}$$

and argued that for $r \to \infty$ this gave $\langle L(T)\rangle \simeq \exp\{-F_Q/T\}$, with $F_Q = F_{Q\bar{Q}}(r = \infty, T)/2$ denoting the energy of an isolated quark. In a confining medium, F_Q becomes infinite; if color screening leads to deconfinement, the range of the potential between static quark and antiquark becomes finite, and with it F_Q. We thus obtained for the temperature dependence of the Polyakov loop expectation value

$$\langle L(T) \rangle = \left\{ \begin{array}{ll} 0 & \text{for } T \leq T_c \\[2ex] \exp\{-F_Q/T\} > 0 & \text{for } T > T_c \end{array} \right\}, \qquad (6.3)$$

with T_c denoting the deconfinement temperature. Therefore the functional form of $\langle L(T) \rangle$ is very similar to that of the magnetisation $m(T)$ for a spin system. At first sight it seems surprising that $m(T)$ vanishes at high, $\langle L(T) \rangle$ at low temperatures. The common feature, however, is that both systems are in the ordered phase at high density. For the gluon gas, that is at high temperature; for the spin system, increasing the density is equivalent to increasing the coefficient K in the Hamiltonian (2.1), and that in turn is the same as decreasing the temperature.

The behavior of $\langle L(T) \rangle$ describes only the potential between two static, i.e., infinitely heavy quarks Q and $\bar{Q}$ in a gluonic medium. We had seen in Sect. 5.5.2 that it breaks down if the medium contains light ("dynamical") quarks q and $\bar{q}$, since string breaking through formation of a pair of light-heavy hadrons $Q\bar{q}$ and $\bar{Q}q$ now allows the test quarks to be separated without expending any further energy. The linear rise of the potential is thus stopped at this point, and $\langle L(T) \rangle$ now becomes non-zero also in the confinement regime, as shown in Fig. 5.8. Comparing the behavior of $\langle L(T) \rangle$ in the cases with and without dynamical quarks to the magnetization effect in the Ising model, i.e., comparing Figs. 5.8 and 5.5 to Fig. 2.2, it becomes evident that the inverse mass of dynamical quarks in the theory must play a role similar to that of the external field in the Ising model. We shall address this aspect in the last section of this chapter.

Here we now return to pure $SU(N)$ gauge theory, determined by the Wilson action (see Eq. (5.14))

$$S \sim \sum \{1 - \frac{1}{N} Re\ Tr\ UUU^+U^+\}. \qquad (6.4)$$

It remains invariant under the gauge transformations

$$U_{r,\tau} \rightarrow V_{r,\tau} U_{r,\tau} V^+_{r,\tau+1}, \qquad (6.5)$$

where $U_{r,\tau}$ is an $SU(N)$ matrix "living" on the link from site $(r.\tau)$ to site $(r, \tau + 1)$, and $V_{r,\tau}$ the $SU(N)$ gauge transformation associated to site (r, τ). The periodicity requirement $A(r, \tau = 0) = A(r, \tau = N_\tau)$ for the gauge fields imposes a periodicity on the gauge transformations as well, with

$$V_{r,\tau=0} = V_{r,\tau=N_\tau} \ \forall\ r, \qquad (6.6)$$

assuring the invariance of the trace in Eq. (6.4). However, the trace in Eq. (6.4) remains in fact invariant for an even more general periodicity condition. If we require

$$V_{x,\tau=0} = C_N\ V_{x,\tau=N_\tau} \ \forall\ x, \qquad (6.7)$$

then all loops in space-time planes – and only those are affected by conditions (6.6)/(6.7) – contain a factor C_N from going "up" in time and a factor C_N^+ from going down. If C_N commutes with all elements of $SU(N)$, we can bring the two together and use $C_N C_N^+ = 1$. Thus the action remains invariant for the more general periodicity condition (6.7), with $C_N \in Z_N \subset SU(N)$, where Z_N is the *center* of $SU(N)$, i.e., the subgroup consisting of those elements of $SU(N)$ which commute with all others. As we saw above, Z_2 consists of ± 1 only, while Z_3 consists of $\exp\{ni(2\pi/3)$, $n = 1, 2, 3$. We thus note that pure $SU(N)$ gauge theory is invariant under global transformations belonging to the center $Z_N \subset SU(N)$.

As in the case of the Ising model, this invariance of the theory need not be shared by the actual state of the system. To check possible spontaneous symmetry breaking in $SU(N)$ gauge field thermodynamics, we note that under the center Z_N transformations just considered, the Polyakov loop operator (see Eq. (5.47))

$$L_N(r) \equiv \frac{1}{N} Tr \prod_{\tau=1}^{N_\tau} U_{r,\tau}, \tag{6.8}$$

does not remain invariant: it is a closed loop only by periodicity and hence contains only one factor C_N, so that transformation (6.7) results in

$$L_N(r) \to C_N L_N(r) \ \forall \ r. \tag{6.9}$$

Hence $\langle L(T) \rangle$ constitutes the order parameter for Z_N symmetry in $SU(N)$ gauge theory; it vanishes when the state shares the symmetry of the Lagrangian, it becomes non-zero when that symmetry is spontaneously broken [28, 31, 32].

It is thus evident that there is much similarity between the Polyakov loops in gauge theory and the spins in the Ising model. In the latter, we have at each lattice site a spin which takes on a certain value (± 1); in gauge theory, we have instead at each spatial lattice site a Polyakov loop with a certain value. In both cases, the average over the lattice tells us if the global Z_N symmetry of the theory is broken or not.

Starting from this similarity, one might try to change variables in the partition function (5.32) from the U_{ij} to the L_i plus whatever remains to be integrated over. For $SU(2)$ gauge theory [34] this leads to a partition function of a structure very much like that of the Ising model. Continuing in this vein, Svetitsky and Yaffe [40] have conjectured that the critical behavior of an $SU(N)$ gauge theory at deconfinement is in the same universality class as the order-disorder transition in the corresponding Z_N spin theory of the same spatial dimension. The conjecture consists of two statements, depending on the nature of the transition.

- If the Z_N spin system shows a discontinuous (first order) transition, then also the corresponding $SU(N)$ gauge theory is expected to lead to a first order deconfinement transition.

- If the corresponding Z_N and the $SU(N)$ transitions are both continuous, then they will be in the same universality class, i.e., they will be governed by the same critical exponents.

In accord with the conjecture, the deconfinement transition in $SU(3)$ gauge theory in three space dimensions is found to be of first order [7, 30], just as the corresponding order-disordered transition in Z_3 spin theory (the so-called three-state Potts model) is of first order. The agreement continues to higher N, where both spin and gauge theories lead to first order transitions. The form of the order-disorder transition in Z_N spin systems is given in Fig. 6.1 as a function of N and of the space dimension d [41].

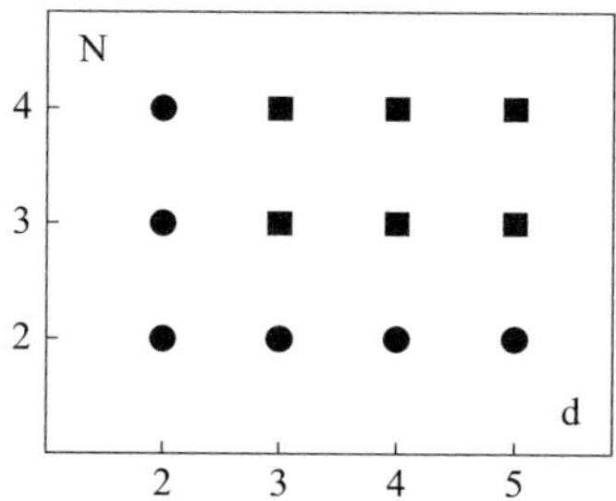

Fig. 6.1 Transition structure of Z_N spin systems in d space dimensions; circles denote continuous, squares discontinuous transitions

For $N = 2$, both spin and gauge theory show continuous transitions. The Svetitsky-Yaffe conjecture thus predicts the critical exponents in $SU(2)$ gauge theory to be those of the Ising model. This prediction has been tested in considerable detail, extrapolating results from finite lattice size studies to infinite volume [13]. We see in Table 6.1 that the conjecture is indeed very well satisfied. The last line in this table provides an additional independent check of the scaling relation $(\gamma + 2\beta)/\nu = 3$ obtained for $d = 3$ by combination of Eqs. (2.23) and (2.26).

The deconfinement transition thus seems well understood in pure $SU(N)$ gauge theory: we can define color deconfinement as the spontaneous breaking of the global Z_N center symmetry. The deconfinement transition is thus structurally related to the disorder-order transitions in Z_N spin systems, and the Svetitsky-Yaffe conjecture specifying the details of this relation is well satisfied.

Exponent	Ising	$SU(2)$
y_t	1.590 ± 0.002	1.587 ± 0.027
y_h	2.482 ± 0.007	2.475 ± 0.008
$(\gamma + 2\beta)/\nu$	3.006 ± 0.018	2.994 ± 0.021

Table 6.1 Critical exponents for the Ising model and SU(2) gauge theory in three space dimensions

In full QCD, the presence of dynamical quarks explicitly breaks the Z_N symmetry. However, as we have seen in the last chapter, the Polyakov loop expectation value nevertheless remains small up to a certain temperature; it then increases rapidly in a very narrow temperature interval, which coincides with that in which the energy density suddenly increases. In other words, some almost critical behavior seems to persist even in the limit of light quarks. We shall later on return to the possible origin of this behavior.

6.3   Chiral Symmetry Restoration

The quark and gluon fields of QCD define a quantum number or "index" space of color, flavor, charge and baryon number, in addition to the usual vector and spinor spaces determined by Poincaré invariance. The breaking of the global center Z_N symmetry just discussed as a basis for deconfinement dealt with transformations of gluon fields in color space. Gluons are flavor-blind, i.e., they interact the same way with all species of quarks, as well as with their different electric charge and baryon number states. This separates the quark and the gluon spaces of internal quantum numbers, and the symmetries to be considered in this section arise in the quark space of flavor, charge and baryon number.

The u and d quarks, with masses below 10 MeV, are very much lighter than the four other species. In the transition regime, they therefore play the dominant role as the quark constituents in QCD thermodynamics, with heavier quarks suppressed by the requirement of pair production (to conserve, e.g., strangeness) and their correspondingly smaller Boltzmann factors. At high temperatures, however, this $u - d$ predominance will weaken and at least strange quarks will come into play. At this point, we are concerned with the transition region and will therefore restrict ourselves for the time being to the two basic flavors. In the Lagrangian (see Eqs. (5.1)/(5.9)),

$$\mathcal{L} = -\frac{1}{4}F_{\mu\nu}F^{\mu\nu} - \sum_f \left\{ \bar{\psi}_f \gamma_\mu \left(i\partial^\mu + gA^\mu \right) \psi_f - m_f \bar{\psi}_f \psi_f \right\}, \tag{6.10}$$

the quark four-spinors ψ_u and ψ_d then form a vector in a two-dimensional flavor space,

$$\psi = \begin{pmatrix} \psi_u \\ \psi_d \end{pmatrix} \equiv \begin{pmatrix} u \\ d \end{pmatrix}; \tag{6.11}$$

we have here suppressed all other indices. Note that the operator $D \equiv (i\gamma^\mu \partial_\mu + g\gamma^\mu A_\mu)$ is flavor-blind. Consider now continuous unitary transformations $V \in U(2)$ in the mentioned flavor space, acting on the two-component vector, $\psi \to \psi' = V\psi$,

$$\psi = \begin{pmatrix} u \\ d \end{pmatrix} \to \psi' = \begin{pmatrix} u' \\ d' \end{pmatrix} = V \begin{pmatrix} u \\ d \end{pmatrix} = V\,\psi. \tag{6.12}$$

Which transformations of this type leave the Lagrangian $\mathcal{L}$ invariant? That depends, as we shall see, on the quark masses. Multiples of the unit matrix, $V = e^{i\alpha}\mathbf{1}$, quite generally do not affect $\mathcal{L}$. Invariance under continuous transformations leads to conserved quantum numbers, as a consequence of the Noether theorem. The invariance under $U(1)$ transformations $\psi \rightarrow e^{i\alpha}\psi$ assures baryon number conservation, keeping the total baryon balance (baryons minus antibaryons) fixed. The remaining transformations of $U(2)/U(1) = SU(2)$ in general do not leave $\mathcal{L}$ invariant; this becomes immediately evident if we rewrite the mass term as

$$\sum_f m_f \bar{\psi}_f \psi_f = m_u \bar{u}u + m_d \bar{d}d = \frac{1}{2}(m_u + m_d)(\bar{u}u + \bar{d}d) + \frac{1}{2}(m_u - m_d)(\bar{u}u - \bar{d}d).$$

$$(6.13)$$

The last term destroys invariance under rotations in $u - d$ space; hence we get invariance under flavor $SU(2)$ only if $m_u = m_d$, i.e., for equal quark masses. The conserved quantity in this case is the isospin: for equal quark masses, the hadrons built out of u and d quarks form isospin multiplets, and in all reactions the overall isospin is conserved. The invariance of $\mathcal{L}$ under $U(1) \times SU(2)$ transformations $u - d$ in flavor space thus provides the isospin structure of the observed hadrons as well as baryon number conservation. What it does not tell us is why the pion mass, in comparison to that of other $u - d$ meson states, is so small. This is where chiral invariance enters.

For massless quarks, the four-spinors can be decomposed into two two-spinors, corresponding to left- and right-handed massless fermions; Lorentz transformations do not mix these different helicity states. We can therefore subdivide the two-dimensional flavor space into two subspaces, one of left- and one of right-handed states, respectively. Each of these is again two-dimensional (for $N_f = 2$), and in it we can then again consider rotations, such as

$$\psi'_L = V_L \psi_L, \quad \text{with} \quad \psi_L = \begin{pmatrix} u_L \\ d_L \end{pmatrix} \quad \text{and} \quad V \in SU(2), \qquad (6.14)$$

and similarly for right-handed states. For massless quarks, the Lagrangian remains invariant under this additional symmetry group. To see that, we introduce the projection operators

$$P_R = \frac{1}{2}(1 + \gamma_5), \quad P_L = \frac{1}{2}(1 - \gamma_5), \qquad (6.15)$$

where $\gamma_5 = \gamma_0 \gamma_1 \gamma_2 \gamma_3$ anticommutes with the Dirac matrices γ_μ, $\gamma_5 \gamma_\mu = -\gamma_\mu \gamma_5$. The operators (6.15) satisfy

$$P_R + P_L = 1 \,; \quad P_L^2 = P_L \,; \quad P_R^2 = P_R \,; \quad P_R P_L = P_L P_R = 0 \,; \quad \gamma_\mu P_R = \gamma_\mu P_L.$$

$$(6.16)$$

In terms of the projection operators and the state ψ of Eq. (6.11), we define

$$\psi_L = P_L\psi \,, \quad \psi_R = P_R\psi, \tag{6.17}$$

and hence get

$$\bar{\psi}_L = \bar{\psi}P_R \,, \quad \bar{\psi}_R = \bar{\psi}P_L. \tag{6.18}$$

Using these relations, we find

$$\bar{\psi}\gamma_\mu\psi = \bar{\psi}_L\gamma_\mu\psi_L + \bar{\psi}_R\gamma_\mu\psi_R, \tag{6.19}$$

so that rotations in the L-part of flavorspace only leave the term $\bar{\psi}D\psi$ of the Lagrangian (5.38) invariant. On the other hand,

$$\bar{\psi}\psi = \bar{\psi}_R\psi_L + \bar{\psi}_L\psi_R, \tag{6.20}$$

so that the mass term, as expected, will not remain invariant: the Lagrangian $\mathcal{L}$ has the additional chiral symmetry only for massless quarks.

In that case, $\mathcal{L}$ still has a further axial $U(1)$ symmetry, obtained by the gauge transformation $V_A = \exp\{i\gamma_5\alpha\}$; it corresponds to independent phase transformations of the left- and right-handed parts of ψ.

We thus find a hierarchy of symmetries. $\mathcal{L}$ is always invariant under $U(1)$ transformations assuring baryon number conservation. If $m_u = m_d$, we have in addition isospin conservation, obtained from the invariance under $SU(2)_{\text{isospin}}$. If $m_u = m_d = 0$, $\mathcal{L}$ is furthermore invariant under $SU(2)_{\text{chiral}}$ and $U_A(1)$, so that in this case the full symmetry group of $\mathcal{L}$ becomes

$$U(1)_{\text{baryon}} \times U(1)_A \times SU(2)_L \times SU(2)_R, \tag{6.21}$$

or, equivalently,

$$U(1)_V \times U(1)_A \times SU(2)_V \times SU(2)_A, \tag{6.22}$$

where the subscripts V and A refer to vector and axial vector operations in flavor space. Generalizing the result to N_f flavors, we obtain the group

$$SU(N_f)_L \times SU(N_f)_R \sim SU(N_f)_V \times SU(N_f)_A; \tag{6.23}$$

it is generally denoted as the chiral symmetry group; it is the product of rotations in left-handed flavor space and rotations in right-handed flavor space. It is equivalent to that obtained as a product of independent rotations of vectors and axial vectors.

The conservation of baryon number and isospin, predicted by the invariance of $\mathcal{L}$ under $U(1)_{\text{baryon}} \times SU(2)_V$, is well satisfied in nature; deviations with respect to isospin conservation are of the size of the difference between proton and neutron masses. The invariance under $SU(2)_A$, on the other hand, would require an axial vector partner of each isospin multiplet, e.g., a negative parity nucleon state. This is not observed, indicating that this symmetry must be spontaneously broken in the hadronic state of matter. The axial symmetry $U_A(1)$ is also broken, because of the two-gluon anomaly of the axial vector current [1, 6], leading to a non-conservation of the axial quark charge.

Since $SU(2)_A$ is a continuous global symmetry, its breaking implies an associated isospin triplet of Goldstone excitations – the pion. In other words: while the vacuum state $|0\rangle$ is invariant under vector operations in flavor space,

$$\mathbf{I}_V|0\rangle = 0, \tag{6.24}$$

this is not the case for axial vector operations

$$\mathbf{I}_A|0\rangle = |\mathbf{G}\rangle \neq \mathbf{0}. \tag{6.25}$$

The pion thus becomes the massless particle associated to the axial isovector $\mathbf{G}$. Such a massless state exists, strictly speaking, only for massless u and d quarks. In the real world, the actual pion has a mass, m_u and m_d are not zero, and hence the chiral symmetry of $\mathcal{L}$ is explicitly broken; the picture of pions as Goldstone particles can therefore be only approximate. Nevertheless, we see how a light axial isovector state appears in QCD as consequence of the spontaneous breaking of the chiral symmetry of the Lagrangian. The pion thus plays a special role among the hadrons: in an idealised QCD of vanishing quark masses, we have conventional hadrons (vector mesons, nucleons) which are quark-antiquark bound states, and pions, which are massless Goldstone excitations. At deconfinement, the quark-antiquark binding is dissolved, so that the conventional hadrons "melt". The pions, however, would still persist beyond this point, unless chiral symmetry is also restored here. A measure of chiral symmetry breaking is provided by the effective quark mass term $\langle \bar{\psi}\psi \rangle$; when it does not vanish, the chiral symmetry of the Lagrangian is spontaneously broken. Let us now consider the critical behavior associated to chiral symmetry breaking and restoration; for a more extensive discussion of chiral quark dynamics and the chiral phase transition, see e.g. [2, 35].

The transformation group $SU(2)$ is specified by three real parameters; it is locally isomorphic to the group $O(3)$ of rotations in a three-dimensional space, so the three parameters correspond to the three Euler angles. The flavor symmetry group $SU(2)_V \times SU(2)_A$ is parametrised in terms of six real parameters, three for each $SU(2)$ factor; it is locally isomorphic to the group $O(4)$ of rotations in a four-dimensional space. Based on this, Pisarski and Wilczek conjectured [33, 36] that the critical behavior of chiral symmetry restoration in QCD with two species of massless quarks and that at the order-disorder point of an $O(4)$ symmetric spin model lie in the same universality class. This relation makes sense only if chiral symmetry restoration in two flavor QCD is a continuous transition, and, as we shall see shortly, that does seem to be the case. This provides a universality relation of direct physical relevance, in contrast to the relation between Z_N symmetry and deconfinement, where only the (unphysical) $SU(2)$ gauge theory provided a continuous transition in three space dimensions. The extension of chiral symmetry to more than two massless quark species (e.g., $m_s = 0$ as well as m_u and m_d) leads to $SU(N)_V \times SU(N)_A$ and hence to $O(2N)$ symmetric spin systems; both show first order transitions.

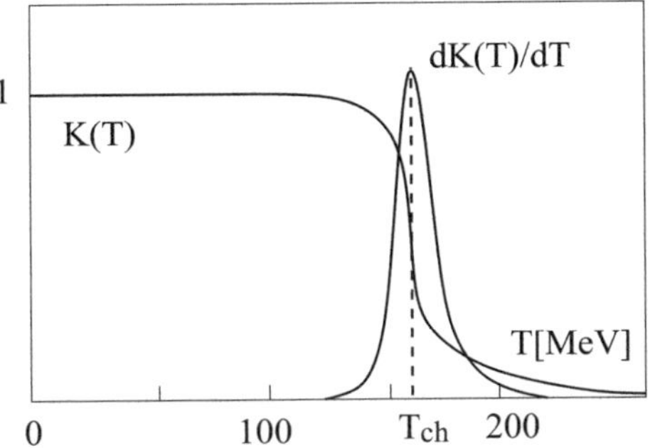

Fig. 6.2 Schemativ view of the temperature dependence of the renormalised chiral condensate $K(T)$ and of the corresponding susceptibility

The chiral order parameter in QCD is defined by

$$\langle \bar{\psi}\psi \rangle = \frac{T}{V} \left(\frac{\partial \ln Z(T, m_q)}{\partial m_q} \right)_{m_q=0}, \tag{6.26}$$

where m_q denotes the quark mass. The calculation of $\langle \bar{\psi}\psi \rangle$ at finite temperature requires both additive and multiplicative renormalisations; for details, see [10]. We normalize it here to its value at $T = 0$, where in the limit $m_q \to 0$, chiral symmetry is spontaneously broken, and denote the resulting condensate as $K(T)$. The behavior obtained for the case of two light (u, d) and one heavy (s) quarks is shown schematically in Fig. 6.2. To determine the resulting critical temperature in the limit of vanishing light quark masses, we can differentiate $K(T)$ either with respect to temperature or with respect to quark mass. The first leads to the "mixed" susceptibility

$$\chi_q^T(T) = \left(\frac{\partial K(T)}{\partial T} \right)_{m_q} \sim \left(\frac{\partial}{\partial T} \left(\frac{\partial \ln Z}{\partial m_q} \right)_T \right)_{m_q}, \tag{6.27}$$

which (see Chap. 2) should diverge as $|t|^{-(1-\beta)}$ for $t \to 0$, with $t = (T - T_c)/T_c$; the corresponding behavior in our case is included in Fig. 6.2. The second gives the chiral susceptibility

$$\chi_q^m(T) = \left(\frac{\partial K(T)}{\partial m_q} \right)_T = \left(\frac{\partial^2 \ln Z(T, m_q)}{\partial m_q^2} \right)_T, \tag{6.28}$$

with a divergence as $m_q^{(1/\delta)-1}$ on the critical isotherm $T = T_c$. Either could be used to locate the most rapid temperature change of the condensate and thus define the critical point. In actual lattice studies, for technical reasons the chiral susceptibility (6.28) is used and the resulting temperature is then employed here as scale.

To specify the resulting critical temperature in physical units, one should have

- calculations on different size lattice, allowing an extrapolation to the continuum limit; and
- calculations providing the continuum limit of a measured observable, allowing the gauging of the lattice scale in physical units.

Such studies have been performed by different groups, and the outcome appears to be converging to a value of

$$T_\chi^0 = 160 \pm 10\,\text{MeV} \tag{6.29}$$

for the critical temperature of chiral symmetry restoration in the limit $m_{u,d} \to 0$ of $N_f = 2+1$ QCD [3–5, 12]. More recently, the pseudo-critical temperature, at which the chiral susceptibility peaks for small but non-zero light quark masses, has been determined in the continuum limit [10], giving

$$T_c^\chi = 154 \pm 9\,\text{MeV}. \tag{6.30}$$

Corresponding lattice results are shown in Fig. 6.3. It is this value which has to be compared to phenomenological determinations of the hadronization temperature; we return to that in Chap. 11.

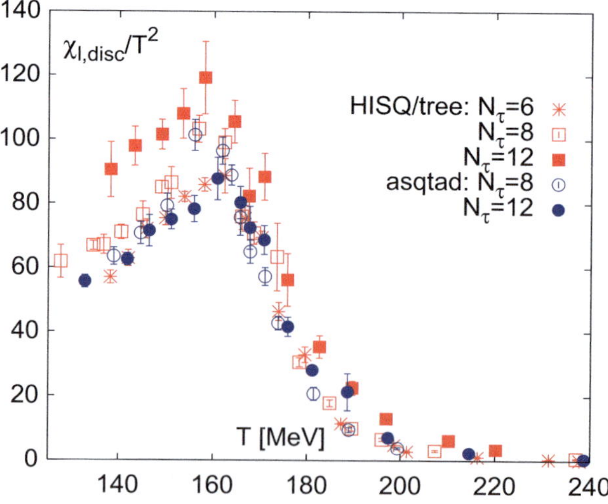

Fig. 6.3 Temperature variation of the chiral susceptibility in (2+1) QCD with physical quark masses [5]

Given these lattice results, we can return to the universality conjecture of Pisarki and Wilczek, which predicts chiral symmetry restoration to be in the universality class of three-dimensional $O(4)$ spin theory, just as the Svetitsky-Yaffe conjecture related deconfinement in $SU(2)$ gauge theory to the Ising model. The critical exponents for the $O(4)$ system are listed in Table 6.2 [24]. A direct comparison with those of two-flavor QCD encounters several problems, however. In early work [25], it was found that while the magnetic chiral exponent y_h agrees very well with its predicted value, the thermal exponent y_t appears to be about 50 % larger than its $O(4)$ prediction, though in both cases with considerable error. One difficulty is the fact that the exponent of the specific heat, α, is negative for $O(4)$, so that the specific heat does not diverge at T_c, but only shows a cusp. As a consequence, the regular part of C_v plays a considerable role there and can modify the temperature dependence. A second aspect to be kept in mind is that the lattice formulation of the quark fields breaks the $O(4)$ symmetry to an effective $O(2)$ symmetry; the full $O(4)$ is recovered only in the continuum limit. The problem is further complicated by the fact that the exponents in the two cases are quite similar. Finally, even now the lattice studies of full QCD are not nearly as precise as they are for pure $SU(N)$ gauge theory, so that a determination of the critical behavior in a perhaps very small region around T_c remains difficult for the moment.

α	β	γ	δ	ν		y_t	y_h
-0.2	0.38	1.5	4.8	0.76		0.45	0.83

Table 6.2 Critical exponents for the O(4) spin system

As alternative approach [22], one has therefore compared the observed behavior, e.g. of the chiral condensate and susceptibilities derived from it, to the scaling functions in the critical region, using the $O(4)$ or $O(2)$ exponents. The crucial variables are the reduced temperature $t = (T - T_c)/T_c$ and the "external field" $h = m_q/T_c$; the latter is here effectively determined by the quark mass, since $m_q \to 0$ leads to a chirally symmetric Lagrangian. In Chap. 2 we have seen that the order parameter $M(t, h)$ should scale for $t = 0$ as

$$M(t = 0, h) = \left(\frac{\partial F_s(t, h)}{\partial h} \right)_{t=0} = h^{1/\delta} f_G(z), \tag{6.31}$$

where $F_s(t, h)$ is the singular part of the free energy. The scaling function $f_G(z)$, with $z = t/h^{1/\beta\delta}$, can be determined in the corresponding spin class, and the exponents β and δ are known (for $O(4)$, see Table 6.2). The so-called magnetic equation of state (6.31) has been analysed in QCD with two light quark flavors, making use of

$$M(t, h) \sim \langle \bar{\psi}\psi \rangle(t, h) \tag{6.32}$$

as the relevant chiral order parameter. It is found to agree well with the $O(4)/O(2)$ predictions [23], so that also the Pisarski-Wilczek conjecture appears to be satisfied for QCD.

6.4 Quark Mass and Transition Structure

We now want to consider how the transition is affected when the mass values for the different quark flavors are varied.

- When $m_q \to \infty$ for all quark flavors, QCD reduces to pure $SU(N)$ gauge theory, which is invariant under a global Z_N symmetry. Hence the critical behavior of $SU(N)$ gauge theory is in the same universality class as that of Z_N spin theory (the N-state Potts model): both are due to the spontaneous symmetry breaking of a global Z_N symmetry [40]. In accord with this, the critical exponents in $SU(2)$ gauge theory are identical to those of the Ising model, and for $SU(3)$ gauge theory, the first order transition of the corresponding Potts model is recovered.
- When $m_q = 0$ for all quark flavors, the Lagrangian is chirally symmetric, so that we now have a phase transition corresponding to chiral symmetry restoration. For three massless quark flavors, the transition is again of first order. It remains of this form for a range of strange quark mass values, until at a certain $m = m_s^{\text{tri}}$, it becomes of second order; the case of two massless quarks is thus also continuous.

In Fig. 6.4, the corresponding overall transition pattern is shown, with m_u, m_d and m_s denoting the mass values for the three quark flavors. Let us look in detail at what this means.

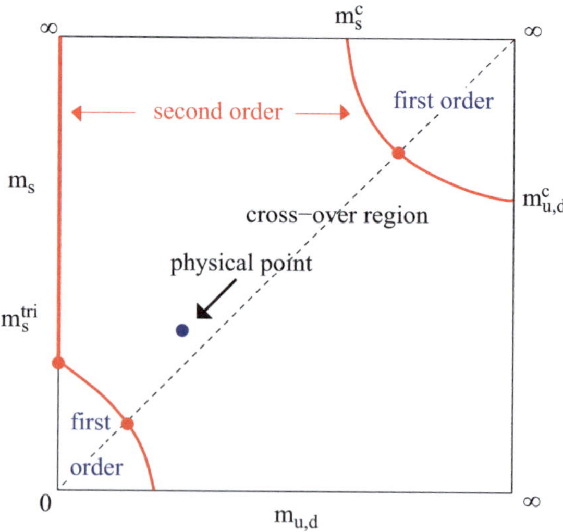

Fig. 6.4 Finite temperature phase structure for three quark flavors

Consider the case of the first order transition in pure gauge theory, located in the upper right-hand corner of Fig. 6.4. With no quarks present, the Polyakov loop is a genuine order parameter, vanishing for $T \leq T_c$ and finite for $T > T_c$; at $T = T_c$, $L(T)$ has a discontinuity (see Fig. 5.5). When quarks of large but finite mass are introduced, this discontinuity does not disappear immediately; it does so only when m_q falls below a certain value m_q^c. We thus obtain a finite region of discontinuous first order behavior, bounded by a line of second order transitions. For still lower quark masses the Polyakov loop continues to vary rapidly in a certain narrow temperature range, but it shows no more singular behavior (we ignore the chiral region for the moment). This is generally referred to a "rapid cross-over".

A similar pattern emerges when we start at the lower left-hand corner. There now is a first order chiral transition when all three quark flavors are massless, and the resulting discontinuity in $\chi(T)$ persists for a while as the quark masses become finite. And again the resulting region is bounded by a line of second order transitions. Between the two first-order regions with their second-order boundaries there is no genuine thermal critical behavior, but "only" the rapid cross-over already mentioned above. On the other hand, for $m_u = m_d = 0$, there is a range in the strange quark mass, $m_s^{\text{tri}} \leq m_s \leq \infty$, for which the transition is again of second order, up to a two-flavor point $m_u = m_d = 0$, $m_s = \infty$.

Let us comment briefly on the different second order transitions in Fig. 6.4. Those bounding a region of first order transitions are generally believed to be (and sometimes shown to be) in the universality class of the three-dimensional Ising model [20, 26]. There does not seem to be any proof of this behavior; it may be related to the "yes-no" nature of the Z_2 symmetry. The two-flavor second-order point (the upper left-hand corner) has been conjectured to be in the O(4) universality class [33], as mentioned above, and this behavior is expected to persist for values $m_s \geq m_s^{\text{tri}}$. The point $m_u = m_d = 0, m_s = m_s^{\text{tri}}$ in the diagram thus becomes a tri-critical point, at which there is a "meeting" of O(4), Z_2 and first order critical behavior.

The location of the actual "physical point", realized in nature for two light u and d quarks and a heavier s quark, has been under study for quite some time. It is now generally thought to be in the cross-over region.

6.5 Deconfinement and Chiral Symmetry Restoration

We have seen that there are two *bona fide* phase transitions in finite temperature QCD at vanishing baryochemical potential. For $m_q = \infty$, $L(T)$ provides a true order parameter which specifies the temperature range $0 \leq T \leq T_c$ in which the Z_3 symmetry of the Lagrangian is present, implying confinement, and the range $T > T_c$, with spontaneously broken Z_3 symmetry and hence deconfinement. For $m_q = 0$, the chiral condensate defines a range $0 \leq T \leq T_\chi$ in which the chiral symmetry of the Lagrangian is spontaneously broken (quarks acquire an effective dynamical mass), and one for $T > T_\chi$ in which $\langle\psi\bar\psi\rangle(T) = 0$, so that the chiral symmetry is restored. Hence here $\langle\psi\bar\psi\rangle(T)$ is a true order parameter.

In the real world, the (light) quark mass is small but finite: $0 < m_q < \infty$. This means that the string breaks for all temperatures, even for $T = 0$, so that $L(T)$ never vanishes. On the other hand, with $m_q \neq 0$, the chiral symmetry of $\mathcal{L}_{QCD}$ is explicitly broken, so that $\langle\psi\bar\psi\rangle$ never vanishes. The temperature behavior of $\langle\bar\psi\psi\rangle$ is shown in Fig. 6.5 together with that of the corresponding Polyakov loop, both for small but finite quark masses [25]. These are rather old results, and we shall return shortly to more recent developments. We show them here simply to illustrate that the average Polyakov loop and the chiral condensate vary rapidly at the same temperature, at which, as we saw in Fig. 5.6, also the energy density undergoes a rapid variation. To underline the point, we also show the corresponding susceptibilities $\chi_L(T)$ and $\chi_m(T)$, obtained by differentiating the corresponding order parameters with respect to the temperature. Note that these results, as all other mentioned lattice results, are obtained for vanishing baryon number density, i.e., for an equal number of baryons and antibaryons. The coincidence of the peaks in Fig. 6.5 was in most earlier studies interpreted as the coincidence of deconfinement and chiral symmetry restoration at vanishing baryon number density.

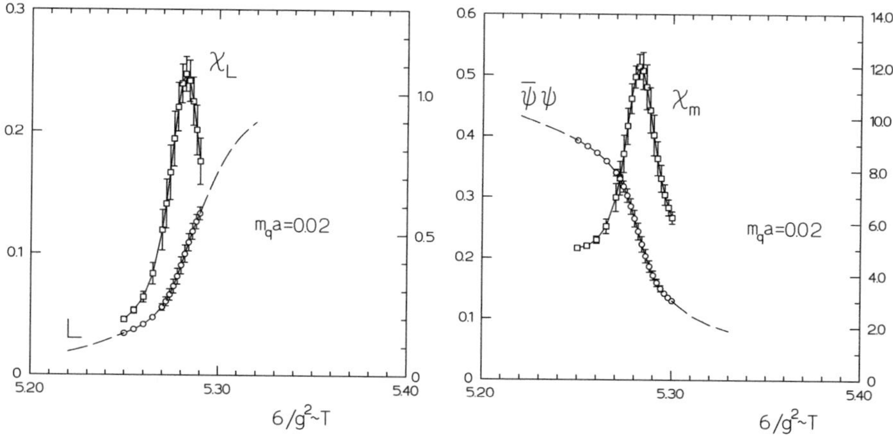

Fig. 6.5 The temperature dependence of the Polyakov loop L and the chiral condensate $\psi\bar\psi$, as well as of the corresponding susceptibilities [25]

In the past few years, the situation concerning the relation between the two expected critical phenomena has become somewhat more clarified. It is found that the peak in the chiral susceptibility indeed increases with decreasing quark mass, indicating a divergence in the chiral limit $m_q \to 0$, which is in accord with the predicted exponents in the two-flavor case. The behavior of $\langle\psi\bar\psi\rangle(T)$ thus appears to be a genuine reflection of chiral symmetry restoration, and hence it is an indicator of the corresponding transition temperature T_χ even for small but finite quark masses. The average Polyakov loop, on the other hand, does not seem to show singular behavior for $m_q \to 0$. This suggests some reconsideration of what deconfinement and chiral symmetry restoration conceptually imply.

At a critical point (for the sake of discussion, let us consider a continuous transition), all thermodynamic observables obtained through derivatives of the partition function must show non-analytic behavior. In other words, derivatives of sufficiently high order will diverge. The chiral condensate is the derivative of $Z(T, m_q)$ with respect to quark mass and hence it shows critical behavior in the chiral limit. The Polyakov loop operator is not present in the QCD action, neither in full QCD nor in pure gauge theory. It can, however, be added as "external" field; observables are then evaluated for a vanishing coupling of this field. In pure gauge theory, this external field breaks the Z_N symmetry of the proper Lagrangian and hence the average value of $L(T)$ serves as order parameter. In particular, the intrinsic critical behavior of the theory is reflected in a non-analytic behavior of $L(T)$ as well as in that of the energy density, the specific heat, etc. In full QCD, in the presence of dynamical quarks, the quark mass term $m_q^2 \langle \psi \bar{\psi} \rangle (T)$ acts as the external field breaking the chiral symmetry, so that the derivative with respect to m_q at $m_q = 0$ serves as the corresponding order parameter. On the other hand, here the Z_N symmetry is explicitly broken by the presence of the quarks, even in the absence of a Polyakov loop external field. Hence adding such a term does not change the Z_N symmetry properties, and so $L(T)$ can no longer be used as genuine order parameter for deconfinement. It is presently not known if it nevertheless reflects the singular behavior of the partition function in the chiral limit. It does continue to change rapidly at the chiral restoration temperature; this can perhaps be interpreted as an effect of the medium on the external field. In the next subsection, we shall elaborate a little on this possibility [11, 37]. In any case, the rapid variation of $L(T)$ indicates a sudden change in the screening pattern seen by quarks in the transition from below to above T_χ. It therefore makes sense to ask for a more general way of specifying deconfinement and chiral symmetry as distinct phenomena, and to check whether in a given theory they coincide.

Deconfinement is basically the transition from bound to unbound quark constituents, from a state of color-neutral hadrons to one of colored quarks. Chiral symmetry restoration is the transition from a state of massive "dressed" constituent quarks to one of massless current quarks. These two phenomena need not coincide, and there exist theories in which they do not, for quarks in the adjoint, rather than the fundamental color group representation [18, 29]. In this case, deconfinement occurs at a much lower temperature than chiral symmetry restoration, i.e., deconfinement leads to a state of colored massive "dressed" quarks. The basic indications of the two phenomena are therefore

- a sudden change in the number of degrees of freedom, from bound to unbound color, for deconfinement, and
- a sudden change in the effective quark mass, from finite to zero, for chiral symmetry restoration.

The relevant measure in the first case is the entropy density $s(T, m_q)$, in the second the chiral condensate; we had denoted the normalized form of the latter as $K(T, m_q)$. If the system shows one common critical point in the chiral limit $m_q \rightarrow 0$, both $s(T, 0)$ and $K(T, 0)$ should be inherently singular at $T = T_c$. The vanishing of the constituent quark mass at the deconfinement point implies that here gluons are deconfined as well as quarks, so that the entropy density increases correspondingly. This appears to be the behavior found for $N_f = 2$ QCD in the chiral limit at zero baryon density. If the system would have two transitions, with distinct deconfinement and chiral symmetry restoration, the entropy density would not immediately increase to the value of ideal quarks and gluons; some of the gluons are "needed" to dress the quarks. If this removes most of them, the entropy density would first increase at $T = T_{\text{dec}}$ towards the ideal quark limit,

$$s_q/T^3 \simeq 21 \frac{4}{3} \frac{\pi^2}{30} \left[1 - \frac{15}{7\pi^2} \left(\frac{M_q}{T} \right)^2 \right] \tag{6.33}$$

with the gluons still confined to the constituent quark mass M_q It is not known what effect this transition would have on the chiral condensate $K(T)$. Subsequently, when the quarks melt, at $T = T_\chi$, $K(T)$ would vanish and the entropy density would increase further toward the ideal quark-gluon plasma limit

$$s_{qg}/T^3 = (16 + 21) \frac{4}{3} \frac{\pi^2}{30}. \tag{6.34}$$

The two cases are illustrated in Fig. 6.6. The two-step scenario would in general lead to different critical exponents for the different transitions, so that it is indeed possible to consider distinct deconfinement and chiral symmetry restoration transitions. A behavior of the second type will be discussed in more detail in Chap. 7, since it remains a possible scenario for the deconfined medium at large baryon number density.

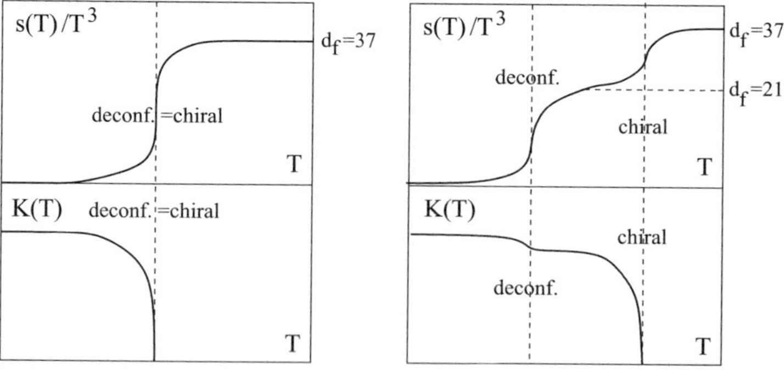

Fig. 6.6 Deconfinement and chiral symmetry restoration for one (left) and two (right) distinct transitions

In Fig. 6.7 we show schematically recent results for the entropy density in the physical case of one heavy and two light quark flavors, with masses tuned to approximate the physical hadron masses [10]. It is seen here that the entropy density also shows its greatest change at the chiral critical temperature $T_\chi = 160 \pm 10\,\text{MeV}$, determined above (see Eq. (6.29)). At this point there is a pronounced peak in the corresponding susceptibility, which is essentially the specific heat. Present results thus indicate that for the physical situation, (2+1) flavor QCD, we do in fact have one transition, at which both deconfinement and chiral symmetry restoration occur in the "cross-over" sense discussed: the actual singular behavior of all thermodynamic observables will occur at a unique "chiral" temperature defined from $K(T, m_q)$ in the limit $m_{u,d} \to 0$.

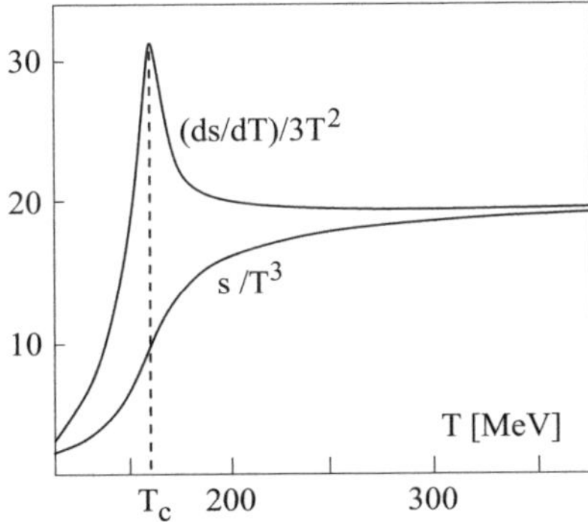

Fig. 6.7 The temperature variation of the entropy density and its derivative as a function of the temperature, for QCD with $N_f = 2 + 1$ and almost physical quark masses [8, 10]

We have here used the chiral definition of the temperature, since a more detailed study of the critical behavior of the entropy density in the chiral limit of $N_f = 2$ QCD encounters one further difficulty. As seen in Table 6.2, the exponent α for the specific heat $C_V(t)$ is negative and hence does not produce a divergence. For this, a derivative of one order more is required, and such studies are in progress.

At this point, a comment seems in order on the "schematic" Figs. 5.9, 6.2, and 6.7, shown for the energy density, the chiral condensate and the entropy density for the case of $N_f = 2 + 1$ QCD. These studies have been and are being performed with different lattice actions, for different lattice sizes, and for different quark mass values. While initially there were some 10% differences in the results obtained, at present they appear to converge to the unique critical temperature of about 160 MeV, in the chiral limit $m_{u,d} \to 0$. The curves shown in the mentioned figures are meant to indicate the general behavior presently observed; they are an interpolation of different results and should not be taken as correctly showing quantitative details.

6.6 Does Chiral Symmetry Restoration Drive Deconfinement?

In this section, we want to consider in some more detail a speculative answer to why the Polyakov loop continues to vary rapidly at the temperature of chiral symmetry restoration.

In the confined phase of pure gauge theory, we have $L(T) = 0$; the Polyakov loop as generalized spin is disordered, so that the state of the system shares the Z_3 symmetry of the Lagrangian. Deconfinement then corresponds to ordering through spontaneous breaking of this Z_3 symmetry, making $L \neq 0$. In going to full QCD, the introduction of dynamical quarks effectively brings in an external field $H(m_q)$, which in principle could order L in a temperature range where it was previously disordered. Since this field is not present for $m_q \to \infty$, H must be inversely proportional to m_q for large quark masses. On the other hand, since $L(T)$ shows a rapid variation signalling an onset of deconfinement even in the chiral limit, the relation between H and m_q must be different for $m_q \to 0$. We therefore conjecture [11, 19, 37] that H is determined by the effective constituent quark mass M_q, setting

$$H \sim \frac{1}{m_q + c\langle \psi \bar{\psi} \rangle}, \tag{6.35}$$

since the value of M_q is determined by the amount of chiral symmetry breaking and hence by the chiral condensate. From Eq. (6.35) we obtain

- for $m_q \to \infty$, $H \to 0$, with deconfinement at T_c^∞;
- for $m_q \to 0$,

$$\langle \psi \bar{\psi} \rangle = \begin{cases} \text{large, } H \text{ small, } L \text{ disordered,} & \text{for } T \leq T_\chi; \\ \text{small, } H \text{ large, } L \text{ ordered,} & \text{for } T > T_\chi. \end{cases} \tag{6.36}$$

In full QCD, it is thus the onset of chiral symmetry restoration that drives the onset of deconfinement, by ordering the Polyakov loop at a temperature value below the point of spontaneous symmetry breaking. In Fig. 6.8 we compare the behavior of $L(T)$ in pure gauge theory to that in the chiral limit of QCD. In both cases, we have a rapid variation at some temperature T_c. This variation is for $m_Q \to \infty$ due to the spontaneous breaking of the Z_3 symmetry of the Lagrangian at $T = T_c^\infty$; for $m_q \to 0$, the Lagrangian retains at low temperatures an approximate Z_3 symmetry which is explicitly broken at T_χ by an external field becoming strong when the chiral condensate vanishes. For this reason, the peaks in the Polyakov loop and the chiral susceptibility coincide and we have $T_\chi = T_c < T_c^\infty$.

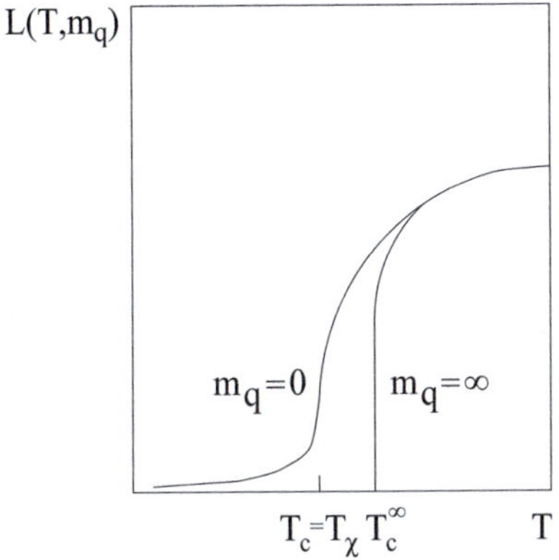

Fig. 6.8 Temperature dependence of the Polyakov loop in the chiral and the pure gauge theory limits

A quantitative test of this picture could be obtained from finite temperature lattice QCD. It is clear that in the chiral limit $m_q \to 0$, the chiral susceptibilities (derivatives of the chiral condensate $\langle \psi \bar{\psi} \rangle$) will diverge at $T = T_\chi$. If deconfinement is indeed driven by chiral symmetry restoration, i.e., if $L(T, m_q) = L(H(T), m_q)$ with $H(T) = H(\langle \psi \bar{\psi} \rangle(T))$ as given in Eq. (6.35), then also the Polyakov loop susceptibilities (derivatives of L) must diverge in the chiral limit. Preliminary lattice studies support this picture. In Fig. 6.9 we see that the peaks in the Polyakov loop susceptibilities as a function of the effective temperature increase as m_q decreases. Further lattice calculations for smaller m_q (which requires larger lattices) would certainly be helpful; in particular, the chiral limit $m_q \to 0$ in the case $N_f = 3$ should result in discontinuous behavior also for the Polyakov loop and its derivatives.

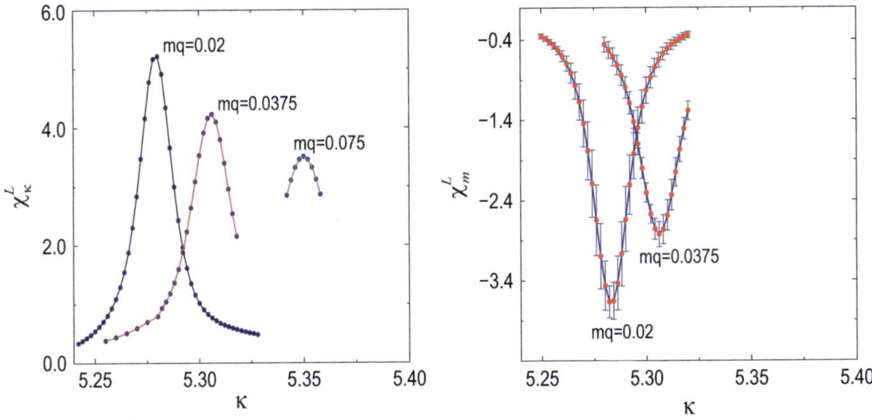

Fig. 6.9 The Polyakov loop susceptibilities with respect to temperature χ_κ^L (left) and to quark mass χ_m^L (right) as function of the temperature variable $\kappa = 6/g^2$ for different quark masses [11]

6.7 Percolation and Rapid Cross-Over

In the quark mass range $0 < m_q < \infty$, neither the Polyakov loop nor the chiral condensate constitute genuine order parameters, since both are non-zero at all finite temperatures. Nevertheless, in the first-order regions, the discontinuity in the order parameter still persists, until it finally vanishes on a boundary line of second order transitions. Between the boundary lines in Fig. 6.4 we have what was referred to as rapid cross-over. It does not correspond to thermal critical behavior in a strict mathematical sense, but nevertheless all thermodynamic variables vary quite sharply at some pseudo-critical temperature (see e.g., Fig. 6.5).

We believe that a deeper understanding of the behavior along this pseudo-critical "transition" line could come from a study of cluster percolation. In Sect. 2.2, we had seen that for spin systems without an external field, the thermal magnetization transition can be equivalently described as thermal transition or as percolation transition of suitably defined clusters [9, 14, 15]. One can thus characterize the Curie point of a spin system either as the point where, with decreasing temperature, spontaneous symmetry breaking sets in, or as the point where the size of suitably bonded like-spin clusters diverges.

In Sect. 2.2, we had also noted that in the presence of a finite external field H, there is no more *thermal* critical behavior; for the 2d Ising model, as illustration, the partition function now is analytic. In a corresponding cluster description, however, the percolation transition continues to persist as *geometric* critical behavior for all H. The critical indices now become those of random percolation and hence differ from the thermal (magnetization) indices. For the 3d three-state Potts model (with a global Z_3 symmetry instead of the Z_2 symmetry of the Ising model) the transition for $H = 0$ is of first order, so that turning on the field does not immediately remove the discontinuity. The resulting phase diagram is shown on the left of Fig. 6.10; here the dashed line, the so-called Kertész line [27], is defined as the line of the geometric critical behavior obtained from cluster percolation. The phase on the low temperature side of the Kertész line contains percolating clusters, the high temperature phase does not. Comparing this result to the $T - m_q$ diagram of two-flavor QCD, one is tempted to speculate that deconfinement for $0 < m_q < \infty$ corresponds to the Kertész line of QCD [38]. This would maintain a relation of deconfinement to critical behavior for all m_q, but in general this would be geometric critical behavior. First studies have shown that in pure gauge theory, one can in fact describe deconfinement through Polyakov loop percolation [16, 17, 39]. It would indeed be interesting to see if this can be extended to full QCD, and if deconfinement in the most general sense is indeed a geometric critical phenomenon, related to percolation, which only for some specific parameter values also leads to thermal critical behavior. The crucial feature in this context is the correct definition of what is to be considered a cluster, given a specific dynamics basis. This was already the essential point in the percolation formulation of spin systems, and in QCD the possibility of longer range interactions (not just next neighbors) enhances the problem.

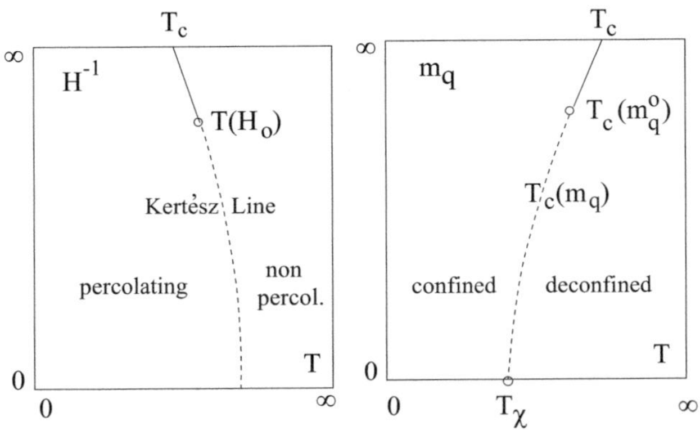

Fig. 6.10 Phase structure of the 3-state Potts model (left) and two-flavor (right) QCD

6.8 Conclusions

We have surveyed the general description of critical behavior in terms of spontaneous symmetry breaking and then applied this formalism to deconfinement and chiral symmetry restoration. In particular, we have seen that

- in pure $SU(N)$ gauge theory, deconfinement can be associated to the spontaneous breaking of a global center Z_N symmetry; in full QCD with massless quarks, the invariance of the Lagrangian under the chiral group $SU(N)_V \times SU(N)_A$ is spontaneously broken in the hadronic phase.
- In both cases, conjectures relate the critical behavior in gauge theories with that in spin models. The Svetisky-Yaffe conjecture relating the critical exponents in $SU(2)$ gauge theory with those of the Ising model was found to be well supported by numerical studies. The Pisarski-Wilczek conjecture connecting chiral symmetry restoration in two-flavor QCD with that in the $O(4)$ spin models now is also in accord with recent lattice studies based on scaling function behavior.
- In full QCD at vanishing baryon number density, deconfinement is signalled by a rapid increase of the entropy density towards its ideal quark-gluon plasma limit. The most rapid increase occurs at the same temperature as that defined by the peak in the susceptibility of the chiral condensate, so that the two critical phenomena here appear to coincide.
- The persistence of almost-critical behavior in the "rapid cross-over region" could be a reflection of a cluster percolation transition, which remains present even when the partition function becomes analytic.

There thus exists a rather well-developed theory of strong interaction thermodynamics for systems of vanishing overall baryon density. A number of interesting open questions remain, to be sure, but the overall picture has become quite clear: above a critical or quasi-critical temperature, strongly interacting matter at $\mu = 0$ finds itself in a new, deconfined state. The next thing to ask is what will happen if we introduce the baryon number density as a further variable of strongly interacting matter. We had already sketched some first answers in Chap. 1, and we will now, in the following chapter, return in more detail to the structure of the QCD phase diagram.

References

1. S.L. Adler, Phys. Rev. **177**, 2426 (1969)
2. R. Alkofer, H. Reinhardt, *Chiral Quark Dynamics* (Springer, Berlin, 1995)
3. Y. Aoki et al., Phys. Lett. B **643**, 46 (2006)
4. Y. Aoki et al., JHEP **0906**, 088 (2009)
5. A. Bazazov et al., (Hot QCD), Phys. Rev. D **85**, 065503 (2012)
6. J.S. Bell, R. Jackiw, Nuovo Cim. A **60**, 47 (1969)
7. T. Celik, J. Engels, H. Satz, Phys. Lett. B **125**, 411 (1983)
8. M. Cheng et al., Phys. Rev. D **81**, 054504 (2010)
9. A. Coniglio, W. Klein, J. Phys. A **13**, 2775 (1980)
10. H.-T. Ding et al., arXiv:1111.0185
11. S. Digal, E. Laermann, H. Satz, Eur. Phys. J. C **18**, 583 (2001)
12. S. Ejiri et al., Phys. Rev. D **80**, 094505 (2009)
13. J. Engels et al., Phys. Lett. B **365**, 219 (1996)
14. C.M. Fortuin, P.W. Kasteleyn, J. Phys. Soc. Jpn. **26**(Suppl.), 11 (1969)
15. C.M. Fortuin, P.W. Kasteleyn, Physica **57**, 536 (1972)
16. S. Fortunato, H. Satz, Phys. Lett. B **475**, 311 (2000)
17. S. Fortunato, H. Satz, Nucl. Phys. A **681**, 466c (2001)
18. K. Fukushima, Phys. Rev. D **68**, 045004 (2003)
19. R.V. Gavai, A. Goksch, M. Ogilvie, Phys. Rev. Lett. **56**, 815 (1886)
20. S. Gavin, A. Goksch, R.D. Pisarski, Phys. Rev. **D49**, 3079 (1994)
21. J. Goldstone, Nuovo Cim. **19**, 380 (1960)
22. Y. Hatta, T. Ikeda, Phys. Rev. D **67**, 014028 (2003)
23. O. Kaczmarek et al., Phys. Rev. D **83**, 014504 (2011)
24. M. Kanaya, M. Kaya, Phys. Rev. D **51**, 2404 (1995)
25. F. Karsch, E. Laermann, Phys. Rev. D **50**, 6954 (1994)
26. F. Karsch, E. Laermann, C. Schmidt, Phys. Lett. B **520**, 41 (2001)
27. J. Kertész, Physica A **161**, 58 (1989)
28. J. Kuti, J. Polónyi, K. Szlachányi, Phys. Lett. B **98**, 199 (1981)
29. J. Kogut et al., Phys. Rev. Lett. **48**, 1140 (1982)
30. J. Kogut et al., Phys. Rev. Lett. **50**, 353 (1983)
31. L.D. McLerran, B. Svetitsky, Phys. Lett. B **98**, 195 (1981)
32. L.D. McLerran, B. Svetitsky, Phys. Rev. D **24**, 450 (1981)
33. R. Pisarski, F. Wilczek, Phys. Rev. D **29**, 338 (1984)
34. J. Polónyi, K. Szlachányi, Phys. Lett. B **110**, 395 (1982)
35. K. Rajagopal, The chiral phase transition in QCD, in *Quark-Gluon Plasma 2*, ed. by R. Hwa (World Scientific, Singapore, 1995)
36. K. Rajagopal, F. Wilczek, Nucl. Phys. B **399**, 395 (1993)
37. H. Satz, Nucl. Phys. A **642**, 130 (1998)
38. H. Satz, Nucl. Phys. A **642**, 130c (1998)
39. H. Satz, Comput. Phys. Commun. **147**, 46 (2002)
40. B. Svetitsky, L. Yaffe, Nucl. Phys. B **210**, 423 [FS6] (1982)
41. F.Y. Wu, Rev. Mod. Phys. **54**, 235 (1982)

Chapter 7
The QCD Phase Diagram

> *And God said, Let the waters under the heaven be gathered*
> *together unto one place, and let the dry land appear: and it*
> *was so.*
>
> The Bible, *Genesis 1. Moses*

Abstract In this chapter, we turn to the phase structure of strongly interacting matter as a function of finite baryon density. We first discuss the different limiting forms for mesonic and baryonic matter, and elaborate the resulting phase pattern. Next, we consider the possible phase structure at low temperature and high baryon density, where deconfinement and chiral symmetry restoration need not coincide. Following this, we survey results obtained on the basis of the Nambu-Jona-Lasinio model, and we conclude with a discussion of the present status of finite density lattice QCD.

7.1 States of Matter in QCD: A Second Look

The essential features of hadron structure are color confinement and spontaneous chiral symmetry breaking. The former binds colored quarks interacting through colored gluons to color-neutral hadrons. The latter brings in pions as Goldstone bosons and gives the essentially massless quarks in the QCD Lagrangian a dynamically generated effective mass. We had already noted that both features will come to an end when hadronic matter is brought to sufficiently high temperatures and/or baryon densities. A priori, they need not end simultaneously; however, rather basic arguments suggest that chiral symmetry restoration occurs either together with or after color deconfinement [1].

In principle, QCD could thus lead to a three-state phase structure as a function of the temperature T and the baryochemical potential μ, as shown in Fig. 7.1 [2]. In such a scenario, color deconfinement would result in a plasma of massive "dressed"

© Springer International Publishing AG 2018

H. Satz, *Extreme States of Matter in Strong Interaction Physics*,
Lecture Notes in Physics 945, https://doi.org/10.1007/978-3-319-71894-1_7

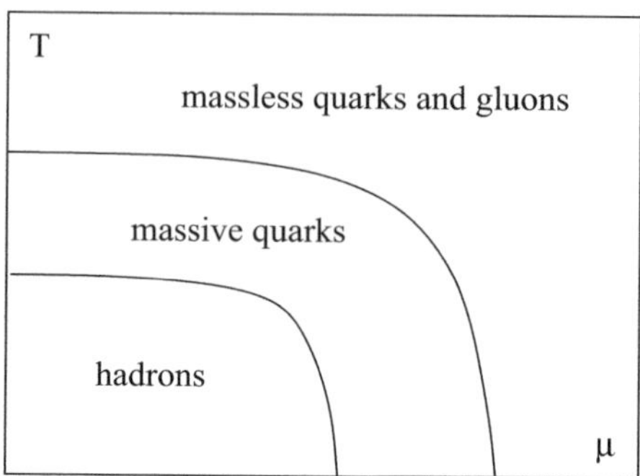

Fig. 7.1 Three-state scenario of QCD matter [2]

quarks; the only role of gluons in this state would be to dynamically generate the effective quark mass, maintaining spontaneous chiral symmetry breaking. At still higher T and/or μ, this gluonic dressing of the quarks would then "evaporate" or "melt", leading to a plasma of deconfined massless quarks and gluons: the conventional QGP, with restored chiral symmetry. Evidently, this view of things ignores the possibility of bosonic diquark binding and condensation as well as that of the color superconductivity states which could result as a consequence (see Fig. 1.5); but we shall return to these aspects later on.

The possible existence of a phase of quarks with a dynamically generated effective mass has been considered quite often [2–10], and in recent years, finite temperature lattice QCD investigations have provided an impressive amount of further information which could be used in determining the properties of constituent quarks [11–14]. However, as we saw in the last chapter, lattice studies at $\mu = 0$ have shown that there deconfinement and chiral symmetry restoration in fact coincide [15]. This is in accord with a scenario in which the constituent quark mass is a polarization cloud excited by the quark in the surrounding gluonic medium. With increasing temperature, this cloud melts away and at T_c, it has "evaporated", leaving pointlike quarks and gluons. In this region, there thus remains only one transition, the simultaneous onset of color deconfinement and chiral symmetry restoration. However, since the melting of the effective quark mass is a consequence of the hot gluonic environment, it is not at all evident that such behavior also occurs at low T and large μ. In that region, an intermediate plasma of massive quarks thus appears quite conceivable, separating hadronic matter from quark-gluon plasma [16].

In the next section, we first return to the limits of hadronic matter, already discussed in Chap. 3. Here we will in particular study in some detail the different limits for mesonic and baryonic matter. Following this, we consider constituent quarks and a possible deconfined phase in which chiral symmetry remains broken. After these rather phenomenological considerations, we summarize some more theoretical proposals for the QCD phase diagram, together with first numerical results from finite density lattice QCD. In the final section, we briefly return to diquarks and color superconductivity.

7.2 Interaction Regimes of Hadronic Matter

At low temperature and density, hadronic matter consists of interacting mesons and baryons. Increasing either temperature or baryon density eventually drives the system towards limits beyond which a description in terms of interacting hadrons breaks down. Since the dominant interactions in a dense meson gas are quite different from those in a dense baryon system, these limits are also of different nature. As a result, the boundary curve of the hadronic matter regime in Fig. 7.1 is expected to correspond to different transition patterns at low and high μ. Let us therefore begin by looking at the underlying interaction dynamics and its thermodynamic consequences.

All hadrons experience short-range attractive interactions. For systems of several mesons, or for systems of one baryon plus mesons, this leads to abundant resonance formation, with two pions combining to make a ρ, or a pion and a nucleon making a Δ, to cite just two examples. The size of these resonances is, as far as we can determine, essentially of the common hadronic scale of about 1 fm; multihadron resonances do not seem to become significantly larger in size. Nevertheless, larger clusters of mesonic matter are certainly possible, and since there is no indication of any repulsion between mesons, such clusters allow arbitrarily much overlap between constituents. The interactions in multiple baryon systems are also strongly attractive at separation distances of about 1 fm; but for distances below 0.5 fm, they become strongly repulsive. The former is what makes nuclei, the latter (together with Coulomb and Fermi repulsion) prevents them from collapsing. The repulsion between a proton and a neutron shows a purely baryonic "hard-core" effect and is connected neither to Coulomb repulsion nor to Pauli blocking of nucleons. As a consequence, the volume of a nucleus grows linearly with the sum of its protons and neutrons. We thus expect a dominantly attractive resonance interaction in the region of low baryon density, while at higher baryon density in addition baryon repulsion comes into play. These features alone already provide some first hints of the phase diagram of strongly interacting matter.

Since both mesons and baryons have an intrinsic spatial size of about 1 fm, the formation of percolating clusters provides a natural limit to the hadronic form of strongly interacting matter [16–19]. Increasing the density eventually leads to a medium in which the distance between a quark constituent from one hadron and an antiquark from another (using mesonic matter as example) is equal to or less than the typical hadronic size, so that defining specific quark-antiquark pairs as hadrons ceases to make any sense. We had already discussed this in the Introduction, and we illustrate it here once more in a schematic form in Fig. 7.2.

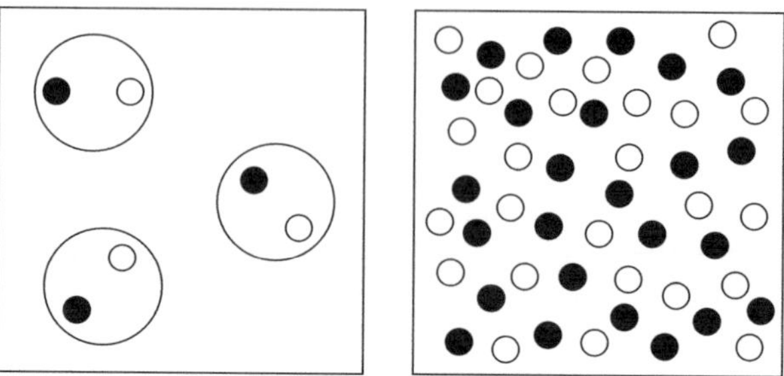

Fig. 7.2 Transition from confined (left) to deconfined (right) matter

In a baryon-rich medium, an increase of density of hard-core nucleons moreover leads to "jamming", i.e., a restriction in the mobility of the nucleons [20, 21], see Fig. 7.3. This is a genuine critical phenomenon, with the accessible volume as an order parameter. Up to a certain density n_j, a given nucleon can move anywhere in the overall volume, while for $n_B \geq n_j$, its range of mobility abruptly becomes finite. The resulting "gas-liquid" transition has been studied extensively in nuclear physics; it results in a first-order transition line starting at $\mu = \mu_j, T = 0$ and ending in a second order point at finite temperature (see Fig. 7.3c). It separates a low density "nucleon gas" from a higher density "nucleon liquid"; we shall here not address this nuclear matter phase structure any further. Turning to the high density form of nuclear matter, beyond the liquid-gas transition, we note that there the mobility of nucleons is restricted, but the percolation limit is not yet attained.

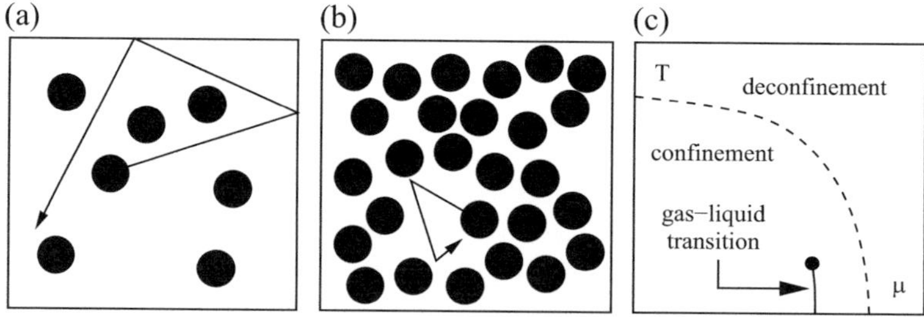

Fig. 7.3 States of matter for hard core baryons: full mobility (**a**), "jammed" (**b**); liquid-gas transition in nuclear matter (**c**)

At very high baryon density $n_B(T)$, the pressure of a gas of nucleons can be calculated, taking into account the reduction of the accessible volume; it will approach the Clausius form

$$P(T) = \frac{P_0(T)}{1 - n_B(T)V_e},\tag{7.1}$$

where P_0 denotes the pressure of pointlike nucleons, and V_e the effective excluded volume. The pressure of hard-core nucleons diverges when the density $n_B(T)$

approaches $1/V_e$, the random dense packing limit.[1] But here as well we reach eventually the formation of a percolating medium [23], in which the overall quark density is too high to define individual nucleons, because any given quark "sees" numerous others within reach. We thus consider the limit of confinement in the $T-\mu$ diagram of strongly interacting matter to be defined in general by the percolation of the relevant hadron species in the given region [16]. It should be emphasized that such a percolation limit is a well-defined geometric form of critical behavior, specified by critical exponents and corresponding universality classes, just as found for thermal critical behavior; see Chap. 2. The essential difference is that geometric singularities (formation of infinite clusters) do not necessarily imply non-analytic behavior for the partition function.

In general, we then have a percolation limit of overlapping mesons for increasing temperature at low baryon density, and a baryonic hard core percolation limit at low temperatures and high baryon density. The two resulting regimes in the $T-\mu$ plane are schematically illustrated in Fig. 7.4.

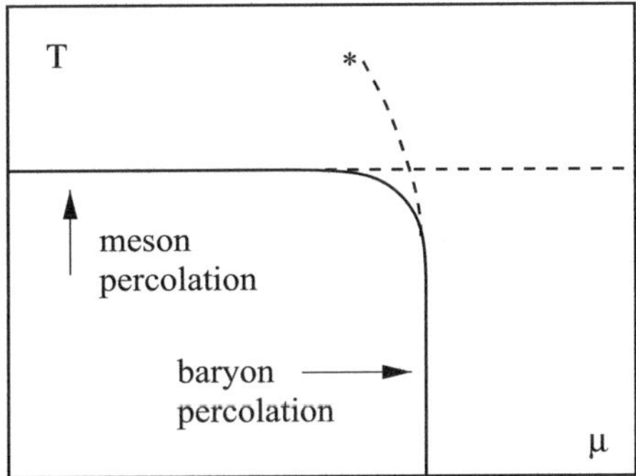

Fig. 7.4 Limits of hadronic matter

Let us estimate some transition parameter values. At $\mu = 0$, the percolation density for permeable hadrons (mesons and low density baryons) of size $V = 4\pi R_h^3/3$ is found to be (see Chap. 2)

$$n_M^{\text{dec}} \simeq \frac{1.2}{V_h} \simeq 0.6\,\text{fm}^{-3}, \tag{7.2}$$

with $R_h \simeq 0.8\,\text{fm}$ for the hadron radius. This is the density of the largest cluster at the onset of percolation, or, equivalently, the density at which the medium no longer allows a spanning vacuum. The corresponding temperature can be obtained once the hadronic medium is specified. At low baryon density, resonance formation is the

---

[1]The close packing limit of spheres, achieved in the orderly packing of oranges in a supermarket display, is not reached when the density of a medium of randomly distributed spheres is increased. This instead leads to the lower "random packing limit" $1/V_e$ [22], specified by the hard core volume $V_{hc} < V_e$.

dominant interaction, and this can be taken into account by replacing the interacting medium of pions and nucleons by an ideal gas of all observed resonance states [24, 25]. For such an ideal gas one finds as deconfinement temperature [16]

$$T_{\mathrm{dec}} \simeq 180\,\mathrm{MeV}, \tag{7.3}$$

which agrees well with the value presently obtained in finite temperature lattice QCD. Assuming mesons to be the dominant constituents, the density (7.2) implies

$$d_q^M \simeq \frac{1}{n_M^{1/3}} \simeq 1.2\,\mathrm{fm}, \tag{7.4}$$

for the average separation between quarks and/or antiquarks at the deconfinement point.

At the other extreme, for $T = 0$, we have to consider the percolation of nucleons with a hard core [16, 23]. Assuming a hard core radius $R_{\mathrm{hc}} = R_h/2$, one obtains for the critical density

$$n_B^{\mathrm{dec}} \simeq \frac{2}{V_h} \simeq 0.9\,\mathrm{fm}^{-3} \simeq 5 n_0, \tag{7.5}$$

a value about 30% higher than that for permeable hadrons, as consequence of the baryon repulsion; $n_0 \simeq 0.17\,\mathrm{fm}^{-3}$ denotes standard nuclear density. The value (7.5) can be used to obtain the percolation value of the baryochemical potential, using the excluded volume approach mentioned above to account for the repulsion in the determination of the density as function of μ (see Sect. 4.4). As result, one obtains

$$\mu_{\mathrm{dec}} \simeq 1.1\,\mathrm{GeV} \tag{7.6}$$

as deconfinement point for μ. The separation between quarks at this point becomes

$$d_q^B \simeq \frac{1}{n_B^{1/3}} \simeq 1.0\,\mathrm{fm}, \tag{7.7}$$

slightly less than at $T = 0$, due to the higher density. We note that since $\mu \geq M$, where M denotes the nucleon mass, this leaves as function of μ a rather small window

$$M \leq \mu \leq 1.2\,M \tag{7.8}$$

for the range of confined baryonic matter at T = 0. This window contains essentially all strongly interacting matter in the real world, from nuclei to neutron stars. The corresponding density range runs from standard nuclear matter to the deconfinement value (7.5) of about $5\,n_0$.

We thus confirm through percolation arguments that color deconfinement is expected to set in at hadron densities for which in general the quark constituents are separated by about 1 fm. At this point, any partitioning into hadrons becomes meaningless, and we have a medium of deconfined quarks.

The percolation picture also gives some indications about the nature of the different limits for hadronic matter. The percolation of permeable hadrons is in general not a thermal phase transition; it could thus correspond to a rapid cross-over [26]. Nevertheless, given specific dynamics, it can also result in singular behavior of the partition function [27–29]. Hadronic size alone does not suffice to decide the issue, and the form of the restoration of chiral symmetry was seen to depend on N_f (see Sect. 6.5). In the case of hard core percolation, the connection to thermodynamic critical behaviour is more evident [23]. If a system of hard-core constituents is subjected to a density-dependent negative background potential (the attractive nuclear interaction), first order critical behavior appears. A classical case is the van der Waals equation. The pressure in our hard core medium is given by Eq. (7.1), and the density n of the hard core constituents by

$$n(T, \mu) = \frac{n_0(T, \mu)}{1 + n_0(T, \mu)V_e}, \tag{7.9}$$

where the subscript 0 again refers to point-like constituents. In addition to the short-range repulsion, the forces between nucleons at a range of about 1 fm are strongly attractive, as already noted: this attraction is what allows the formation of nuclei. In the equation of state, it causes a reduction of the pressure, due to a decrease of the momentum of the nucleons and to a decrease in the number of nucleons hitting the surface of the volume. The attraction is thus generally argued to depend on n_B^2, leading to the well-known van der Waals equation

$$P(T) = \frac{P_0(T)}{1 - n_B(T)V_e} - a n_B^2(T), \tag{7.10}$$

where a is a constant measuring the strength of the attractive nucleon-nucleon interaction. In this case, we have a first order phase transition ending in a second order critical point, specified by the parameters a and V_e. We have included such an end point in Fig. 7.4, although its position so far remains open.

7.3   Constituent Quarks and Constituent Quark Plasma

A region of particular interest in Fig. 7.4 is that below the mesonic limit, but beyond the baryon boundary. Here matter is deconfined and thus cannot consists of baryons, so the binding of quarks to color-singlet triplets must be dissolved. On the other hand, quarks could still retain their gluonic dressing, leading to a plasma of deconfined but massive quarks. Thus here deconfinement and chiral symmetry restoration need not coincide. Let us look at this "terra incognita" in more detail [30]; it is (presently) accessible neither to lattice calculations nor to experiment, so it provides a tempting ground for speculation.

There are two different regimes for the quark infrastructure of hadrons, depending how we probe. Relatively hard probes, such as deep inelastic lepton-hadron scattering or hadron-hadron interactions at large momentum transfer, lead to massless pointlike quarks and gluons. In this regime, the parton model with hadronic quark and gluon distribution functions provides a suitable description. On the other hand, soft interactions, as seen in minimum bias proton-proton or pion-proton interactions, suggest that mesons/nucleons consist of two/three "constituent" quarks having a size of about 0.3 fm and a mass of about 0.3–0.4 GeV. Here many features are well accounted for by the additive quark model [31–33]. We can thus imagine that inside a hadron, a quark polarizes the gluon medium in which it is held through color confinement, and the resulting gluon cloud forms the constituent quark mass M_q [34, 35].

This picture is today found to be quite compatible with heavy quark correlation studies in finite temperature lattice QCD at vanishing baryon density. By evaluating Polyakov loop correlations in a QCD medium of two or three light quark flavors below deconfinement ($T < T_c$), one obtains the free energy $F(r, T)$ as a function of the quark separation distance r. In the low temperature limit, $F(r, T = 0)$ saturates beyond a separation of $r \simeq 1.5$ fm, converging to the value $F(\infty, T = 0) \simeq 1.2 \pm 0.1$ GeV [11]. This result is quite universal; it is obtained by separating a heavy quark-antiquark pair, where the separation requires the formation of a light $q\bar{q}$ pair to assure color neutrality. It is obtained as well also if we separate a heavy quark-quark pair, where the formation of a light antiquark-antiquark pair is necessary [12]. Moreover, it is reached for any color channel (singlet or octet for $Q\bar{Q}$, antitriplet or sextet for QQ). The large r behavior of all these cases coincides. To create an isolated heavy-light system, we thus need a gluonic energy input F_g, with

$$F(\infty, T = 0) = 2F_g \simeq 1.2 \, \text{GeV}. \qquad (7.11)$$

This result can also be checked experimentally, comparing the mass values of the open charm and open bottom mesons to those of the charm and bottom quarks. This yields

$$M_D - m_c = F_g = 0.60 \pm 0.10 \, \text{GeV}$$

$$M_B - m_b = F_g = 0.53 \pm 0.15 \, \text{GeV}, \qquad (7.12)$$

using the relevant mass values as given by the PDG listing [36], and thus confirms that the resulting value does not depend on the quark source mass. The sizes of the heavy-light mesons are expected to be of hadronic scale, so that a gluonic polarization system of spatial extent makes sense.

Light-light quark systems are also in accord with such a value of F_g, giving reasonable estimates of the vector meson masses, for both non-strange and strange ground states, as well as for the different ground state baryon masses. In the case of hadrons containing strange quarks, the relevant number of strange quark masses ($m_s \simeq 100\,\mathrm{MeV}$) has to be added, with

$$M_\rho \simeq F_g, \quad M_{K^*} \simeq F_g + m_s \tag{7.13}$$

as illustration.

What is the meaning of F_g? One possible and rather widely accepted interpretation is that it is the "mass" or the energy content of the gluonic string connecting quark and antiquark. With

$$F_g \simeq \sigma R_h \simeq 0.6 - 0.8\,\mathrm{GeV} \tag{7.14}$$

and using $\sqrt{\sigma} = 0.4\,\mathrm{GeV}$ and $R_h = 0.8 - 1.0\,\mathrm{fm}$, this does lead to the correct value of F_g, at least in the case of mesons; baryons are not so easily dealt with.

We want to consider here instead an alternative scenario in which F_g is the sum of the gluonic dressing masses of two constituent quarks. Then both mesons and baryons can be treated on equal footing, giving us

$$M_q = \frac{F_g}{2} + m_q \rightarrow 0.3 - 0.4\,\mathrm{GeV}, \tag{7.15}$$

where m_q denotes the bare quark mass; the last term thus corresponds to the light quark limit. We emphasize that the constituent quarks retain their intrinsic quantum numbers; the gluon cloud thus is color-neutral and without any spin.

Such an interpretation is, as noted, supported by the additive quark model [31–33]. In a collision energy range of about $\sqrt{s} \simeq 5 - 20\,\mathrm{GeV}$, in which hard processes do not yet play a significant role, the total cross sections for proton-proton and pion-proton collisions are given as

$$\sigma_{pp} = 3 \times 3\sigma_{qq} \simeq 38 \text{ mb}, \quad \sigma_{\pi p} = 2 \times 3\sigma_{qq} \simeq 25 \text{ mb}. \tag{7.16}$$

The predicted ratio 3/2 between proton and pion projectiles is seen to be in accord with the data; moreover, $\sigma_{pp} = \pi R_h^2$ leads to $R_h \simeq 0.9\,\mathrm{fm}$ for the hadronic radius. Using Eq. (7.16) and $\sigma_{qq} = \pi R_q^2$, we obtain

$$\sigma_{qq} \simeq 3.3 \text{ mb} \rightarrow R_q \simeq 0.33\,\mathrm{fm} \tag{7.17}$$

for the corresponding constituent quark sizes in the case of light bare quarks; we return to the more general case of $m_q \gg 0$ shortly. A similar constituent quark radius was also obtained through partonic arguments [35].

As mentioned, we consider the constituent quark to be made up of the bare quark and the gluonic polarization cloud surrounding it. This means that as we move a distance r away from the pointlike quark, we find an effective quark mass $M_q^{\mathrm{eff}}(r)$, depending on how much of the cloud we include at a given r. String breaking limits the size of the cloud, so that beyond $r_0 \simeq 0.3\,\mathrm{fm}$, the cloud mass saturates, with the constituent quark mass M_q as saturation limit. The resulting behavior [3, 4] is illustrated in Fig. 7.5, with $R_h \simeq 1\,\mathrm{fm}$ denoting the radius of color confinement.

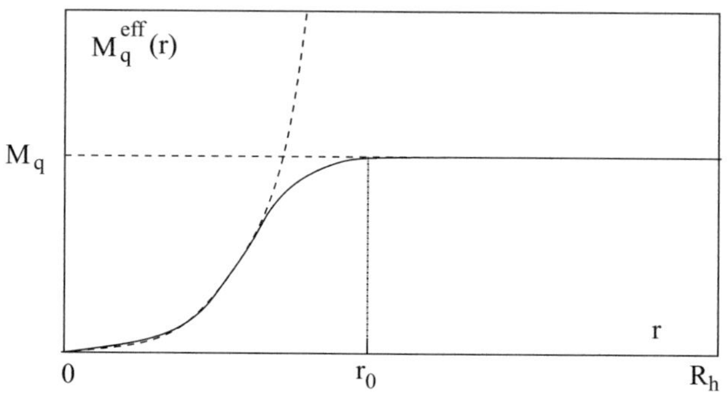

Fig. 7.5 Effective quark mass $M_q^{\mathrm{eff}}(r)$ as seen from a distance r [3, 4]

The conceptual scenario just discussed is supported by perturbative QCD estimates [37]. In the chiral limit ($m_q \rightarrow 0$), the effective quark mass $M_q^{\mathrm{eff}}(r)$ at scale r is determined by the (non-perturbative) chiral condensate $\langle \bar{\psi}\psi \rangle$ and renormalization factors,

$$M_q^{\mathrm{eff}}(r) = 4\, g^2(r)\, r^2 \left[\frac{g^2(r)}{g^2(r_0)} \right]^{-d} \langle \bar{\psi}\psi(r_0) \rangle, \tag{7.18}$$

where r_0 denotes a reference point for the determination of $\langle \bar{\psi}\psi(r_0) \rangle$; it is something like the meeting point of perturbative and non-perturbative regimes. For the case of massless quarks with $N_c = 3$, $N_f = 3$, the anomalous dimension is $d = 4/9$, and

$$g^2(r) = \frac{16\pi^2}{9} \frac{1}{\ln[1/(r^2 \Lambda_{\mathrm{QCD}}^2)]} \tag{7.19}$$

is the running coupling at scale r. The constituent quark mass M_q is defined as the solution of Eq. (7.18) at the scale $r = 1/2M_q$ [37]. This allows us to rewrite Eq. (7.18) in the form

$$M_q^{\mathrm{eff}}(r) = M_q \left[4M_q^2 r^2 \right] \left[\frac{g^2(1/2M_q)}{g^2(r)} \right]^{d-1}, \tag{7.20}$$

showing how the effective quark mass decreases at short distances $r < 1/2M_q$, starting from the constituent quark value. The value itself is determined by the non-perturbative chiral condensate at some reference point r_0. From Eq. (7.20) we have

$$\frac{M_q^3}{\langle \bar{\psi}\psi(r_0)\rangle} = \left(\frac{16\pi^2}{9}\right)\left(\ln[1/(r_0^2\Lambda_{QCD}^2)]\right)^{-4/9}\left(\ln[4M_q^2/\Lambda_{QCD}^2]\right)^{-5/9}. \tag{7.21}$$

As reference scale we use $r_0 = 1/2M_q$, since that is where the perturbative evolution stops and the non-perturbative regime starts. We thus obtain

$$\frac{M_q^3}{\langle \bar{\psi}\psi(r_0)\rangle} = \frac{16\pi^2}{9}\frac{1}{\ln(4M_q^2/\Lambda_{QCD}^2)}, \tag{7.22}$$

showing that the constituent quark mass indeed provides a scale parameter for chiral symmetry breaking. To get an estimate for its value, we use $\Lambda_{QCD} = 0.2\,\text{GeV}$ and $\langle \bar{\psi}\psi(r_0)\rangle^{1/3} = 0.2\,\text{GeV}$; this yields as solution of Eq. (7.22) $M_q = 375\,\text{MeV}$; the corresponding constituent quark radius becomes $R_q = r_0 = 0.26\,\text{fm}$. At the point $r_0 = 1/2M_q$, Eq. (7.19) gives for the strong coupling

$$\alpha_s(r_0) = \frac{g^2(r_0)}{4\pi} \simeq 0.5; \tag{7.23}$$

we have assumed a perturbative evolution in r up to this point, which may be subject to some doubt. The value (7.23) is in approximate agreement with the value obtained in static quark lattice studies [38], which indicate, however, remaining non-perturbative contributions at this r. Nevertheless, it seems remarkable that the results agree rather well with the estimates obtained above from lattice calculations as well as from experiment.

The constituent quark radius is thus found to be $R_q = 1/2M_q$ in the chiral limit. More generally, we then expect

$$R_q \simeq \frac{1}{2(M_q + m_q)}, \tag{7.24}$$

leading to a decrease in size with increasing bare mass. This is in accord with the decreasing cross sections for the interaction of strange or charm mesons [39].

The effective constituent quark mass is thus determined by the size and energy density of the gluon cloud, or equivalently, by the chiral condensate value in the non-perturbative region. How do these quantities change with temperature in a hadronic medium at vanishing baryon density?

This can again be deduced from heavy quark correlation studies as a function of the temperature of the medium. They show that the effective mass of the gluon cloud of an isolated static color charge (obtained by separating a static $Q\bar{Q}$ pair), starting from confinement values around 300 MeV, drops sharply at $T \simeq T_c$ [15]. This is accompanied by a corresponding drop of the screening radius. We thus expect the effective quark mass to show the temperature dependence illustrated in Fig. 7.6.

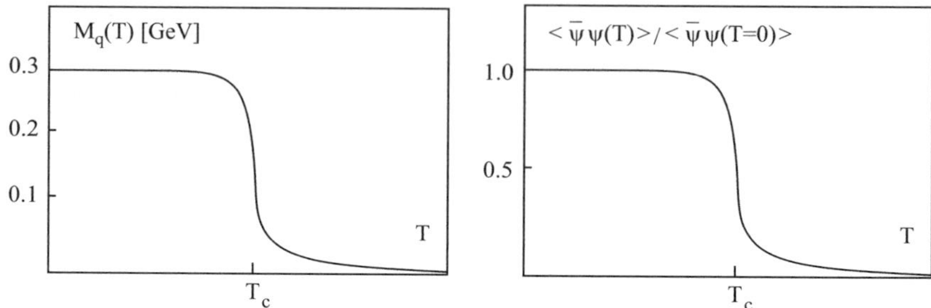

Fig. 7.6 Constituent quark mass $M_q(T)$ (left) and chiral condensate $\langle \bar{\psi}\psi(T)\rangle$ (right) as function of temperature T

Complementary to this, the temperature dependence of the chiral condensate is determined directly in finite temperature lattice QCD. Its behavior is also shown in Fig. 7.6; at the deconfinement point, the chiral condensate vanishes as well, as discussed in the previous chapter. This is in accord with the idea that at the deconfinement point, the gluon cloud around the given quark has essentially evaporated.

These considerations show that there are two distinct ways to reach chiral symmetry restoration. On the one hand, even for an interquark distance $2R_h$ well above $2R_q$, a sufficiently hot medium will through gluon screening cause the effective quark mass to vanish, as shown in Fig. 7.6. On the other hand, when a cold medium becomes so dense that the average interquark distance is $2R_q$ or less, the quarks form a connected cluster containing pointlike bare quarks.

We note in passing that the two scales, R_h and R_q, have also been considered as the quark and gluon confinement scales, respectively. This implies that color-neutral hadrons have size R_h, whereas color-neutral glueballs have the much smaller intrinsic size R_q, and the spatial ground state glueball size is indeed in most calculations found to be about $R_q \simeq 0.3$ fm.

Percolation arguments thus suggest that also at $T = 0$, color deconfinement sets in at hadron densities for which in general the quark constituents are separated by about 1 fm. At this density, any partitioning into hadrons becomes meaningless, and we have a medium of deconfined quarks of mass $M_q \simeq 0.4$ GeV and size $r_0 \simeq 0.3$ fm, separated by a distance $r \simeq 1$ fm $> r_0$. Hence at low temperatures in the density range corresponding to $r_0 \leq r \leq 1$ fm, the quarks can retain their effective constituent mass, so that the deconfined medium now is a plasma of quarks of finite mass and spatial extent, with continued chiral symmetry breaking. A sufficient further increase in density will eventually lead to overlap and percolation of the constituent quarks. We assume that beyond this percolation point, chiral symmetry is effectively restored. Let us see what density the above obtained value of R_q leads to.

At $T = 0$, the density for a system of quarks of mass M_q is given by

$$n_q(\mu_q) = \frac{2}{\pi^2}(\mu_q^2 - M_q^2)^{3/2},\tag{7.25}$$

with $\mu_q = \mu/3$ for the quark chemical potential. The percolation condition for quarks of radius R_q (see Eq. (7.2)),

$$n_q^{\rm ch} = \frac{1.2}{(4\pi R_q^3/3)} \simeq \frac{0.29}{R_q^3},\tag{7.26}$$

then defines the onset of chiral symmetry restoration. With $R_q \simeq 0.3\,\rm fm$, we obtain

$$n_B^{\rm ch} \simeq 3.5\,\rm fm^{-3} \simeq 3.9 n_B^{\rm dec} \simeq 20 n_0,\tag{7.27}$$

indicating that the baryon density threshold for chiral symmetry restoration is about four times higher than that for color deconfinement. The corresponding value for the baryochemical potential is found to be $\mu_B^{\rm ch} \simeq 2.2\,\rm GeV$, to be compared to $\mu_B^{\rm dec} \simeq 1.1\,\rm GeV$. Using the μ counterpart of the two-loop form (7.19), the strong coupling α_s has now dropped to the value $\alpha_s(\mu_c^{\rm ch}) \simeq 0.5$. The resulting phase structure is schematically illustrated in Fig. 7.7.

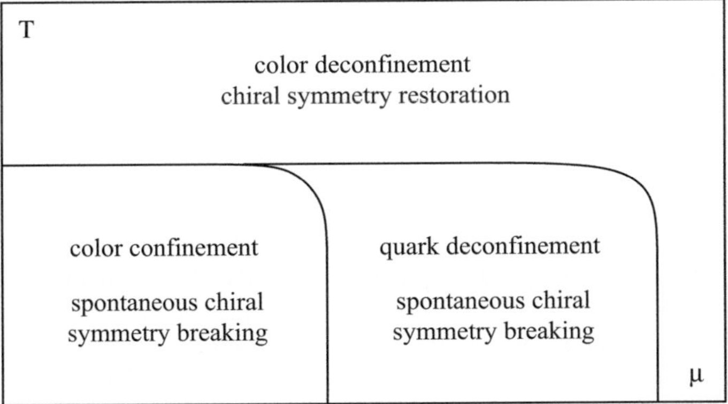

Fig. 7.7 Phase structure of strongly interacting matter

Our considerations thus suggest the existence of a plasma of massive deconfined quarks between the hadronic matter state and the quark-gluon plasma. In this state, quarks are deconfined; gluons, however, are "bound" into the constituent quark mass and thus remain in a sense confined. The quark dressing is made up of gluons which form a color-neutral cloud, so that the massive quarks retain their fundamental color state as well as their other intrinsic quantum numbers. The effective degrees of freedom in the resulting quark plasma thus are just those of massive quarks. Its lower limit in baryon density is defined by the onset of vacuum formation, forcing the quarks to bind into color neutral nucleons. The corresponding high density limit is given by the percolation point of the (spatially extended) quarks, beyond which there is a connected medium containing bare quarks and gluons. Finally, increasing the temperature at fixed μ leads here, just as in the hadronic phase, to an evaporation of the gluonic dressing of the quarks and thus to a restoration of chiral symmetry.

The constituent quarks in the deconfined medium will in general be interacting with each other. Of particular interest here is the presence of a qq attraction, which at sufficiently low temperatures could lead to the formation of bound colored bosonic qq states ("diquarks"). Baryons have in fact often been considered in terms of two quarks bound in a color antitriplet state, which then in turn binds with the remaining quark to form a color singlet. For $T \simeq 0$, extensive studies have addressed the formation of QCD Cooper pairs; their condensation would then lead to a color superconductor [40]. In contrast to electromagnetic superconductors, where only the global background field (phonons) of the medium can result in a binding of electrons into Cooper pairs, we have in QCD a local qq color anti-triplet attraction, provided by gluon exchange. Hence diquarks could here exist as local "particle" bound states and not just as momentum state pairs of undefined spatial separation. The thermodynamics of a medium of such "localizable" diquarks was in fact considered some time ago [41–43].[2] Moreover, heavy quark studies [12] indicate that the interquark potential for a color anti-triplet QQ pair in a deconfined medium is attractive and essentially identical to that for a color singlet $Q\bar{Q}$ pair.

The resulting picture of the new medium thus parallels somewhat that of hadronic matter at $\mu \simeq 0$, where resonance interactions lead to a gas of basic hadrons (pions and nucleons) plus their resonance excitations. Here we have instead a gas of basic constituent quarks, together with the diquark excitations formed through their interaction.[3] The essential difference is that the basic "particles" now are massive colored fermions, which can exist only in the colored background field provided by a sufficiently dense strongly interacting medium.

An interesting open aspect of a massive quark phase is the role of pions. As long as chiral symmetry remains broken, pions will be present as Goldstone bosons, even though their hadronic features as $q\bar{q}$ states of standard hadronic size no longer exist.

We have argued that nuclear matter, with color confinement and chiral symmetry breaking, is separated from the canonical plasma of massless quarks (plus some gluons and antiquarks at $T \neq 0$) by an intermediate phase of massive quarks as basic constituents. Recent arguments dealing with strongly interacting matter in the large N_c limit have introduced "quarkyonic" matter as an intermediate phase [44, 45]. The relation of these considerations to our scenario remains an interesting, though still

---

[2] It should be noted, however, that possible observable consequences (H-dibaryon, diquark contributions to hadronic structure functions from deep inelastic scattering) in the low baryon density region have not been found.

[3] We restrict ourselves here to diquarks. In principle, the formation of multi-quark clusters seems also conceivable.

rather open question. We comment here only briefly and refer to Ref. [30] for further comparisons.

The limit of large N_c results in several rather drastic modifications of the phase diagram. The state of color deconfinement and chiral symmetry restoration becomes a gluon plasma, since the gluonic degrees of freedom ($N_c^2 - 1$) dominate of the N_c quark states. The baryon density in the hadronic regime is given by

$$n_B \sim \exp\{(\mu - M)/T\}, \tag{7.28}$$

where M again denotes the nucleon mass. Since both M and μ are linear in N_c, the nuclear matter region is in the large N_c limit contracted to $\mu = M$, so that the hadronic regime becomes purely mesonic. The resulting phase diagram thus features mesonic matter for $T \leq T_c$, $\mu < M$, and a gluon plasma for $T \geq T_c$ and all μ. The remaining section, with $T \leq T_c$ and $\mu \geq M$, is the regime of the proposed quarkyonic matter. Its effective color degrees of freedom are $d_{\text{eff}}^c \sim N_c$ and it has non-vanishing baryon density; it must thus consist of deconfined quarks and confined gluons, either as quark dressing or as glueballs.

7.4   The Nambu–Jona-Lasinio Model

The phase diagram according to QCD is obviously one of the central topics of strong interaction thermodynamics, and hence over the years it has been addressed in many studies. Nevertheless, there exists so far no firm theoretical basis. The phase and transition structure is inherently non-perturbative, and the only ab initio tool we have in that domain is lattice QCD. Unfortunately, as we had already pointed out in Chap. 6, the canonical method for evaluating lattice QCD, numerical simulation based on Monte Carlo techniques, breaks down for finite baryon density. For this reason, much of what is presently discussed for the QCD phase structure is based on effective field theory models, in particular on the Nambu–Jona-Lasinio (NJL) model [46, 47], its application to QCD [48–50] and its further extensions [51–59]. The basic idea here is to model the Lagrangian of QCD by one in which the interaction is given in terms of quarks alone.

The Nambu–Jona-Lasinio model, as applied in QCD, addresses the chiral properties of quark interactions.[4] The QCD Lagrangian (5.1)/(5.2) is replaced by

$$\mathcal{L}_{NJL} = \bar{\psi}(i\gamma^\mu \partial_\mu - m_q)\psi + G\left\{(\bar{\psi}\psi)^2 + (\bar{\psi}i\gamma_5\psi)^2\right\}, \tag{7.29}$$

where we have suppressed the color and flavor indices. It contains no gluon fields; the interaction between quarks is instead assumed to be described by a point-like

---

[4]The model was initially proposed for meson-nucleon interactions and then extended to QCD, with quarks replacing nucleons.

chirally symmetric four-fermion term. As a consequence, it contains neither color confinement nor asymptotic freedom. The interaction term is just the bilinear form $\bar{\psi}\psi$ leading to the chiral condensate and its chiral transform, rendering the sum invariant under chiral transformations. More recently, there have been attempts to add confinement features a posteriori [55–59]. The constant G is of dimension $(\text{mass})^{-2}$ and thus brings in the hadronic scale through an adjustable parameter.

The interesting feature for our considerations is that the interaction in Eq. (7.29) leads to a dynamically generated effective or constituent quark mass M_q, determined by the "gap equation"

$$M_q = m_q + 4N_c N_f G M_q \int \frac{d^3 p}{(2\pi)^3} \frac{1}{\sqrt{p^2 + M_q^2}}. \tag{7.30}$$

Its non-trivial solution leads to an effective mass M_q much larger than the current quark mass m_q, so that Eq. (7.30) indeed specifies the gap $M_q - m_q$ between the effective and the current quark mass. The integral in Eq. (7.30) is in turn related to the chiral condensate, with

$$\langle \bar{\psi}\psi \rangle = -2N_c M_q \int \frac{d^3 p}{(2\pi)^3} \frac{1}{\sqrt{p^2 + M_q^2}}, \tag{7.31}$$

per flavor degree of freedom, giving

$$M_q = m_q - 2G N_f \langle \bar{\psi}\psi \rangle. \tag{7.32}$$

This expression is evidently the NJL counterpart of the perturbative constituent quark mass obtained in Eq. (7.18). This becomes even more explicit in the evaluation of M_q and $\langle \bar{\psi}\psi \rangle$. The integrals in Eqs. (7.30) and (7.31) are ultra-violet divergent and have to be regularized. This is generally achieved by introducing a momentum cut-off $|p| \leq \Lambda$, and the removal of the large momenta effectively corresponds to the elimination of the short-distance region which was described perturbatively in Eq. (7.18). The formulation of the NJL model thus contains three open parameters, the bare quark mass m_q, the dimensionful coupling constant G, and the cut-off Λ. On the other hand, the value of the chiral condensate, together with that of the bare quark mass m_q, must satisfy the Gell-Mann–Oakes–Renner relation

$$f_\pi^2 m_\pi^2 = -m_q \langle \bar{\psi}\psi \rangle. \tag{7.33}$$

Hence, given the pion decay constant, $f_\pi \simeq 90$ MeV, the pion mass, $m_\pi \simeq 140$ MeV, and the vacuum chiral condensate per flavor degree of freedom, $\langle \bar{\psi}\psi \rangle^{1/3} \simeq 250$ MeV, the three open quantities G, Λ and m_q can be determined.[5] For the noted

---

[5]Note that Eq. (7.30) is a self-consistency relation for M_q, and thus it need not have a solution for all parameter values, nor need the solution be unique.

parameter values, one finds [54] $G \simeq 5\,\mathrm{GeV}^{-2}$, $\Lambda \simeq 0.65\,\mathrm{GeV}$ and $m_q \simeq 10\,\mathrm{MeV}$, leading to $M_q \simeq 320$ MeV as constituent quark mass, quite in accord with what we found above.

At finite temperature and density, the gap equation (7.30) becomes

$$M_q(T, \mu) = m_q + 4N_c N_f G \int \frac{d^3p}{(2\pi)^3} \frac{M_q}{E} \left(1 - N_q^+(T, \mu) - N_q^-(T, \mu)\right), \qquad (7.34)$$

where

$$N_q^{\pm} = \frac{1}{1 + \exp\{E \pm \mu\}/T} \qquad (7.35)$$

specifies the quark/antiquark occupation numbers, with $E = \sqrt{p^2 + M_q^2}$. Using these,

$$n_q(T, \mu) = 2N_c N_f \int \frac{d^3p}{(2\pi)^3} \frac{M_q}{\sqrt{p^2 + M_q^2}} \left(N_q^+(T, \mu) - N_q^-(T, \mu)\right), \qquad (7.36)$$

gives the overall quark number density $n_q(T, \mu)$. In vacuum, for $T = \mu = 0$, $N_q^{\pm} = 0$, so that Eq. (7.34) reduces to (7.30). For increasing T and/or μ, the weight factor $(1 - N_q^+(T, \mu) - N_q^-(T, \mu))$ decreases, and so does M_q. In particular, in the chiral limit $m_q \to 0$, the gap equation has solutions only for temperatures $T \le 200$ MeV at $\mu = 0$ and for baryon densities $\rho_B \le 2\rho_0$ at $T = 0$, with $\rho_B = 3n_q$ and $\rho_0 \simeq 0.17\,\mathrm{fm}^{-3}$ for standard nuclear matter density [51–54]. These points thus specify the onset of chiral symmetry restoration in the NJL model; the two limiting cases are illustrated in Fig. 7.8.

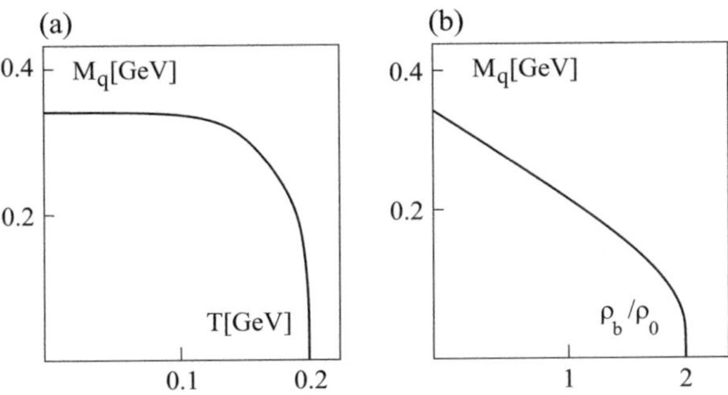

Fig. 7.8 Constituent quark mass at $\mu = 0$ as a function of the temperature (**a**), and at $T = 0$ as a function of the baryon density (**b**); both figures are for the chiral limit $m_q = 0$

The general phase boundary in the chiral limit is given by [54]

$$\mu^2 + \left(\frac{\pi^2}{3}\right) T^2 = \Lambda^2 \left(1 - \frac{\pi^2}{G\Lambda^2 N_c N_f}\right) \qquad (7.37)$$

and illustrated in Fig. 7.9a. Using expression (4.20), we can translate the baryochemical potential into the baryon density $\rho_b = 3\,n_q$ and thus obtain the corresponding phase diagram as function of T and ρ_b; it is shown in Fig. 7.37b. The increase of the critical baryon density with T, starting from the $T = 0$ limit, is an effect of Fermi statistics: for $T \neq 0$, the occupation of momenta above the limiting $T = 0$ Fermi momentum becomes possible.

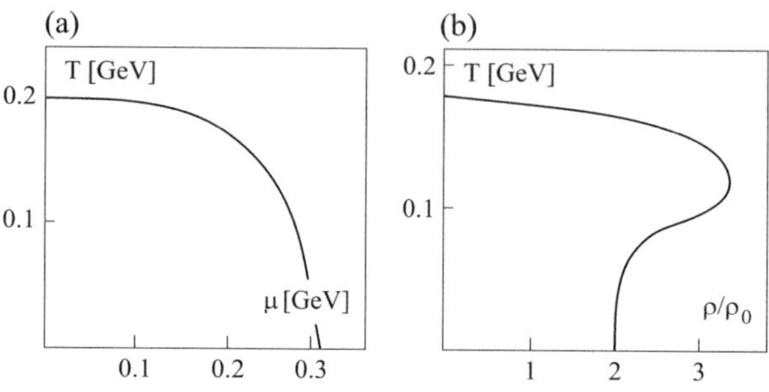

Fig. 7.9 Phase boundary for the NJL model in the chiral limit, (**a**) as a function of temperature and baryochemical potential, and (**b**) as a function of temperature and baryon density

From Eq. (7.34) it follows that in the chiral limit $m_q \to 0$ the constituent quark mass, and thus (see Eq. (7.32)) also the chiral condensate, vanish outside the region defined by Eq. (7.37). In this limit, the NJL model thus leads to a two-state phase structure defined by the vanishing of the chiral order parameter $\langle \bar{\psi}\psi \rangle$. At low T and/or μ, the system is in a "hadronic" phase of broken chiral symmetry, in which the quarks have a finite, dynamically generated effective mass; beyond the phase boundary specified by $\langle \bar{\psi}\psi \rangle$, it is in a "quark" phase, in which chiral symmetry is restored and the effective quark mass is zero.

The chiral order parameter $\langle \bar{\psi}\psi \rangle$ vanishes smoothly along the entire phase boundary in T and μ. This seems to suggest a second-order chiral phase transition everywhere. More detailed studies [48–50] show, however, that this need not be so. The crucial point is that the self-consistency nature of the gap equation allows more than one solution for a given set of parameters; in particular $M_q = 0$ is always a trivial solution. To determine the order of the transition, the grand thermodynamic potential per volume V,

$$\Omega(T, \mu) = -\frac{T}{V} \ln Tr \, \exp\left(-\frac{1}{T}\right) \int d^3x \, \{\mathcal{H} - \mu \mathcal{N}_B\} \qquad (7.38)$$

has to be evaluated in terms of T and μ as a function of the effective quark mass M_q; here the trace is over all physical states. The system will then choose the state of lowest Ω, and as long as the choice is unique, the transition is continuous. For the actual calculations, see e.g. Refs. [48–54]; here we only indicate the main result, which is illustrated in Fig. 7.10 for the specific choice of parameters already noted above, but in the chiral limit $m_q = 0$. For $\mu = 0$, we have at low temperatures a unique minimum of Ω, defining the constituent quark mass $M_q \simeq 320$ MeV. With increasing T, the maximum at $M_q = 0$ decreases, and at a specific $T = T_{ch}$, the maximum is gone and the minimum is at $M_q = 0$. It remains there for all higher T. As a result, for $\mu = 0$ the chiral restoration transition is continuous (second-order). For $T = 0$, the result for low μ is similar, the thermodynamic potential has again a well-defined minimum at $M_q(0) \simeq 320$ MeV. Here, however, with increasing μ a further relative minimum develops at $M_q = 0$, and at some μ_{ch}, the value of the potential at the two minimum points becomes equal. The presence of two equal minima corresponds to a discontinuity in M_q and hence also in $\langle \bar{\psi}\psi \rangle$, leading to the conclusion [48–50] that in the chiral limit at $T = 0$, the NJL model predicts a first-order phase transition at $\mu = \mu_{ch}$. The two limiting cases are illustrated in Fig. 7.10. For the result as a function of baryon density, it is assumed that the lower density corresponds to the vacuum density value of M_q, the upper one to the chiral restoration value $M_q = 0$ [51–53].

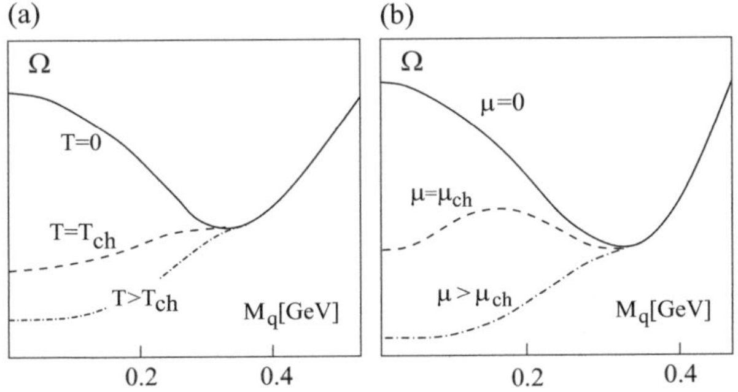

Fig. 7.10 Thermodynamic potential as function of the constituent quark mass M_q (**a**) for different values of T at $\mu = 0$ and (**b**) for different μ at $T = 0$

Increasing the temperature will weaken the first-order transition, until it eventually ends in a second-order critical point $T = T^*, \mu = \mu^*$. For values of μ below μ^*, the thermodynamic potential has a unique minimum, and so the transition is from then on continuous. As a result of these considerations, Fig. 7.8 has a fine structure schematically illustrated in Fig. 7.11. At low T, a first-order transition separates the phases of broken and restored chiral symmetry, which terminates at a critical point, beyond which the transition is of second-order. The position of the critical end point is very much dependent on the details of the model, the number of flavors and the parameters chosen. Moreover, for physical quark masses, chiral symmetry is broken explicitly, so that now M_q and chiral condensate never vanish exactly. This implies that continuous transitions will be washed out; there is no longer any singular behavior of thermodynamic observables, only a possible "rapid cross-over" of regimes. The first-order transition, however, will persist for a range of (small) quark masses, since the discontinuity cannot disappear instantaneously when m_q is "turned on".

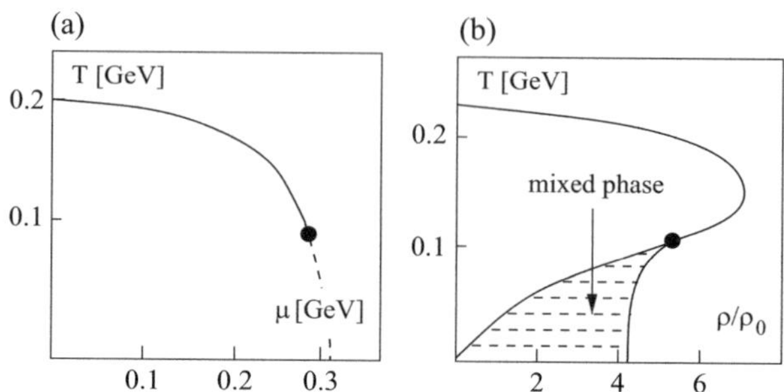

Fig. 7.11 NJL phase structure (**a**) as function of T and μ, solid line: continuous transition, dashed line: first-order transition; (**b**) as function of T and baryon density relative to standard nuclear density

We note that this conclusion is quite similar to the one obtained in our conceptual considerations above. There, the short-range nuclear repulsion in baryonic matter, competing with a longer-range nuclear attraction, led to a Van der Waals scenario and hence to a first order percolation transition. In contrast, meson percolation provides in general only a geometric transition and hence no singular behavior for thermodynamic quantities.

For the interpretation of the NJL-model results, a caveat seems appropriate. As mentioned, the model contains neither gluons nor confinement, and, at least for increasing temperature at $\mu = 0$, that is a rather drastic approximation. The gluon interactions lead to a string tension, and this in turn provides color confinement. Screening of the gluon interaction at higher temperatures in turn results in color deconfinement. It is thus rather remarkable that all these features can somehow be approximated by a chirally symmetric four-fermion interaction alone. And in particular, it is evident that such a description cannot lead to different forms of chiral behavior at $\mu = 0$ and $T = 0$, as we had considered in Sect. 7.3. The mentioned extensions of the NJL model to include confinement features [55–59] are patterned according to the lattice results at $\mu = 0$ and thus by construction cannot result in a qualitatively different behavior at $T = 0$.

The result of these considerations is thus that in the case of two light and one heavy quark species, the phase diagram has the structure shown in Fig. 7.12 [60, 61]. We concentrate here on the regions of confinement and deconfinement, ignoring for the moment the possibilities of a massive quark plasma and a colored diquark state. The diagram shows a non-singular "cross-over" region for $0 \leq \mu < \mu_{\text{crit}}$, a critical point (continuous transition) at μ_{crit}, and then a line of first order transitions down to $T = 0, \mu_c$. It is evidently of great interest to check in lattice simulations whether this picture is correct.

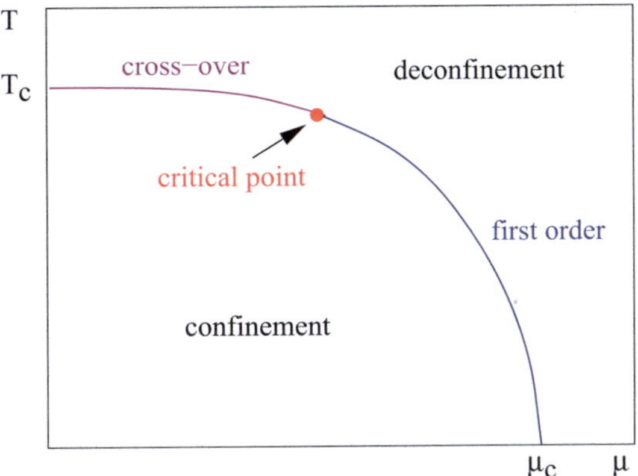

Fig. 7.12 Expected QCD phase structure for one heavy and two light quarks

7.5   QCD at Finite Baryon Density

Given the various model proposals for the phase diagram of QCD as function of temperature and baryon density, the interest in ab initio QCD calculations became more and more intensive, in spite of the known difficulties. Could one find evidence for a critical point at some μ and T, at which the "rapid cross-over" would terminate in a critical point and then turn into a discontinuous transition?

The basic problem for lattice evaluations of finite baryon density QCD is that the weight $P(U) \sim (\det Q)^{N_f}$ in the integral (5.39)/(5.40) is no longer positive-definite, since the Fermion determinant Q becomes complex when $\mu \neq 0$. The standard Metropolis procedure for reaching equilibrium configurations is thus no longer applicable, so that some new approach is required.

First such attempts have appeared in the past few years.

- In the *reweighting method* [62], the lattice configurations are calculated at $\mu = 0$ and extended to finite μ with the help of the ratios of the fermion determinants at $\mu = 0$ and $\mu > 0$, which are then used to establish the weight of a given configuration.
- Using *analytic continuation* [63, 64], the calculations are performed for imaginary μ and then analytically continued to real values.
- In the *power series approach* [65], the fermion determinant is expanded in powers of μ/T and the resulting expressions are evaluated term by term.

In all schemes, one obtains an extension from $\mu = 0$ to small but finite μ. Applied to the $(2 + 1)$ flavour case, the results obtained by the different methods applied to the $(2 + 1)$ flavour case were initially all consistent with the phase diagram shown in Fig. 7.12.[6] The position of the critical end-point of the line of first order transition

---

[6]Recent studies based on analytic continuation [66] have modified the conclusions of this approach: they do not indicate critical behavior for any value of μ and thus suggest for the $(2 + 1)$ case a cross-over behavior in the entire $T - \mu$ plane.

depends on the position of the physical point in the $m_{u,d} - m_s$ plane, see Fig. 6.4. Thus the cross-over region disappears and the critical point moves to $\mu = 0$ for $m_{u,s} = 0, m_s > m_s^{\text{tri}}$. The expected general phase structure in the three-dimensional space defined by $m_{u,d}, m_s$ and μ is outlined in [67].

All methods so far have in common at least two basic short-comings. They start from $\mu = 0$ calculations and then extend these in some approximative way. This does not really permit a quantitative error determination for the results obtained. Moreover, in view of the more or less analytic extension procedure used, it appears likely that they will fail when a critical point is reached. Whether such a failure is indicative of the existence of such a point is not evident; but it certainly prevents a study in the region of large μ and low T.

To consider the situation in some detail, we use the power series approach. In order to address the critical behavior of QCD at finite temperature and finite baryon density, we recall the general pattern discussed in Chap. 2 for spin systems. The thermodynamic potential $\Omega(T, H) = T \ln Z(T, H)$, depends on two variables:

- a "temperature" T, which describes the energy dependence of the undisturbed system and which for $H = 0$ leaves $\Omega(T, 0)$ invariant under the symmetry in question;
- an "external field" H, which shows the reaction of the system to an outside intervention breaking this symmetry.

In the case of QCD at finite baryon density, the "temperature" variable becomes two-dimensional, with T and μ as variables, while the quark mass m_q plays the role of the external field [68–70]; for simplicity, we restrict ourselves for the moment to the case of two light quark flavors, where in the chiral limit the transition at $\mu = 0$ is second order and presumably in the universality class of the three-dimensional $O(4)$ spin system. Instead of a critical point, we now get a critical line in the $T - \mu$ plane, and in the vicinity of this line, the temperature variable is in leading order near $\mu = 0$ given by

$$\tau = \frac{1}{t_0}\left[\left(\frac{T - T_c(\mu)}{T_c(\mu)}\right) + \kappa\left(\frac{\mu}{T}\right)^2\right] = \frac{1}{t_0}\left[t + \kappa\,\bar{\mu}^2\right] \tag{7.39}$$

where κ gives the relative position in the plane and t_0 the scale. For spin systems, we had $\tau = t = (T - T_c)/T$ as variable for the critical behavior; Eq. (7.39) extends this to finite μ. The baryochemical potential enters through the fugacity in the form $\bar{\mu} \equiv \mu/T$, and the dependence is quadratic, since the system is invariant under baryon-antibaryon interchange. As external field variable, we have

$$h = \frac{1}{h_0}\frac{m_q}{T_c}; \tag{7.40}$$

in view of the open scale parameter h_0, the normalization by T_c is arbitrary.

The thermodynamics of the system can now be derived from the pressure

$$P(t, \bar{\mu}, m_q) = (T/V) \ln Z(t, \bar{\mu}, m_q) \tag{7.41}$$

as the basic function. Suppressing the m_q dependence for the moment and setting $t_0 = 1$, we obtain at fixed μ

$$\epsilon(t, \bar{\mu}) = (t + 1) \left(\frac{\partial P}{\partial t} \right)_{\bar{\mu}} - P(t, \bar{\mu}) \tag{7.42}$$

for the energy density and

$$C_V(t, \bar{\mu}) = \frac{t + 1}{T_c} \left(\frac{\partial^2 P}{\partial t^2} \right)_{\bar{\mu}} \tag{7.43}$$

for the specific heat. Similarly, we obtain

$$n_B(t, \bar{\mu}) = \frac{1}{T_c} \left(\frac{\partial P}{\partial \bar{\mu}} \right)_t = \frac{1}{T_c} \left(\frac{\partial P}{\partial \tau} \right)_{\bar{\mu}} \left(\frac{\partial \tau}{\partial \bar{\mu}} \right)_t = \frac{2\kappa \bar{\mu}}{T_c} \left(\frac{\partial P}{\partial \tau} \right)_{\bar{\mu}} = 2\kappa \bar{\mu} \frac{s(t, \bar{\mu})}{(t + 1)} \tag{7.44}$$

for the net baryon density, which vanishes for $\mu = 0$; here $s = (\epsilon + P)/T$ denotes the entropy density. The baryon-number susceptibility is given by

$$\chi_B(t, \bar{\mu}) = \frac{1}{T_c} \left(\frac{\partial n_B(t, \bar{\mu})}{\partial \bar{\mu}} \right)_t = \frac{2\kappa}{T_c^2} \left[\left(\frac{\partial P}{\partial \tau} \right)_{\bar{\mu}} + 2\kappa \bar{\mu}^2 \left(\frac{\partial^2 P}{\partial \tau^2} \right)_{\bar{\mu}} \right]; \tag{7.45}$$

with $\chi(t, \bar{\mu}) = \langle n_B^2 \rangle - \langle n_B \rangle^2$, it determines the fluctuations of the baryon number density. Considering the expansion around $\bar{\mu} = 0$ we thus find

$$\chi_B(t, \bar{\mu}) = \frac{2\kappa}{T_c^2} \left[\frac{s(t, 0)}{(t + 1)} + 2\kappa \bar{\mu}^2 \frac{T_c}{(t + 1)} C_v(t, 0) \right]; \tag{7.46}$$

I.e., at $\mu = 0$, the baryonic susceptibility is essentially given by the entropy density, and with increasing μ, there is an additional term determined by the specific heat. For a second order transition, $C_V(T)$ is expected to diverge at $T = T_c$; in the specific case of $O(4)$ symmetry, the corresponding exponent is negative (see Table 6.2 in Chap. 6), so that it only has a cusp-like peak there. For a rapid cross-over, there will also only be a pronounced peak.

The behavior of $\chi_B(t, \mu)$ has been studied extensively in $(2 + 1)$ flavor QCD [71]. In Fig. 7.13, we show results based on a second order Taylor expansion in μ [67]. It is seen that $\chi_B(T, \mu)$ for $\mu = 0$ indeed shows the form expected for the entropy density, and with increasing μ, a peak develops at $T = T_c$.

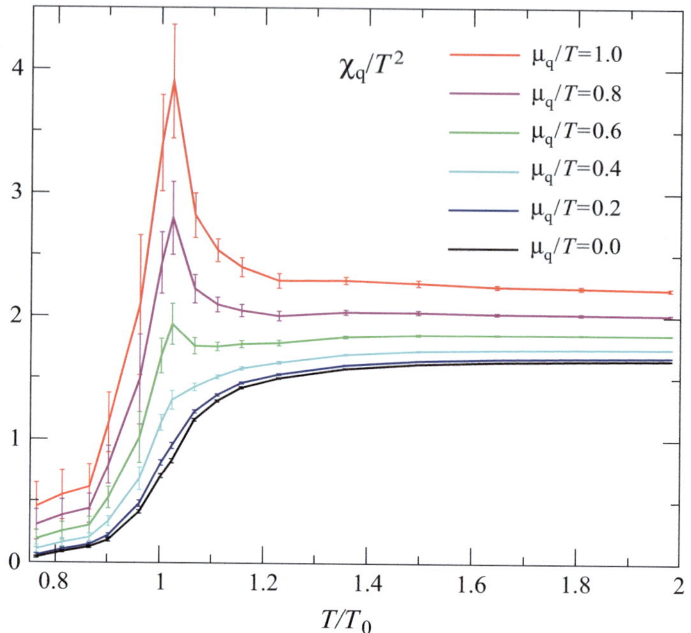

Fig. 7.13 Baryon number susceptibility in two-flavour QCD [71]

We see from Eq. (7.45) that the baryon susceptibility, i.e., second order derivative of $P(t, \bar{\mu})$ with respect to $\bar{\mu}$, leads to a form containing the second order derivative with respect to t; this is a consequence of the general variable structure (7.39). Consider then the higher order moments,

$$\chi_B^n(t, \bar{\mu}) = \frac{1}{T_c^{(n-1)}} \left(\frac{\partial^n P(t, \bar{\mu})}{\partial \bar{\mu}^n} \right)_t ; \tag{7.47}$$

for $n = 1$, they give the baryon density, for $n = 2$ the baryon number susceptibility. Their general structure is such that at $\mu = 0$ they vanish for odd n, while even n lead to higher derivatives of P with respect to t. For $n = 2$, $\chi_B^{(2)}$ gives at $\mu = 0$ the first derivative, $n = 4$ the second (i.e., the specific heat), and $n = 6$ the third. With the form

$$P(t, 0) \sim (t - 1)^{(2-\alpha)} \tag{7.48}$$

for the critical part of the pressure, this means in the case of $O(4)$ universality (negative critical exponent $\alpha \simeq -0.2$) that the specific heat has a finite peak, while all (even) moments for $n \geq 6$ diverge at $T = T_c$ and $\mu = 0$ [72, 73]. In other words, the $T - \mu$ structure implies even at $\mu = 0$ a specific form for the moments of the baryon number distribution. If such moments can be studied experimentally with sufficient precision, this will provide a possible quantitative test of statistical QCD (see Chap. 11 for more details).

In closing, we note that the dependence on the variable which we have so far suppressed, the quark mass m_q, can provide further information on the extension to finite μ. The derivative of P with respect to m_q defines the chiral condensate

$$\langle \bar{\Psi}\Psi \rangle \sim \left(\frac{\partial P}{\partial m_q} \right)_{t,\bar{\mu}}. \tag{7.49}$$

The first derivative of $\langle \bar{\Psi}\Psi \rangle$ with respect to $\bar{\mu}$ vanishes at $\bar{\mu} = 0$, as all odd such derivatives, due to the baryon symmetry. The second derivative, however, remains finite and is proportional to κ, again in the same way as found above (see Eq. (7.45)). It can thus be used to determine κ in lattice studies, leading to first estimates $\kappa \simeq 0.06$. From Eq. (7.39) we have

$$\frac{T_c(\mu)}{T_c} = 1 - \kappa \left(\frac{\mu}{T} \right)^2 ; \tag{7.50}$$

hence the critical curve in the $T - \mu$ plane has a very small downward curvature starting from $\mu = 0$ [70].

7.6 Conclusions

The phase structure of strongly interacting matter at finite baryon density remains today still largely a *terra incognita*. In the confined region at low baryon density, we have mainly mesonic matter, with resonance-dominated interactions. At high baryon density, nucleons attract each other at large and repell each other at short distance; the former leads to the formation of nuclei, the latter to their size growing with the number of nucleons. The different interactions in different baryon density regions will presumably also result in different deconfinement transition patterns. Rather general arguments suggest that the transition is of first order at low temperature and high baryon density. What follows "on the other side" of this transition is presently one of the main playgrounds for theorists in this field. Our favorite version features a deconfined state of broken chiral symmetry, i.e., a plasma of colored massive quarks; their masses will then melt at much higher density, to eventually restore chiral symmetry.

References

1. T. Banks, A. Casher, Nucl. Phys. B **169**, 103 (1980)
2. G. Baym, in *Quark Matter and Heavy Ion Collisions*, ed. by H. Satz (World Scientific, Singapore, 1982)
3. E.V. Shuryak, Phys. Lett. B **107**, 103 (1981)
4. E.V. Shuryak, Phys. Rep. **115**, 151 (1984)
5. R. Pisarski, Phys. Lett. B **110**, 155 (1982)
6. J. Cleymans, K. Redlich, H. Satz, E. Suhonen, Z. Phys. C **33**, 151 (1986)
7. H. Schulz, G. Röpke, Z. Phys. C **35**, 379 (1987)
8. A. Drago, M. Fiolhais, U. Tambini, Nucl. Phys. A **588**, 801 (1995)
9. H. Toki, K. Suzuki, Prog. Theor. Phys. **87**, 1435 (1992)
10. P. Guo, A.P. Szczepaniak, arXiv:0902.131 [hep-ph]
11. O. Kaczmarek, F. Zantow, Phys. Rev. D **71**, 114510 (2005)
12. M. Döring, K. Hübner, O. Kaczmarek, F. Karsch, Phys. Rev. D **75**, 05404 (2007)
13. F. Karsch, E. Laermann, A. Peikert, Nucl. Phys. B **605**, 579 (2001)
14. K. Hübner, F. Karsch, O. Kaczmarek, O. Vogt, Phys. Rev. D **77**, 074504 (2007)
15. For the most recent study and references to earlier literature, see A. Bazavov et al., Phys. Rev. D **80**, 014504 (2009)
16. P. Castorina, K. Redlich, H. Satz, Eur. Phys. J. C **59**, 67 (2009)
17. G. Baym, Physica A **96**, 131 (1979)
18. T. Celik, F. Karsch, H. Satz, Phys. Lett. B **97**, 128 (1980)
19. V. Magas, H. Satz, Eur. Phys. J. **32**, 115 (2003)
20. F. Karsch, H. Satz, Phys. Rev. D **21**, 1168 (1980)
21. For a more general discussion, see e.g., G. Biroli, Nat. Phys. **3**, 222 (2007)
22. G.Y. Onoda, E.G. Liniger, Phys. Rev. Lett. **22**, 2727 (1990)
23. K.W. Kratky, J. Stat. Phys. **52**, 1413 (1988)
24. E. Beth, G.E. Uhlenbeck, Physica **4**, 915 (1937)
25. R. Dashen, S.-K. Ma, H.J. Bernstein, Phys. Rev. **187**, 345 (1969)
26. See e.g., H. Satz, Nucl. Phys. A **681**, 3c (2001)
27. C.M. Fortuin, P.W. Kasteleyn, J. Phys. Soc. Jpn. **26**(Suppl), 11 (1969)
28. C.M. Fortuin, P.W. Kasteleyn, Physica **57**, 536 (1972)
29. A. Coniglio, W. Klein, J. Phys. A **13**, 2775 (1980)
30. P. Castorina, R.V. Gavai, H. Satz, Eur. Phys. J. C **69**, 169 (2010)
31. E.M. Levin, L.L. Frankfurt, JETP Lett. **2**, 65 (1965)
32. H.J. Lipkin, F. Scheck, Phys. Rev. Lett. **16**, 71 (1966)
33. H. Satz, Phys. Rev. Lett. **19**, 1453 (1967)
34. For an early reference to such a picture, see e.g. V.A. Novikov, M.A. Shifman, A.I. Vainshtein, V.I. Zakharov, Ann. Phys. **105**, 276 (1977)
35. E.V. Shuryak, A.I. Vainshtein, Nucl. Phys. B **199**, 451 (1982)
36. C. Amsler et al., (Particle Data Group), Phys. Lett. B **667**, 1 (2008). For m_c, we use the $\overline{MS}$ value at scale 2 GeV, which agrees with that found in potential model studies; for m_b, we use instead the $1S$ value, which is closer to that found in potential model studies than the $\overline{MS}$ value at scale 2 GeV
37. H.D. Politzer, Nucl. Phys. B **117**, 397 (1976)
38. O. Kaczmarek, F. Karsch, P. Petreczky, F. Zantow, Phys. Rev. D **70**, 074505 (2004)
39. B. Povh, J. Hüfner, Phys. Rev. Lett. **58**, 1612 (1987)
40. See e.g. M. Alford, K. Rajagopal, T. Schäfer, A. Schmitt, Rev. Mod. Phys. **80**, 1455 (2008)
41. S. Ekelin, in *Strong Interactions and Gauge Theories*, ed. by J. Tran Thanh Van (Editions Frontières, Gif-sur Yvette, 1986)
42. J.F. Donohue, K.S. Sateesh, Phys. Rev. D **38**, 360 (1988)
43. M. Anselmino, S. Ekelin, D.B. Lichtenberg, E. Predazzi, Rev. Mod. Phys. **65**, 1199 (1993)
44. L. McLerran, R. Pisarski, Nucl. Phys. A **796**, 83 (2007)

45. See A. Andronic et al., arXiv:0911.4806 [hep-ph], for a recent summary and further references
46. Y. Nambu, G. Jona-Lasinio, Phys. Rev. **122**, 345 (1961)
47. Y. Nambu, G. Jona-Lasinio, Phys. Rev. **124**, 246 (1961)
48. T. Hatsuda, T. Kunihiro, Phys. Lett. B **145**, 7 (1984)
49. V. Bernard, R.L. Jaffe, U.-G. Meissner, Nucl. Phys. B **308**, 753 (1988)
50. M. Asakawa, K.Yazaki, Nucl. Phys. A 668 (1989)
51. For surveys, see e.g., U. Vogl, W. Weise, Prog. Part. Nucl. Phys. **27**, 196 (1991)
52. For surveys, see e.g., S.P. Klevansky, Rev. Mod. Phys. **64**, 649 (1992)
53. For surveys, see e.g., M. Buballa, Phys. Rept. **407**, 205 (2005)
54. T.M. Schwarz, S.P. Klevansky, G. Papp, Phys. Rev. C **60**, 055205 (1999)
55. P.N. Meisinger, M.C. Ogilvie, Phys. Lett. B **379**, 163 (1996)
56. K. Fukushima, Phys. Lett. B **591**, 277 (2004)
57. A. Mocsy, F. Sannino, K. Tuominen, Phys. Rev. Lett. **92**, 182302 (2004)
58. C. Ratti, W. Weise, Phys. Rev. D **70**, 054013 (2004)
59. C. Ratti, M.A. Thaler, W. Weise, Phys. Rev. D **73**, 0140 (2006)
60. M. Halasz et al., Phys. Rev. D **58**, 096007 (1998)
61. M. Stephanov, K. Rajagopal, E. Shuryak, Phys. Rev. Lett. **81**, 4816 (1998)
62. Z. Fodor, S. Katz, JHEP **0203**, 014 (2002)
63. P. de Forcrand, O. Philipsen, Nucl. Phys. B **642**, 290 (2002)
64. M.-P. Lombardo, Phys. Rev. D **67**, 014505 (2003)
65. C.R. Allton et al., Phys. Rev. D **68**, 014507 (2003)
66. P. de Forcrand, O. Philipsen, arXiv:0811.3858
67. F. Karsch et al., Nucl. Phys. Proc. Suppl. **129**, 614 (2004)
68. Y. Hatta, T. Ikeda, Phys. Rev. D **67**, 0140128 (2003)
69. S. Ejiri, F. Karsch, K. Redlich, Phys. Lett. B **633**, 275 (2006)
70. O. Kaczmarek et al., Phys. Rev. D **83**, 014504 (2011)
71. S. Ejiri et al., Phys. Rev. D **80**, 094505 (2009)
72. F. Karsch, K. Redlich, PL B **695**, 136 (2011)
73. B. Friman et al., arXiv:1103.3511

Chapter 8
The Quark-Gluon Plasma

Hic sunt leones
(Here there are lions)

Warning of unchartered territories
on old Roman maps

Abstract In this chapter, we summarize the main properties of the quark-gluon plasma at vanishing overall baryon number density, as so far obtained in finite temperature lattice studies. The defining plasma property, charge screening, will be addressed by studying the effect of introducing test charges into the medium. Subsequently, we note that in a temperature range up to several T_c, the plasma exhibits strong, non-perturbative interaction effects, which can be described in terms of an ideal gas of massive quasi-particles.

8.1 Introduction

Perhaps the most striking result of strong interaction thermodynamics, as it is obtained with QCD as dynamical basis, is the existence of a new state of matter, a plasma of deconfined quarks and gluons. In the hadronic state, the colored quarks and gluons couple such as to form color-neutral bound states, mesons and baryons, and hadronic matter is made up of these as constituents. In the quark-gluon plasma (QGP), the specific color-neutralizing binding is dissolved, so that the medium now consists of colored constituents: it is a plasma. This does not imply the existence of isolated quarks, which would of course violate color confinement. The constituents of the QGP can exist with "open" color only because they always find in their immediate surroundings many other colored constituents, so that we never encounter a spatially isolated colored object.

The fundamental property of a medium of unbound electric charges is charge screening. With increasing density, the presence of many other charges effectively reduces the intrinsic dynamical range of the Coulomb force. Each charge is through polarization surrounded by a cloud of net opposite charge, thereby reducing its

145

H. Satz, *Extreme States of Matter in Strong Interaction Physics*,
Lecture Notes in Physics 945, https://doi.org/10.1007/978-3-319-71894-1_8

interaction range. We expect the same in a plasma of colored constituents, and so color screening will be our first topic. Subsequently we note that the deconfinement of quarks and gluons in the QGP, i.e, the fact that they are no longer being restricted to color-neutral groupings, does not mean that they do not interact. Only in the limit of very high temperature can we expect that asymptotic freedom, i.e., the vanishing of the running coupling constant α_s of strong interaction physics, leads to an ideal gas of quarks and gluons. We shall see that such an "ideal" situation is achieved at best at temperatures which are truly extreme. At all temperature values of present interest both to theoretical and experimental studies, the QGP is very much an interacting medium, and the nature of these interactions is the next topic to be addressed.

8.2   Color Charge Screening and String Breaking

The screening properties of a given medium can be tested by introducing a test charge into the medium, thus producing polarization effects. This phenomenon is of quite general nature in statistical physics, where it is treated in Debye-Hückel theory [1]. Screening effects will in turn also modify the interaction between a given pair of charges in the medium. In finite temperature QCD, this can be studied by considering the behavior of a heavy quark-antiquark ($Q\bar{Q}$) pair placed in hot deconfined medium.

In vacuum, such a $Q\bar{Q}$ pair is connected by a gluonic flux tube or string. If we try to pull the pair apart, the string will break when the separation energy surpasses that needed to create a light $q\bar{q}$ pair, which then produces two heavy-light mesons; these can be pulled apart without requiring any further energy (see Fig. 8.1). String breaking is not exactly color screening, although its effect for the $Q\bar{Q}$ pair is quite similar, since the newly formed light $\bar{q}$ "screens" the Q from its former partner $\bar{Q}$, and vice versa. But it requires an energy input to bring a virtual $q\bar{q}$ pair from its Dirac sea on-shell, and so string breaking measures effectively the sea level of the vacuum, or of the confined medium at temperature T. In a hot deconfined medium, we have unbound color charges, without any outside energy input; nevertheless, the

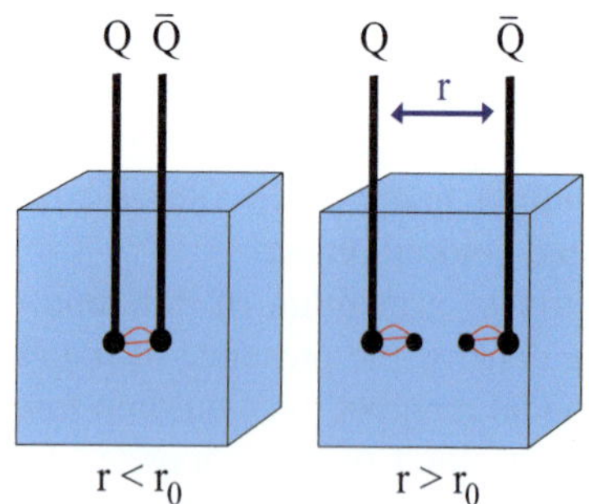

Fig. 8.1   Schematic view of string breaking for a static $Q\bar{Q}$ pair

presence of such charges will modify the $Q\bar{Q}$ binding, and these modifications can provide information both about the medium and about the in-medium behavior of heavy quark bound states ("quarkonia").

The free energy of a static $Q\bar{Q}$ pair at zero temperature can be described in terms of the so-called "Cornell" potential [2, 3],

$$F(r, T = 0) \simeq \sigma r - \frac{\alpha}{r}, \tag{8.1}$$

where r is the $Q\bar{Q}$ separation, σ the string tension and α the color coupling. The first term accounts for the formation of the string between the Q and the $\bar{Q}$ when they are pulled apart. The string free energy thus increases linearly with the separation of the $\bar{Q}Q$, and the string tension is generally assumed to have a universal value $\sigma \simeq 0.2$ $\mathrm{GeV}^2 = 1$ GeV/fm. The second term accounts for the Coulomb interaction between the static charges as well as for transverse string oscillations. String theory predicts $\alpha = \pi/12$, and this value is found to agree with lattice studies until the separation becomes so small that asymptotic freedom renders any string picture inapplicable.

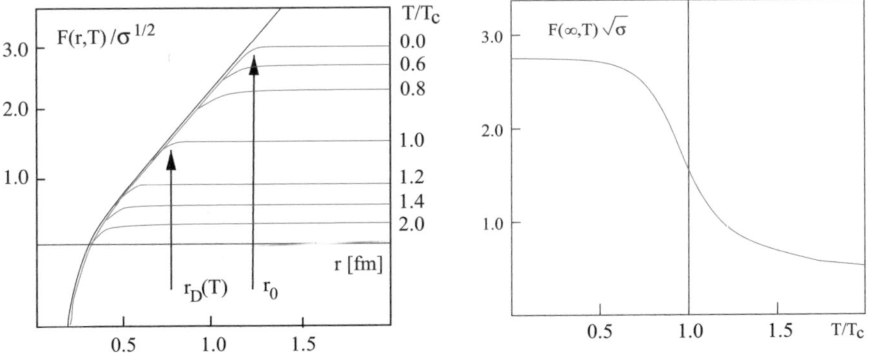

Fig. 8.2 Schematic view of the temperature dependence of the in-medium free energy $F(r, T)$

The increase of the free energy $F(r, T = 0)$ with r eventually comes to a crucial point r_0, beyond which it is energetically favorable for the string to break into two light-heavy mesons (D, B) rather than stretching further (see Fig. 8.1). This leads to a behavior of the form illustrated on the left of Fig. 8.2: After the string breaks at r_0, the free energy remains constant for all separations r larger than r_0. Since string breaking occurs when a light $q\bar{q}$ pair is excited out of the vacuum, it should be a medium effect, independent of the static quarks being pulled apart. This can in fact be tested by looking at the mass of light-heavy quark meson states. For charm quarks, the value of the free energy $F(r_0, T = 0)$ at this point is determined by

$$2M_D = 2m_c + F(r_0, T = 0), \tag{8.2}$$

where M_D denotes the mass ($M_D \simeq 1.9$ GeV) of the lowest open charm meson, the D. One can write a similar equation for the bottomonium system, with $M_B \simeq$

5.3 GeV. Rearranging the terms and using the quark mass values from quarkonium spectroscopy ($m_c \simeq 1.3$ GeV, $m_b \simeq 4.5$ GeV), this leads to

$$F(r_0, T = 0) \simeq 2(M_D - m_c) \simeq 2(M_B - m_b) \simeq 1.2 \,\text{GeV}. \tag{8.3}$$

The fact that the two quarkonium systems with very different quark masses lead to the same value of $F(r_0, T = 0)$ indicates that string breaking is indeed a property of the medium, here the vacuum, not of the heavy quarks. It specifies the energy needed to excite a pair of light quarks from the virtual sea in the vacuum and provide the gluonic "dressing" for both light and heavy quarks. The resulting $F(r_0, T = 0)/4 \simeq$ 300 MeV per quark agrees well with the usual constituent quark mass.[1] With the universal string tension $\sigma \simeq 0.2 \,\text{GeV}^2$, we obtain in turn a universal string breaking length

$$r_0 \simeq \frac{F(r_0, T = 0)}{\sigma} \simeq 1.2 \,\text{fm}. \tag{8.4}$$

This result is also in accord with light hadron sizes as confinement radii.

This phenomenological description of $Q\bar{Q}$ binding in vacuum provides a good starting point for in-medium studies. We now consider a static quark-antiquark pair in a QCD medium of temperature T and consider the difference in free energy between a medium with and one without a static $Q\bar{Q}$ pair; that gives us what we had above called "the free energy of the static pair".

Up to the deconfinement point, the string breaking must produce a color-neutral pair, and at finite temperature, this does not necessarily require the excitation of a (dressed) $q\bar{q}$ pair from the vacuum. The presence of further light-light mesons, in addition to the heavy quark pair, allows a re-arrangement of the coupling, as shown in Fig. 8.3. This constitutes an additional effective screening mechanism for the interquark potential, and near the deconfinement point, when the in-medium hadron density increases rapidly, it makes the $Q\bar{Q}$ separation much easier than at low temperature. Hence the large distance limit $F(r \to \infty, T)$ decreases sharply as $T \to T_c$.

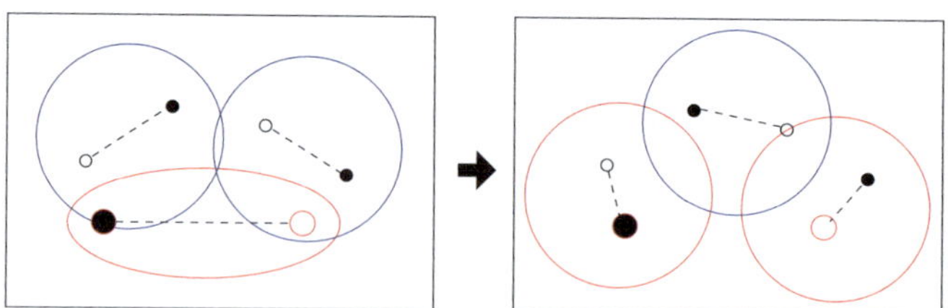

Fig. 8.3 In-medium string breaking through recoupling

---

[1]The two heavy quarks in the $Q\bar{Q}$ pair at short distance do not have such a dressing, see the discussion in Sect. 7.3.

Beyond T_c, we have a colored medium with genuine color screening, which we can try to describe in terms of a color screening radius $r_D(T)$ or mass $\mu(T) = 1/r_D(T)$. We can determine this either operationally as the value of r at which the free energy becomes r-independent, or in the framework of a charge screening theory of the Debye-Hückel type. For this, the basic input is a potential with a given r dependence, e.g., $\sim r^a$, acting in a d-dimensional space. The familiar case of electric Debye screening has $a = -1$, $d = 3$ and leads to the well-known form already shown in the introduction,

$$F_{\text{e−m}}(r, T) \simeq -e_0^2 \left[\frac{1}{r} \exp\{-\mu r\} + \mu \right];\tag{8.5}$$

the T-dependence is contained in $\mu(T)$, and the last term assures that at infinite distance, the correct limit $-e_0^2 \mu$ is obtained for a pair of opposite color charges [1]. Here we want the corresponding screening function for the confining part of Eq. (8.1), i.e., for $a = +1$. For $d = 1$, the result is the so-called Schwinger form and describes screening in one-dimensional electrodynamics,

$$\tilde{F}_s(r, T) = \sigma r \left\{ \frac{1 - \exp\{-\mu r\}}{\mu r} \right\},\tag{8.6}$$

which already shows a behavior of the form seen in Fig. 8.2 (left), with $V(\infty, T) = \sigma/\mu$ for the constant large distance limit. In the more realistic $d = 3$ case, we have instead [4]

$$F_s(r, T) = \frac{\sigma}{\mu} \left[\frac{\Gamma(1/4)}{2^{3/2}\Gamma(3/4)} - \frac{\sqrt{\mu r}}{2^{3/4}\Gamma(3/4)} K_{1/4}[(\mu r)^2] \right].\tag{8.7}$$

The first term again gives the constant value for $r \to \infty$, while the second, through the modified Hankel function, gives a Gaussian cut-off, instead of the exponential in Eq. (8.6). To see how adequate such a screening model is, we compare the free energy

$$F(r, T) = F_s(r, T) + F_w(r, T)\tag{8.8}$$

to lattice calculations in two-flavor QCD [5, 6]. In the first term, we use $\sigma = 0.2$ GeV^2 and $\alpha = \pi/12$, and the second term is the Coulomb form (8.5) with the strong coupling α replacing e_0^2. As shown in Fig. 8.4 (left), the resulting functional description is excellent [7]; the only remaining parameter, the screening mass $\mu(T)$, is shown in Fig. 8.4 (right), including also the values obtained for $T = 0$ and in the confinement region. We see that μ indeed increases rapidly at T_c, indicating a sharp drop in the effective range of the binding force. At high temperatures, μ approaches the linear behavior expected from perturbation theory [8]. The quark-gluon plasma thus indeed shows the color screening pattern given in the usual Debye-Hückel screening theory. The linear increase of $\mu(T)$ with T implies that for sufficiently high temperatures, only $F_w(r, T)$ survives; this is the form obtained in a weak-coupling (perturbative) limit of QCD.

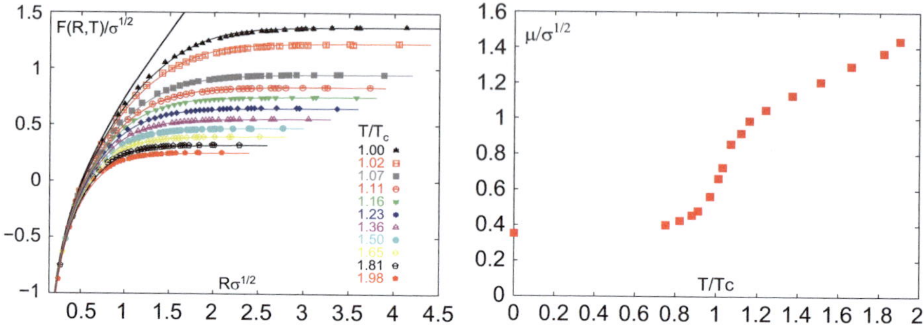

Fig. 8.4 Screening fits to the $Q\bar{Q}$ free energy $F(r, T)$ for $T \geq T_c$ (left), and the corresponding screening mass $\mu(T)$ (right) [7]

From Eqs. (8.7) and (8.8) we obtain as large r limit of the free energy

$$F(T) \equiv F(\infty, T) \simeq \frac{\sigma}{\mu(T)} - \alpha\mu(T), \tag{8.9}$$

and such a functional behavior is indeed found in lattice studies, as illustrated in Fig. 8.5, where $F(T)$ is shown out to much higher temperatures then in Fig. 8.2. For sufficiently high T, $F(\infty, T)$ approaches the conformal limit linear in T, modified by the running coupling. It is clear from this, however, that one cannot simply consider $F(\infty, T)$ as the mass of two dressed color charges at infinite distance, since above a certain T, it becomes negative.

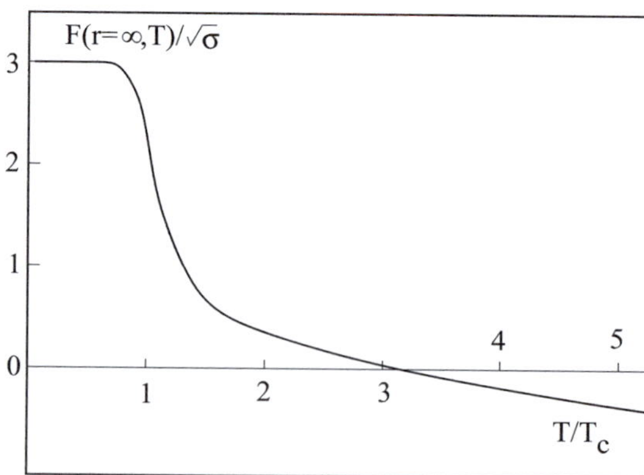

Fig. 8.5 Temperature dependence of the large distance limit of the free energy of a static $Q\bar{Q}$ pair

To understand this behavior, we recall the thermodynamic relation between free energy F, internal energy U and entropy S,

$$F(T) = U(T) - TS(T). \tag{8.10}$$

As mentioned, all thermodynamic quantities under consideration here always give the differences induced by the presence of the $Q\bar{Q}$ pair, i.e., they measure the difference found for the system with and without the pair. The free energy thus contains not only the effective mass acquired by the separated color charges through polarization, but it also measures the change in the entropy of the medium induced by the presence of the charges. In particular, the negative high temperature limit of the free energy is essentially determined by the entropy effect, assuming the change in internal energy due to the $Q\bar{Q}$ pair to be positive.

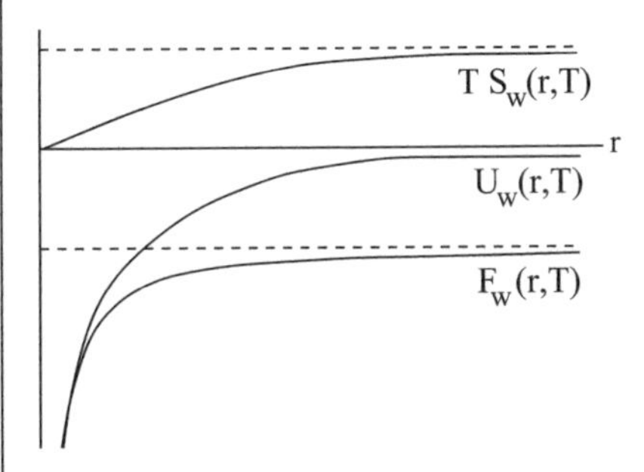

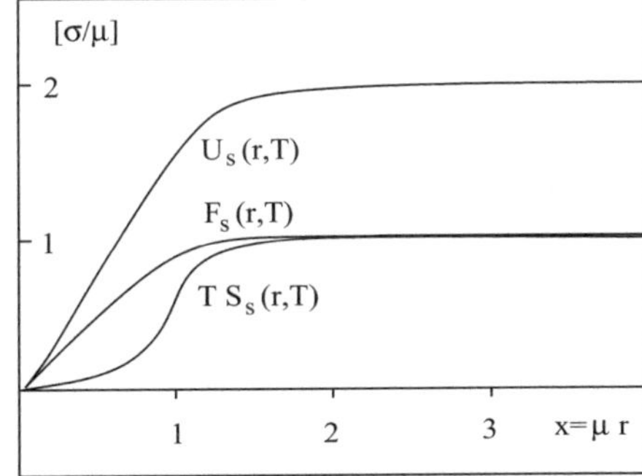

Fig. 8.6 The r-dependence of weak-coupling (left) and strong-coupling (right) behavior of thermodynamic observables

To study this in more detail, we consider separately the r-dependence of the string term and the Coulomb term; they are illustrated in Fig. 8.6. In the weak-coupling form F_w, it is seen that the pair can be separated with no change in internal energy; the work done on the $Q\bar{Q}$ pair is simply converted into entropy. For the strong-coupling form F_s, in contrast, the separation work and the entropy increase are equal but additive, leading to a corresponding change in internal energy. The effect of screening is thus quite different in the strong coupling region $T_c \leq T \leq 3 - 5\,T_c$ and in the weak-coupling high temperature limit.

8.3 Interaction Regimes of the Plasma

In Chap. 5 it was shown that when strongly interacting matter is brought close to the deconfinement temperature, its energy density changes quite abruptly, increasing from a low hadronic value to the much higher one for a medium consisting of quarks and gluons. The amount by which it changes is just the latent heat of deconfinement.

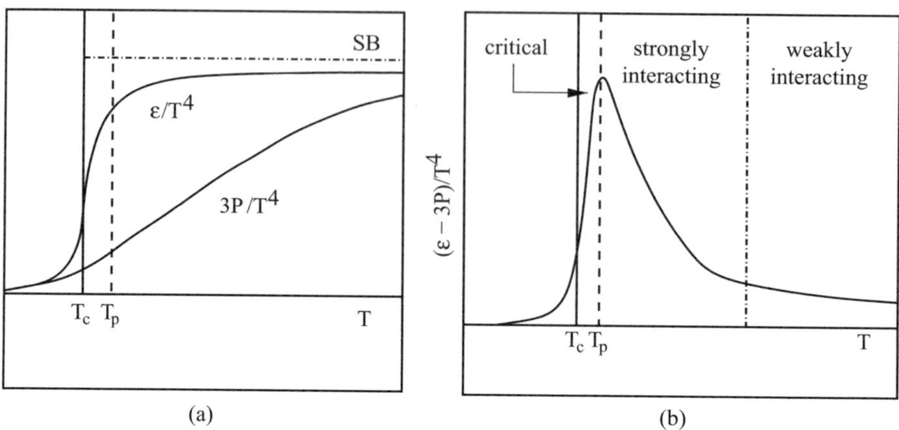

Fig. 8.7 Schematic view of energy density and pressure (**a**) and of the interaction measure (**b**) in the deconfined medium; in (**a**), SB denotes the Stefan-Boltzmann limit

In Fig. 8.7, we show a schematic view of the energy density and the pressure around T_c, as obtained in pure gauge theory as well as in full QCD. In the transition region just above T_c, the pressure increases towards its Stefan-Boltzmann limit much more slowly than the energy density. In Chap. 3 we had already seen a similar effect on the hadronic side, where strong resonance interactions caused the energy density to increase more rapidly than the pressure. The resulting interaction effects are indicated quite clearly by the interaction measure $\Delta(T) \equiv (\epsilon - 3P)/T^4$ [9], which vanishes for an ideal gas of massless constituents (see Eq. (4.17)); in contrast, in all lattice studies it remains quite sizeable up to $T/T_c \simeq 3 - 5$. And even at very high temperature, when ϵ/T^4 does approach $3P/T^4$, the Stefan-Boltzmann limit is not yet reached, so that some form of interaction must still persist in that region.

The nature of these interaction effects has been the subject of much interest and speculation. Let us try to address the topic in a systematic fashion. The schematic picture shown in Fig. 8.7 indicates that the temperature variation of $\Delta(T)$ defines three relatively distinct interaction regimes of the medium. In the "critical region" around T_c, there is a very sharp increase of $\Delta(T)$. It then peaks, and above a temperature $T_p \simeq 1.1 \, T_c$, there is a "strongly interacting plasma" region, in which $\Delta(T)$ decreases quite rapidly. Finally, for $T \gtrsim 5 T_c$, the decrease becomes much slower as we enter the "weakly interacting plasma" regime. The justification of this nomenclature will become clear shortly.

The critical region is most easily identified in the case of pure $SU(2)$ gauge theory, where it is defined by the critical behavior of the Z_2 universality class and the corresponding critical exponents. The singular parts of the pressure and of the energy density in the plasma state, i.e., for $t \equiv (T - T_c)/T_c \geq 0$, are given by

$$P_{\text{sing}} = cT^4 t^{2-\alpha}, \quad \epsilon_{\text{sing}} = 3cT^4 \left\{ t^{2-\alpha} + \frac{2-\alpha}{3}(t+1)t^{1-\alpha} \right\}, \tag{8.11}$$

where c is a constant; in addition, both quantities have non-singular contributions remaining at T_c. From Eq. (8.11) it follows that the specific heat $C_v = (\partial \epsilon / \partial T)_V$ diverges at T_c as $C_v \sim t^{-\alpha}$; the critical exponent $\alpha = 0.11$ is the same as that for the three-dimensional Ising model. In the critical region of the gluon plasma, the interaction measure thus has the singular form

$$\Delta(t) - \Delta(1) = (2 - \alpha)c(t + 1)t^{1-\alpha}, \tag{8.12}$$

where the constant term $\Delta(1)$ removes the non-singular parts of energy density and pressure remaining at T_c. The behavior predicted by Eq. (8.12) is well confirmed by lattice studies [9], though in a very narrow window above T_c, as shown in Fig. 8.8.

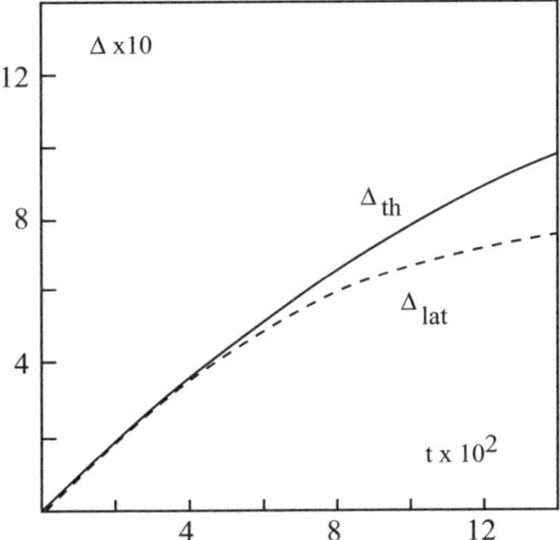

Fig. 8.8 The interaction measure for $SU(2)$ gauge theory in the critical region, comparing lattice calculations (Δ_{lat}, dashed line) to the form (8.12) obtained for the Z_2 universality class (Δ_{th}, solid line)

We see that increasing the temperature 10% above T_c leads to an increase of the interaction measure by about a factor three. As already noted, this increase is due to the fact that in the critical region, the energy density increases much more strongly than the pressure. For large T, the two converge, and so there must be some point T_p where the two exchange roles: above T_p, the pressure grows more strongly than the energy density. We have illustrated this schematically in Fig. 8.7; the effect of the "interchange" is the peak observed in $\Delta(T)$.

As already noted, the form (8.12) holds only in a very narrow temperature range above T_c; in particular, it does not reproduce the peak observed for $\Delta(T)$ at T_p. Instead, it continues to increase further, with a turnover at very much higher temperature. Hence the decrease of $\Delta(T)$ in the range above T_p must have some other origin.

Beyond the peak at T_p, in the region of rapidly decreasing $\Delta(T)$, we have a "strongly interacting plasma", whose behavior, as we shall see, is definitely of non-perturbative nature. Eventually, at high enough temperature, asymptotic freedom could be expected to result in a "weakly interacting plasma", in which the behavior of $\Delta(T)$ is given by perturbation theory. Before discussing this further, we summarize in Fig. 8.9 the results obtained for the interaction measure in lattice studies of pure $SU(N)$ gauge theory [10–13] and of full QCD [14, 15]. In the case of pure gauge theory, the interaction measure is found to scale with the gluonic degrees of freedom, so that $\Delta(T)/(N^2 - 1)$ leads to a unique curve down to the onset of the critical region.

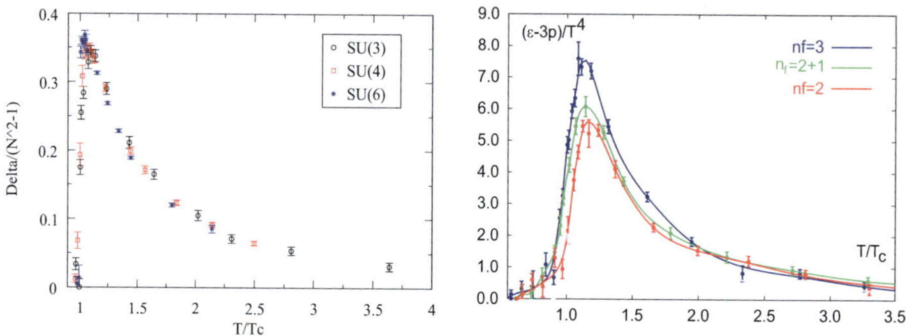

Fig. 8.9 Interaction measure for $SU(N)$ gauge theory [12, 13], scaled by $N^2 - 1$ (left), and for full QCD [14], with $N_f = 2, 2 + 1, 3$ (right)

8.4 Weak Coupling Approaches

In leading order perturbation theory, the interaction measure for pure $SU(N)$ gauge theory is given by [11, 16]

$$\Delta_{pert} = \frac{(N^2 - 1)}{288} \frac{11N^2}{12\pi^2} g^4(T), \qquad (8.13)$$

with

$$g^2(T) = \frac{24\pi^2}{11N \ln(T/\Lambda_T)}. \qquad (8.14)$$

The perturbative interaction measure thus shows the observed scaling in $N^2 - 1$ just mentioned, assuming Ng^2 is kept constant. In Eq. (8.13), Λ_T defines the lattice scale, which in the mentioned $SU(3)$ lattice studies [10] was found to be determined by $T_c/\Lambda_T = 7.16 \pm 0.25$. In this case, we thus obtain

$$\Delta_{pert} = \frac{11}{48\pi^2} g^4 = \frac{4\pi^2}{33} \frac{1}{\{\ln[7.16(T/T_c)]\}^2} \simeq \frac{1.2}{\{\ln[7.16(T/T_c)]\}^2}. \tag{8.15}$$

At $T/T_c = 3$, we thus have $\Delta_{pert} \simeq 0.13$, which is still about a factor 3 below the (continuum extrapolated) lattice result $\Delta_{lat} \simeq 0.4$. Hence at this temperature, leading order perturbation theory cannot yet reproduce the plasma interaction. Nevertheless, we have here $\alpha_s = g^2/4\pi \simeq 0.19$ for the strong coupling α_s, so that in principle perturbation theory seems to be applicable, and we could expect that at somewhat higher temperatures, above $T/T_c \simeq 5 - 10$, the perturbative form might account for the lattice result.

The evaluation of higher order perturbative terms has, however, shown that this is not the case. Divergences in finite temperature field theory limit calculations to a finite order in the coupling g [17, 18]; for the pressure, the limit is $O(g^5)$, and calculations of such corrections have now been extended to this order [19, 20]. In Fig. 8.10 we show the result of expansions in different order g^n for energy and pressure in $SU(3)$ gauge theory, normalized to the Stefan-Boltzmann limit [21]. It is seen that in the temperature region of interest here, $T \le 10\ T_c$, the different orders lead to strong fluctuations; the final form, up to and including $O(g^5)$, still considerably undershoots the lattice results.

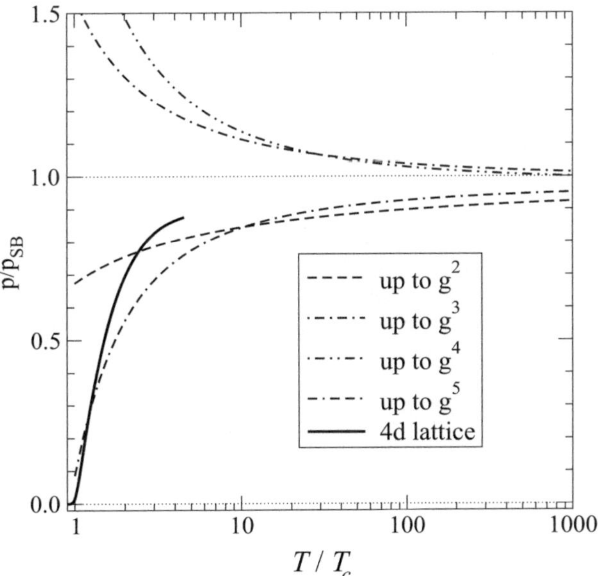

Fig. 8.10 Perturbative expansions of the pressure in $SU(3)$ gauge theory [21], compared to the finite temperature lattice results [10]

Moreover, for an understanding of the interaction effects, a comparison of lattice and perturbation theory results for the pressure is in fact quite misleading, since the major part of the pressure is given by the ideal gas component. To concentrate on just the interaction effects, we return to the interaction measure $\Delta(T)$, and here perturbation theory breaks down completely. The two-loop form for $SU(3)$ gauge theory,

$$\Delta_{pert} = \left(\frac{99}{12\pi^2}\right)\left[\frac{1}{36}g^4 - \frac{1}{6\pi}g^5\right],\tag{8.16}$$

remains negative until $g^2 \simeq 0.27$, which, with the two-loop form of the coupling (see Eqs. (5.24)/(5.25)),

$$g^{-2} = \frac{11}{8\pi^2}\ \ln(T/\Lambda_T) + \frac{51}{88\pi^2}\ \ln[2\ \ln(T/\Lambda_T)],\tag{8.17}$$

requires inconceivably high temperatures, above $10^6\ T_c$. We conclude that the interaction of the plasma in the region of interest here, up to some $10\ T_c$, must definitely require some non-perturbative features.

This situation has triggered numerous efforts to modify the perturbative treatment. In one approach [21], the $O(g^6)$ term is evaluated by a non-perturbative scale determination, using lattice results for the magnetic screening. Another possibility is given by including sums over certain graph classes, thus effectively shifting the point about which the perturbation expansion is performed [22–26]. In particular, this is studied for the terms dominating at high temperature (hard thermal loops, HTL) and leads to a much improved convergence of the perturbation series. Both approaches have in common

- a rather good description of the pressure for high temperatures, but
- the range below about 5 T_c is still not accessible, generally leading to interaction measures which are far too small or even negative.
- Moreover, the strong order-by-order fluctuations raise some doubts that the last order considered is really close to a "final" result.

To illustrate this, we show on the left side of Fig. 8.11 the behavior obtained with the help of a partially non-perturbative $O(g^6)$ term [21], and on the right side the corresponding results from modified HTL calculations [22, 23], in both cases compared to the form obtained in $SU(3)$ lattice QCD. The latter shows for $\Delta(T)$ a decrease as $1/T^2$, so that $T^2\Delta(T)$ becomes approximately constant above T_c. We see in Fig. 8.11 that in leading order (LO) and next-to-leading order (NLO) the breakdown of perturbation theory persists also in a HTL approach, and even the inclusion of a partially non-perturbative NNLO contribution cannot reproduce the lattice result, neither in magnitude nor in functional form.

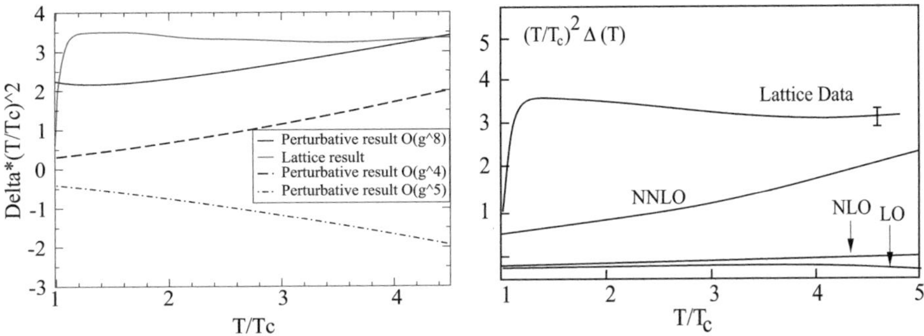

Fig. 8.11 $(T/T_c)^2 \Delta(T)$ as predicted in a systematic effective perturbation theory (left) [21] and in HTL resummed perturbation theory (right) [22, 23], compared to the continuum extrapolation of lattice studies [11]

The convergence of lattice data toward a weak-coupling limit thus remains a rather open issue. Recent lattice studies based on rather small volumes [27] have extended the temperature range in $SU(3)$ gauge theory up to above $100\ T_c$, and as seen in Fig. 8.12, one finds that higher order perturbative results (with a fitted g^6 contribution [21]) approach the lattice curve.

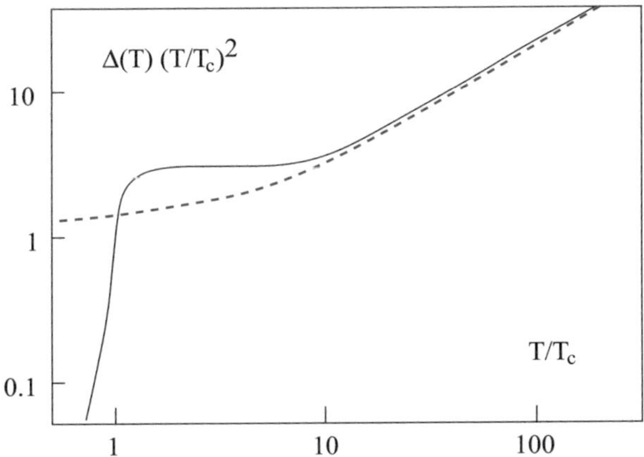

Fig. 8.12 Continuum extrapolation of small volume lattice results for the trace anomaly (solid curve) compared to $O(g^6)$ perturbative calculation [27]

We have here considered pure gauge theory. In recent studies of QCD with one heavy and two light quarks it was shown that in HTL resummed perturbation theory, the agreement with lattice data can be extended down to about $2\ T_c$ [28]; below that, critical behavior sets in. Moreover, effective field theory studies contain an intrinsic renormalisation scale, and variations of this scale lead to a rather wide band of predictions. This, however, can never produce the drop of the interaction measure as T approaches the critical point.

In general, the breakdown observed in any perturbative treatment as we enter the transition region is of course not surprising. Critical or even pseudo-critical behavior with an increasing correlation range is simply not a perturbative phenomenon. We therefore have to find a non-perturbative approach to address the behavior of the plasma in this region.

8.5   Bag Pressure and Gluon Condensate

We had seen in Chap. 4 that one way to implement confinement in an ideal gas picture was to introduce a bag pressure, measuring the "level difference" between the physical vacuum and the ground state in the colored world of QCD. To the ideal gas partition function $Z_0(T, V)$ a bag term was added,

$$T \ln Z_B(T, V) = T \ln Z_0(T, V) - BV \tag{8.18}$$

and this simulates a form of interaction [29], as best seen by the resulting interaction measure,

$$\Delta(T) = \frac{4B}{T^4}. \tag{8.19}$$

We want to check here to what extent this is a viable description of the QGP interaction in the region above T_c.

The thermal expectation value of the trace of the energy-momentum tensor, $\langle \Theta^\mu_\mu \rangle = \epsilon - 3P$, is related to the gluon condensate, i.e., to the expectation value of the gluon term in the QCD Lagrangian [30],

$$G^2 \equiv \frac{\beta(g)}{2g^3} G^a_{\mu\nu} G^{\mu\nu}_a = \frac{11 N_c}{96\pi^2} G^a_{\mu\nu} G^{\mu\nu}_a, \tag{8.20}$$

where $G^a_{\mu\nu} = g F^a_{\mu\nu}$ is given by the gluon field of color a in the QCD Lagrangian. The last term of Eq. (8.20) is obtained using the leading order perturbative beta function,

$$\beta(g) = \frac{11 N_c}{48\pi^2} g^3 + O(g^5). \tag{8.21}$$

The value of $\langle G^2 \rangle = G_0^2$ at $T = 0$ has been estimated numerically, with $G_0^2 = 0.012 \pm 0.006 \, \text{GeV}^4$ as "canonical" value [31]. In both analytical and lattice studies, $\epsilon - 3P$ is normalized to zero at $T = 0$, so that

$$\langle \Theta^\mu_\mu \rangle = \epsilon - 3P = G_0^2 - G_T^2, \tag{8.22}$$

where G_T^2 is the temperature-dependent gluon condensate. In the temperature range below T_c, we expect $G_T^2 = G_0^2$, so that $\epsilon - 3P = 0$. If the gluon condensate melts above T_c, the interaction measure becomes

$$\frac{\epsilon - 3P}{T^4} = \frac{G_0^2}{T^4}, \qquad (8.23)$$

so that $B = G_0^2/4$ relates bag pressure and gluon condensate. The value for the latter given above leads to a bag pressure $B^{1/4} \simeq 230 \pm 30$ MeV, which is in reasonable agreement with that obtained from hadron spectroscopy as given by the bag model.

The color summation in Eq. (8.20) runs over the $N^2 - 1$ gluonic color degrees of freedom, so that we can write

$$G^2 = \frac{11Ng^2}{96\pi^2} \langle F_{\mu\nu}^a F_a^{\mu\nu} \rangle = \frac{11Ng^2}{96\pi^2} (N^2 - 1)\langle \bar{F}_{\mu\nu} \bar{F}^{\mu\nu} \rangle, \qquad (8.24)$$

where $\langle \bar{F}_{\mu\nu} \bar{F}^{\mu\nu} \rangle$ denotes the gluon field contribution per color degree of freedom. The scaling of the interaction measure in $N^2 - 1$ observed for different $SU(N)$ theories is thus in accord with the bag model dependence, if we keep $g^2 N$ constant.

If we assume that $G_T^2 = 0$ for $T \geq T_c$, we obtain for the interaction measure

$$\Delta(T) = \frac{G_0^2}{T^4} = \frac{G_0^2}{T_c^4} \left(\frac{T_c}{T}\right)^4 \simeq \frac{2.3}{(T/T_c)^4}, \qquad (8.25)$$

using $T_c \simeq 0.27$ GeV for the $SU(3)$ deconfinement temperature. The lattice data are found to decrease more slowly and are in fact in accord with a $1/T^2$ dependence. We therefore compare in Fig. 8.13 the results for $T^2 \Delta(T)$ given by the lattice and by the bag model forms. The bag model naturally cannot account for the structure immediately around T_c (the rise to the peak of $\Delta(T)$), but it also fails in the

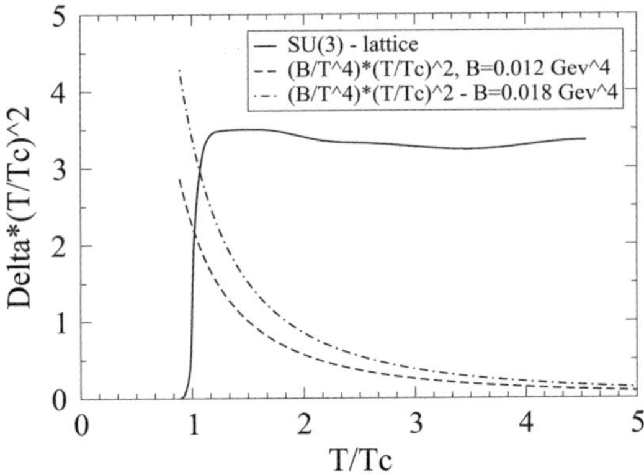

Fig. 8.13 The temperature variation of $\Delta(T)(T/T_c)^2$ obtained from the bag pressure, compared to the corresponding lattice data [11]

temperature region above T_c. There exist different and quite interesting attempts to correct the T-dependence in that region [32–35], including also a combination of bag model and weak coupling expansion; none of these, however, can address the critical behavior.

8.6 The Quasi-particle Approach

There thus remains the task to find a non-perturbative approach which takes into account the critical features arising in the temperature region in the range above T_c, as they were obtained in lattice studies. To illustrate one such possibiity, we stay in pure $SU(3)$ gauge theory, where an extrapolation to the continuum limit is available [11]. The basis for our considerations here is the study of an ideal gas of constituents ("quasi-particles") having dynamically or thermally generated masses [36–39]. The behavior of an ideal gas of such massive gluon modes provides automatically the observed N_c^2 scaling and also leads to other features in accord with the functional behaviour found in $SU(N)$ gauge theories. Let us consider this in more detail.

The partition function of an ideal gas of constituents of mass $m(T)$ is in the Boltzmann limit for $SU(N)$ gauge theory given by

$$\ln Z(T) = \frac{(N^2 - 1)V}{2\pi^2} \int_0^\infty dp\, p^2 \exp(-\frac{1}{T}\sqrt{p^2 + m^2}) = \frac{(N^2 - 1)VTm^2}{2\pi^2} K_2(m/T),$$

(8.26)

where $K_i(x)$ denotes the Hankel function of ith order and imaginary argument. The resulting pressure becomes

$$P(T) = T\left(\frac{\partial \ln Z}{\partial V}\right)_T$$

$$= \frac{(N^2 - 1)T}{2\pi^2} \int_0^\infty dp\, p^2 \exp(-\frac{1}{T}\sqrt{p^2 + m^2}) = \frac{(N^2 - 1)T^2 m^2}{2\pi^2} K_2(m/T)$$

(8.27)

while the energy density is found to be

$$\epsilon(T) = \frac{T^2}{V}\left(\frac{\partial \ln Z(T)}{\partial T}\right)_V$$

$$= \frac{(N^2 - 1)}{2\pi^2} \int_0^\infty dp\, p^2 \exp(-\frac{1}{T}\sqrt{p^2 + m^2})\left\{\sqrt{p^2 + m^2} - T\frac{m}{\sqrt{p^2 + m^2}}\left(\frac{dm}{dT}\right)\right\}$$

$$= \frac{(N^2 - 1)m^2 T^2}{2\pi^2}\left\{3K_2(m/T) + \left[\frac{m}{T} - \left(\frac{dm}{dT}\right)\right]K_1(m/T)\right\}.$$

(8.28)

In these expressions, we have maintained two spin degrees of freedom for the "massive" gluons; we return to this point shortly. Both energy density and pressure thus fall below the Stefan-Boltzmann limit, as is illustrated in Fig. 8.7; the lattice results for the $SU(3)$ case were shown in Fig. 5.3. The resulting interaction measure is given by

$$\Delta = \frac{(N^2-1)}{2\pi^2} \int_0^\infty dp\, p^2 \exp(-\frac{1}{T}\sqrt{p^2+m^2})$$
$$\left\{ \sqrt{p^2+m^2} - 3T - T\frac{m}{\sqrt{p^2+m^2}}\left(\frac{dm}{dT}\right)\right\}$$
$$= \frac{(N^2-1)m^2}{2\pi^2 T^2}\left[\frac{m}{T} - \left(\frac{dm}{dT}\right)\right] K_1(m/T). \tag{8.29}$$

If m is N_c-independent, the scaling in N^2-1 is evident. Moreover, if the effective mass m is linear in T, as in any conformal theory, $\Delta(T)$ vanishes. Given a running coupling, with $m = g(T)T$, we get

$$\Delta(T) = \frac{(N^2-1)m^2}{2\pi^2 T^2} T\left(\frac{dg}{dT}\right). \tag{8.30}$$

We note, however, that such a "naive" quasi-particle description with finite masses seems to encounter a conceptual problem. Physical constituents of non-vanishing mass should have three, rather than two spin degrees of freedom, since a longitudinal polarization is excluded only for massless particles. The resulting changes in all thermodynamic quantities – e.g., the increase of the ideal gas energy density ϵ/T^4 (see Eq. (5.41)) from $8\pi^2/15$ to $12\pi^2/15$ – are definitely in disagreement with the observed high-temperature lattice results. A simple shift to massive gluons thus cannot satisfactorily explain the interactions apparently still present in the high temperature gluon gas. More generally, a gauge invariant theory does not allow massive physical gluons; the mechanism leading to effective thermal masses must thus be more subtle. The mentioned modified HTL perturbation theory approach, in which each order already includes some aspects of gluon dressing, not only leads to a rather rapid convergence of the expansion; in addition, the contribution of longitudinal gluons vanishes in the limit $g \to 0$, so that one also obtains the right number of degrees of freedom for the Stefan-Boltzmann form [24–26]. Moreover, it has recently been argued [40] that masssive gluons should in fact be transversely polarized, since two massless gluons cannot combine to form a longitudinally polarized massive gluon [41]. It thus seems justified to use the thermal mass scenario outlined above to address the temperature behaviour of the quark-gluon plasma.

The form of the effective mass entering in a quasi-particle approach description has been an enigma for quite some time. At sufficiently high temperature, T remains as the only scale, so that there we expect $m \sim T$. From perturbation theory one obtains in leading order for $SU(N)$ gauge theory a thermal screening mass $\sim N\,g(T)\,T/3$, but in view of the above mentioned difficulties, it seems best to leave the proportionality open. As we approach the critical point, perturbation theory in whatever form ceases to be applicable. We now have a medium of strongly interacting gluons, and the range of the forces between them becomes larger and larger as we approach the critical point. This range is governed by the correlation length, or in other words, by the distance up to which a given color charge can "see" other color charges. This distance is the QCD counterpart of the Debye screening radius in QED; we write it as $r_D(T) = 1/\mu(T)$, where $\mu(T)$ denotes the corresponding screening mass. It corresponds to the shift from $1/k^2$ to $1/(k^2 + \mu^2)$ experienced by the gluon propagator due to the presence of the medium.

The perhaps simplest view thus is to consider as mass of the quasi-gluon in the strongly coupled region the energy contained in a volume V_{cor} of the size defined by the correlation range,

$$m_{crit}(T) \sim \epsilon(T) V_{\mathrm{cor}}(T). \tag{8.31}$$

In the case of a continuous transition, the critical part of the energy density becomes

$$\epsilon \sim (t-1)^{1-\alpha}, \tag{8.32}$$

where $t = T/T_c$, and α is the critical exponent for the specific heat. The correlation volume (for three space dimensions) can be written as

$$V_{\mathrm{cor}} = 4\pi \int dr\, r^2\, \Gamma(r, T), \tag{8.33}$$

where

$$\Gamma(r, T) \sim \frac{\exp-\{r/\xi(T)\}}{r^{1-\eta}}, \quad \xi(T) \sim (t-1)^{-\nu} \tag{8.34}$$

specifies the correlation function Γ in terms of the critical exponents ν for the correlation length $\xi(T)$ and η for the anomalous dimension. Combining these expressions, we have

$$m_{\mathrm{crit}}(T) \sim (t-1)^{1-\alpha-2\nu-\eta}; \tag{8.35}$$

for $SU(2)$ gauge theory in three space dimensions, the exponents are given by the corresponding exponents of the 3d Ising model, $\alpha \sim 0.11, \nu \simeq 0.63, \eta \simeq 0.04$, suggesting

$$m_{\text{crit}}(T) \sim (t-1)^{-0.41}. \tag{8.36}$$

This form is correct only very near the critical point $t = 1$; for large temperatures, $\xi(t) \sim t$, so that the overall form expected for the mass of the quasi-gluon becomes

$$m(t) \simeq a(t-1)^{-0.41} + bt, \tag{8.37}$$

where a and b are constants. The resulting behavior is illustrated in Fig. 8.14 (left). It would certainly be of interest to apply this form in an analysis of the thermodynamics of $SU(2)$ gauge theory; unfortunately, there seem to exist no lattice studies allowing an extrapolation to the continuum. Older studies of $\epsilon(T)$ and $P(T)$ in terms of a gluon mass $m(T)$ did in fact lead to the form shown in Fig. 8.14 [36].

For $SU(3)$, the transition is of first order, so that all quantitites remain finite at T_c and an equivalent form cannot be given. Nevertheless, in all cases we have a strong increase of both $\epsilon(T)$ and $\mu(T)$ in some range above T_c, and so we shall maintain the functional dependence (8.35)/(8.36) with an open exponent c. The resulting quasi-particle mass is thus expected to have the form

$$m(T) = \frac{a}{(t-1)^c} + bt, \tag{8.38}$$

with constants a, b, c.

Using this mass, we now determine the parameters a, b, c by calculating $\Delta(T)$ from Eq. (8.29) and the energy density from Eq. (8.28). The resulting mass is shown in Fig. 8.14 (right). The fits to interaction measure and energy density are given in Fig. 8.15 and are seen to reproduce both quantities very well. We can thus conclude that the gluon plasma in $SU(3)$ gauge theory in the temperature region above T_c indeed behaves like a medium of quasi-particles with masses generated through non-perturbative thermal effects.

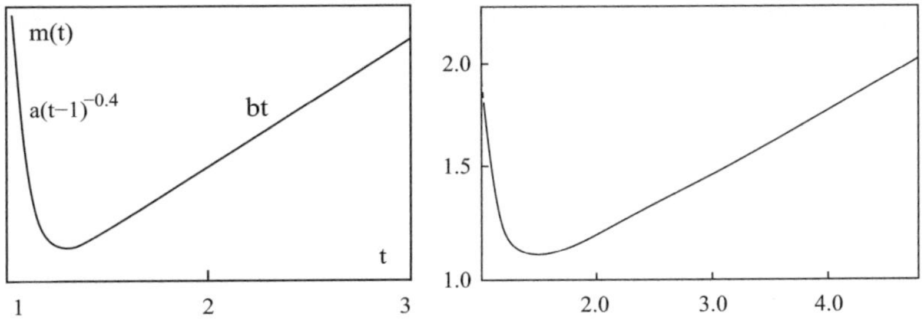

Fig. 8.14 Expected behavior of the effective quasi-particle mass $m(t)$ in $SU(2)$ gauge theory (left), and as obtained from a fit of lattice data in $SU(3)$ gauge theory (right)

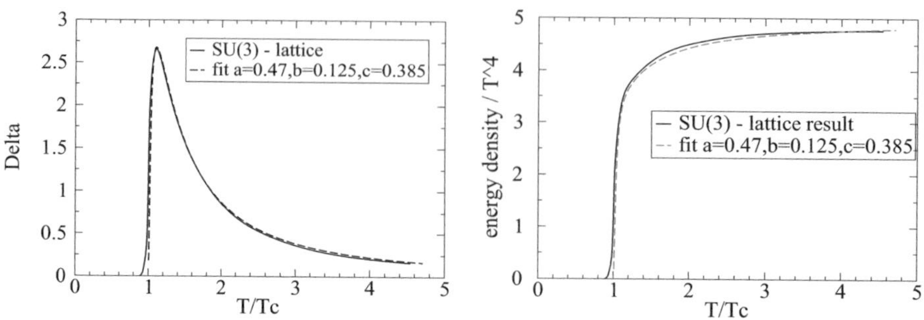

Fig. 8.15 Interaction measure (left) and energy density (right) for $SU(3)$ gauge theory, compared to a quasi-particle description

The same approach can now be used to address the corresponding problem in two-flavor QCD, using the energy density and the screening mass obtained in lattice studies to determine the quasi-particle mass, and then evaluate the interaction measure. In this case, the lattice results are not yet of the same degree of precision, but the general pattern remains the same as for $SU(3)$ gauge theory.

8.7   The Speed of Sound in the QGP

In the hadronic resonance gas study of Chap. 3, we had seen that the speed of sound drops to zero at the critical point defined by the limit of hadronic matter. It does so because any further energy increase goes into making more massive resonances, not into momentum and pressure. On the deconfined side, in the quasi-particle approach just discussed, the behavior is very similar. As we lower the plasma temperature towards the confinement point, the increase of the quasi-particle mass has the same effect. Seen the other way around, a temperature increase above T_c lowers the mass and thus provides more momentum and pressure, causing an increase in the speed of sound.

More specifically, the speed of sound, defined as

$$c_s^2 = \left(\frac{\partial p}{\partial \epsilon} \right)_V = \frac{s(T)}{C_V(T)}, \tag{8.39}$$

vanishes at T_c for a continuous transition, because the specific heat $C_V(T)$ diverges there, while the entropy density $s(T)$ remains finite. In the ideal gas limit, $s(T) \simeq 4\,c_0\,T^3$ and $C_V(T) \simeq 12\,T^3$, so that $c_s^2 \to 1/3$. For the temperature-dependent mass (8.38), the speed of sound can be evaluated numerically, using Eqs. (8.17) and (8.18). It is found [39] that the behavior obtained in such a quasi-particle approach agrees very well with that found in SU(3) lattice calculations.

A similar behavior arises in full QCD, with dynamical quarks. For two massless quark flavors, the specific heat will diverge; otherwise, it will only increase strongly near the cross-over temperature T_c. The general pattern for the speed of sound, both below and above T_c, thus has the form shown in Fig. 8.16. As mentioned above, the dip is on both sides a consequence of the mass increase as the temperature becomes (almost) critical.

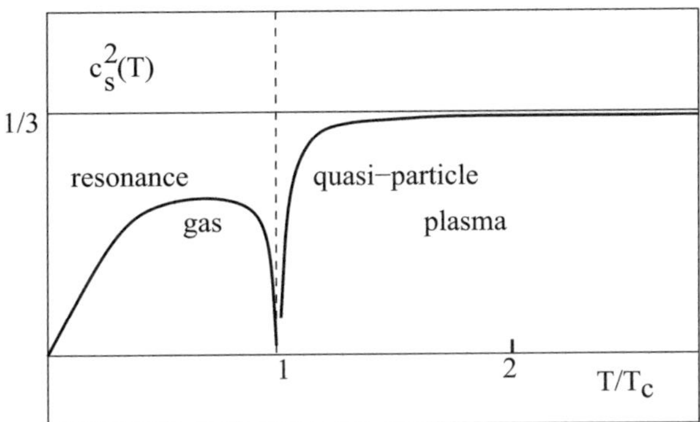

Fig. 8.16 The speed of sound in strongly interacting matter

References

1. See e.g. Landau and Lifshitz, *Statistical Physics*, p. 229ff. (Pergamon Press, London/Paris, 1958)
2. E. Eichten et al., Phys. Rev. D **17**, 3090 (1978)
3. E. Eichten et al., Phys. Rev. D **21**, 203 (1986)
4. V.V. Dixit, Mod. Phys. Lett. A **5**, 227 (1990)
5. O. Kaczmarek et al., Phys. Lett. B **543**, 41 (2002)
6. O. Kaczmarek et al., Phys. Rev. D **70**, 074505 (2004)
7. S. Digal et al., Eur. Phys. J. C **43**, 71 (2005)
8. S. Nadkarni, Phys. Rev. D **33**, 3738 (1986) and D **34**, 3904 (1986)
9. J. Engels et al., Z. Phys. C **42**, 341 (1989)
10. G. Boyd et al., Phys. Rev. Lett. **75**, 4169 (1995)
11. G. Boyd et al., Nucl. Phys. B **469**, 419 (1996)
12. M. Panero, Phys. Rev. Lett. **103**, 23001 (2009)
13. S. Datta, S. Gupta, Phys. Rev. D **82**, 114505 (2010)
14. A. Peikert, F. Karsch, E. Laermann, Nucl. Phys. B **83/84**, 390 (2000)
15. B. Beinlich et al., Eur. Phys. J. C **6**, 133 (1999)
16. See e.g., J.I. Kapusta, *Finite-Temperature Field Theory* (Cambridge University Press, Cambridge, 1989)
17. A.D. Linde, Phys. Lett. B **96**, 289 (1880)
18. D.J. Gross, R.D. Pisarski, L.G. Yaffe, Rev. Mod. Phys. **53**, 43 (1981)
19. P. Arnold, C.-X. Zhai, Phys. Rev. D **50**, 7609 (1994)
20. B. Kastening, C.-X. Zhai, Phys. Rev. D **52**, 7232 (1995)
21. K. Kajantie et al., Phys. Rev. D **67**, 105008 (2003)

22. J.O. Anderson et al., Phys. Rev. D **66**, 085016 (2002)
23. J.O. Andersen, M. Strickland, N. Su, JHEP **1008**, 113 (2010)
24. For reviews, see J.-P. Blaizot, Nucl. Phys. A **702**, 99c (2002)
25. For reviews, see A. Rebhan, Nucl. Phys. A **702**, 111c (2002)
26. For reviews, see A. Peshier, Nucl. Phys. A **702**, 128c (2002)
27. Sz. Borsanyi et al., arXiv:1104.0013 [hep-ph]
28. J.O. Andersen et al., Phys. Lett. B **696**, 468 (2011)
29. M. Asakawa, T. Hatsuda, Nucl. Phys. A **610**, 470c (1996)
30. H. Leutwyler, in *QCD – 20 Years Later*, ed. by P.M. Zerwas, H. Kastrup (World Scientific, Singapore, 1993)
31. M.A. Shifman, A.I. Vainshtein, V.I. Zakharov, Nucl. Phys. B **147**, 385 (1979)
32. D. Zwanziger, Phys. Rev. Lett. **94**, 182301 (2005)
33. R. Pisarski, Phys. Rev. D **74**, 121703(R) (2006)
34. O. Andreev, Phys. Rev. D **76**, 087702 (2007)
35. E. Megias, E. Ruiz, L.L. Salcedo, Phys. Rev. D **80**, 056005 (2009)
36. V. Goloviznin, H. Satz, Z. Phys. C **57**, 671 (1993)
37. A. Peshier et al., Phys. Rev. D **54**, 2399 (1996)
38. F. Brau, F. Buisseret, Phys. Rev. D **79**, 114007 (2009)
39. P. Castorina, D.E. Miller, H. Satz, Eur. Phys. J. C **71**, 1673 (2011)
40. V. Mathieu, PoS QCD-TNT09:024 (2009)
41. C.-N. Yang, Phys. Rev. **77**, 242 (1950)

Chapter 9
The Little Bang

> *Si parva licet componere magnis.*
>
> Publius Virgilius Maro, *Georgica IV,* ~ 30 B. C.
>
> *(If it is allowed to compare the small with the large.)*
>
> *(Virgil, on bee-keeping vs. cattle-raising.)*

Abstract Here we give a short general introduction to the experimental study of strongly interacting matter. After briefly noting applications in cosmology and astrophysics, we turn to the use of high energy nuclear collisions as a tool for such an endeavor. Following a discussion of the conceptual basis, we present the parton basis of high energy nuclear collisions and its limitation through saturation.

9.1 Applying Strong Interaction Thermodynamics

We had already mentioned in the Introduction that the topic of this book deals with a rather esoteric subject: dense strongly interacting matter, i.e., matter of a density beyond that found in heavy nuclei, and in particular the quark-gluon plasma. Such a medium of deconfined quarks and gluons must have been one of the states through which our universe passed in its evolution from the big bang to the present, and quark matter could exist in the core of neutron stars. The great interest this subject holds in today's physics research, however, is to a very large extent based on the hope, the expectation, the certainty (which of these, is in the eye of the beholder) that we will be able to study strongly interacting matter in the laboratory, under controlled conditions, using high energy nuclear collisions. Moreover, increasing the collision energy is expected to increase the initial energy density of the produced matter, so that at sufficiently high energy, we reach the realm of deconfinement and of the quark-gluon plasma. Before turning to nuclear collisions, we briefly comment on the cosmological and astrophysical connections.

The expansion of our universe is described in terms of the Hubble "constant" $H(t)$; it is a constant only in space, but varies with time. It relates the velocity

© Springer International Publishing AG 2018

H. Satz, *Extreme States of Matter in Strong Interaction Physics,*

Lecture Notes in Physics 945, https://doi.org/10.1007/978-3-319-71894-1_9

of a distant galaxy to its position and thus measures the expansion time. The conservation of kinetic expansion and gravitational potential energy requires

$$H^2(t) = \frac{8\pi}{3} G\epsilon(t),$$ (9.1)

where G is the gravitation constant and $\epsilon(t)$ the cosmic mass density. The expansion time is thus given by

$$t_{\mathrm{exp}} = \frac{1}{H(t)} = \sqrt{\frac{3}{8\pi G\epsilon(t)}}.$$ (9.2)

Considering the instant of the big bang as starting point, we can determine the time needed to reach, on a cosmic level, the deconfinement energy density $\epsilon \simeq 1\,\mathrm{GeV/fm}^3$. With the value $G = 6.708 \times 10^{-39}\,\mathrm{GeV}^{-2}$, it is found to be

$$t_{\mathrm{exp}} \simeq 10^{-5}\,\mathrm{s},$$ (9.3)

so that in its first ten microseconds, our universe was in a deconfined state ("quark era"). Only at the end of this era hadrons formed, and there appeared for the first time the background for today's complex cosmic structure, the physical vacuum.

A neutron star is one final evolution stage of sufficiently heavy stars (black holes are the other). In a gigantic supernova explosion, the star collapses from a state of atomic to one of nuclear density. It has a mass of more than 1.5 solar masses, with a radius of around 10 km. Its density varies from about $(1/3)\,n_0$ in the crust to 2–3 n_0 in the core, where $n_0 \simeq 0.17\,\mathrm{fm}^{-3}$ denotes standard nuclear density. The pressure of the collapse has, through inverse beta-decay $(p+e^- \rightarrow n+\nu)$, turned all constituents into neutrons, releasing energy by neutrino radiation. The neutrons are contracted by gravitation, while the exclusion principle provides stability. It is not known if the deep core of such a star reaches conditions which make a state of deconfined quarks energetically favorable. On one hand, one does not have reliable theoretical predictions, since so far finite density lattice studies are restricted to very low baryon density. On the other hand, there are also no clear cut empirical signatures known, which could be used to identify a neutron star with a quark core.

9.2   High Energy Collisions and the Vapour Trail

Normal nuclear matter consists of nucleons of mass 0.94 GeV and has, as just noted, a density of 0.17 nucleons/fm^3; hence its average energy density is 0.16 GeV/fm^3. How can we surpass this value? In a high energy proton-proton collision, the two colliding nucleons do not just stop each other, even if they hit "head-on"; instead, they pass through each other. The reason for this is essentially their finite size [1]. A nucleon has a spatial extension of about 1 fm, and in a nucleon-nucleon collision it therefore takes a certain time τ_0 before the entire nucleon knows that it hit something. Consider a fixed-target experiment, in which a proton beam hits a stationary proton target. In the rest-frame of the projectile, we have $\tau_0 \simeq 1\,\mathrm{fm}$ for the time required to pass the information about the collision from one side to the other. In the rest frame of the target, this time is dilated to

$$t_0 = \tau_0 \left(\frac{P_0}{m}\right) \simeq \tau_0 \left(\frac{\sqrt{s}}{2m}\right), \qquad (9.4)$$

where $P_0 = \sqrt{\mathbf{P}^2 + m^2}$ denotes the energy of the projectile, $\mathbf{P}$ its momentum, m its mass, and $\sqrt{s}$ the center-of-mass collision energy. For large $\sqrt{s}$, the time in the target frame $t_0 \gg 1\,\mathrm{fm}$, so that the projectile has long left the region of the target nucleon before it fully realizes that it was hit.

Hence the colliding nucleons pass through each other, and after the little bang of the collision, they leave behind a "vapour trail" of deposited energy, of disturbed vacuum. It consists of droplets of deposited energy, some almost at rest in the overall center of mass, and then faster and faster droplets moving along the trail, up to the excited target and projectile fragments (see Fig. 9.1).

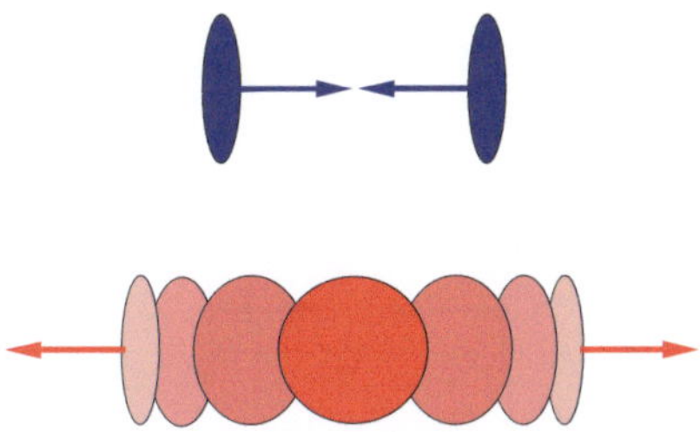

Fig. 9.1 A head-on nucleon-nucleon collision at high energy, as seen in the overall center-of-mass system; top: before, bottom: after the collision

In a boost-invariant picture of such collisions [2], all droplets far enough away in rapidity from the target or projectile regime will be very similar; each will, in its own rest frame, contain a certain amount of deposited energy. Each droplet expands, cools and eventually materializes by forming a number of hadrons of different species. At a center-of-mass collision energy $\sqrt{s}$ and at mid-rapidity, one has on the average per droplet $(dN_h/dy)_0$ mesons of about $0.5\,\mathrm{GeV}$ energy each. We can now "let the film run backwards" and note that initially, this energy was contained in a volume of roughly hadronic size ($r_h \simeq 1\,\mathrm{fm}$). The average initial energy density thus must have been

$$\epsilon_{pp} \simeq \left(\frac{dN_h}{dy}\right)_0 \left\{ \frac{0.5\,\mathrm{GeV}}{(4\pi/3)(1\,\mathrm{fm})^3} \right\} \simeq 0.12 \left(\frac{dN_h}{dy}\right)_0 \,[\mathrm{GeV/fm}^3]. \qquad (9.5)$$

For $\sqrt{s} = 20\,\mathrm{GeV}$, one observes $(dN_h/dy)_0 \simeq 3$, leading to $\epsilon_{pp} \simeq 0.36\,\mathrm{GeV/fm}^3$, which is about twice that of standard nuclear matter. One might here argue that the interaction region in the center of mass frame is Lorentz-contracted in the longitudinal direction, so that instead of a sphere of radius 1 fm, one should consider a much thinner pancake, which would lead to correspondingly higher energy densities. For very "early" phenomena, this may well be correct; but if we consider hadronic consequences of the collision, which take a typical hadronic time to form, a longitudinal extension of about 1 fm is reasonable. Hence in their local rest system, the droplets can presumably be taken as spherical of radius 1 fm.

The energy densities just obtained had initially excluded proton-proton collisions from QGP searches. In the meantime, however, such collisions at the LHC with a collision energy of 7 TeV (7000 GeV) produce on the average 9 secondaries per unit rapidity, and one can study high multiplicity bins with up to more than 30 hadrons. The resulting energy densities can thus be extended to $3.6 \, \text{GeV/fm}^3$, so that the deconfinement regime is attainable here as well.

Let us now look at a high energy head-on collision of two identical heavy nuclei of mass number A; the geometry of the process is schematically illustrated in Fig. 9.2. In the overall center-of-mass system, the Lorentz contraction of the nuclei forces all the nucleons contained in a tube of hadronic radius (1 fm) along the beam axis to interact in essentially the same space-time region. Integrating over the transverse collision area, we find on the average for each nucleus

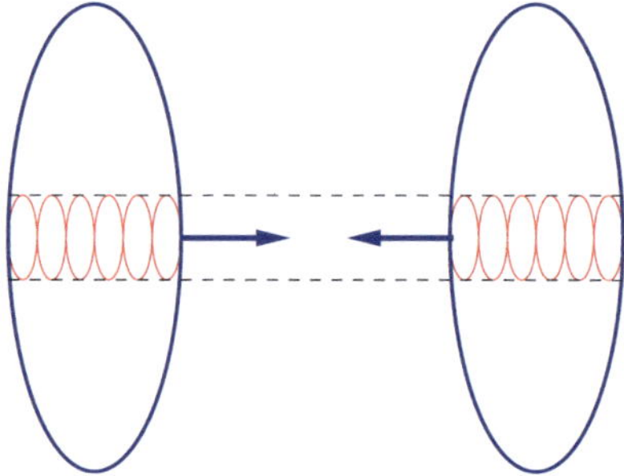

Fig. 9.2 A head-on nucleus-nucleus collision at high energy, as seen in the overall center-of-mass system

$$N_A \simeq \frac{3}{4}[2R_A \pi (1 \, \text{fm})^2] \, n_0 \simeq A^{1/3} \tag{9.6}$$

nucleons in this tube, with $R_A = 1.14 \, A^{1/3}$ fm denoting the radius of a nucleus of mass number A; $n_0 \simeq 0.17 \, \text{fm}^{-3}$ is again standard nuclear density, and the factor 3/4 comes from averaging over the nuclear profile for a central nucleus-nucleus collision. Such a collision will therefore lead to the superposition of about $A^{1/3}$ nucleon-nucleon interactions, resulting in an average initial energy density [2]

$$\epsilon_{AA} \simeq 0.14 \, A^{1/3} \left(\frac{dN_h}{dy} \right)_0, \tag{9.7}$$

which for heavy nuclei ($A \simeq 200$) and $\sqrt{s} = 20\,\mathrm{GeV}$ leads to about $2.5\,\mathrm{GeV/fm}^3$ in a typical central collision. Moreover, the overall interaction region has now become of nuclear size. This means a transverse nuclear area of πR_A^2; in the longitudinal direction, the initial interaction region is at high energies Lorentz-contracted to a thickness of about a fermi (as mentioned above, this is needed for the formation of any hadronic collision products). For a head-on collision of two heavy nuclei we thus obtain an initial volume of about $170\,\mathrm{fm}^3$, decidedly larger than a typical hadronic volume.

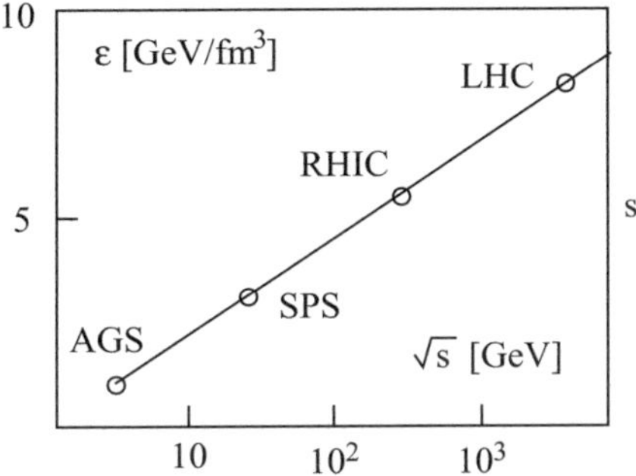

Fig. 9.3 Average initial energy density as a function of collision energy, for central *AA* collisions with $A \simeq 200$

As function of energy, one finds roughly $(dN_h/dy)_0 \simeq \ln\sqrt{s}$ for the multiplicity of produced hadrons; the resulting behavior of the energy density is illustrated in Fig. 9.3, with indications for AA collisions at the top collision energy of the past and present nuclear collision facilities. Experimentally it is indeed found that the hadron multiplicities at top RHIC and LHC energies are by a factor two (RHIC) and a factor 4 (LHC) higher than the SPS value for $\sqrt{s} = 20\,\mathrm{GeV}$, with corresponding increases in the energy density. It thus appears not unreasonable to expect that high energy nuclear collisions bring within our reach the direct experimental study of strongly interacting matter, and if the estimates are correct, the state of this matter when it is initially formed should be a quark-gluon plasma. Let us now consider in some more detail how the formation of the collision medium could occur.

9.3 Parton Interactions and Thermalization

The quark infrastructure of hadrons implies that what really happens in a high energy hadron-hadron collision is an interaction between the sub-hadronic constituents of target and projectile. These are for mesons a quark-antiquark pair, for baryons a three quark state, and for both there are in addition gluons to achieve the binding. These gluons, in turn, lead to a sea of fluctuating quark-antiquark pairs. A simple hadron thus becomes a rather complex entity, consisting of the quarks which determine its quantum numbers, i.e., of the valence quarks, of the gluons and of their quark-antiquark fluctuations, the "sea" quarks. All three types together are denoted as partons, and the momentum of an incident hadron is shared among all of its partonic constituents. This partonic substructure is not a "static" feature, in the way an atom consists of $A = N_p + N_n$ protons and neutrons forming a nucleus surrounded by N_p electrons. Rather, it is a description of the interaction of an energetic hadron with a probe, and the parton substructure of the hadron depends both on the resolution of the probe and on the momentum of the hadron. In other words, a fixed probe "sees" in a specific hadron a different parton content at different momenta of the hadron.

Inelastic lepton-hadron scattering allows us to determine experimentally how many partons of each species there are in a hadron, and how much of its energy is carried by each species of parton. In such reactions, an energetic (virtual) photon hits the hadron, and if the wavelength of the photon is short enough, it sees not the hadron as a whole, but rather one of its partonic constituents (see Fig. 9.4).

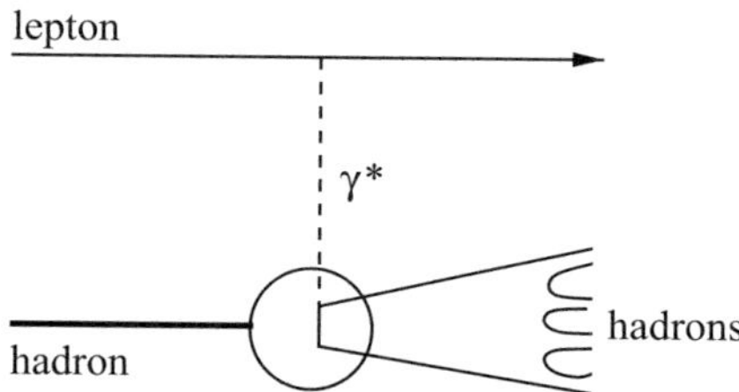

Fig. 9.4 Deep inelastic lepton-hadron scattering

The resulting data therefore allows us to specify how the hadron's momentum $P \simeq \sqrt{s}/2$ is distributed among the partons. The relevant quantity here is the parton distribution function, $f_i(x)$, which specifies the number of partons of species i carrying a fraction $x_{\min} < x < 1$ of the incident momentum of the hadron. Here $x_{\min} \simeq \Lambda_{QCD}/\sqrt{s}$ specifies the low-momentum bound for constituents. Things must add up, of course, so that the different f_i have to satisfy the energy conservation sum rule

$$\sum_i \int dx \, x f_i(x) = 1, \tag{9.8}$$

where the sum runs over gluons, valence and sea quarks. Corresponding relations hold for the conservation of electric charge and of baryon number. A generic illustration of the parton substructure of a proton is shown in Fig. 9.5. We note that the valence quarks provide the dominant contribution for partons carrying a sizeable fraction of the hadron momentum. On the other hand, for increasing hadron energy $\sqrt{s}$ there is an ever increasing number of soft gluons, so that at low x a hadron consists mainly of gluons.

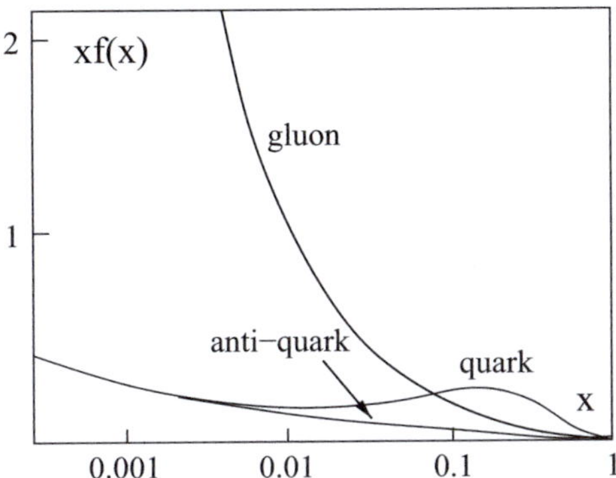

Fig. 9.5 Schematic parton distribution $xf(x)$ for protons, as a function of parton fractional momentum x

An incident energetic hadron thus is "really" a beam of partons. Besides their longitudinal momentum, determined by how much of the hadron momentum they carry, they also have an intrinsic transverse momentum k_T. By the uncertainty relation, this means that they have a transverse size $r_T \sim 1/k_T$, so that a cut through the transverse plane of an incoming hadron looks like a set of small discs of different sizes distributed over a larger disc of hadronic size (Fig. 9.6).

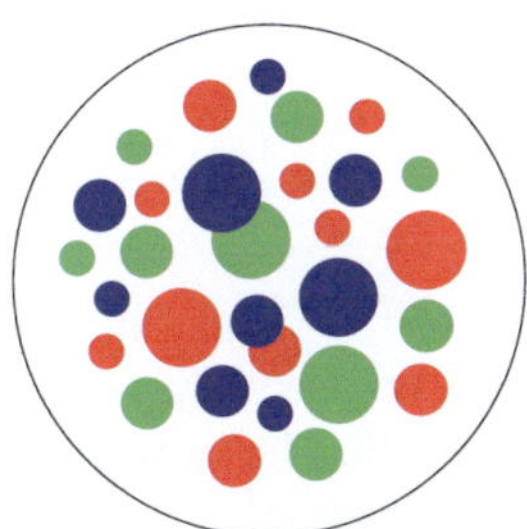

Fig. 9.6 Parton distribution in the transverse plane of an incident hadron

Probing the discs in this picture is possible only if the photon used has a sufficiently short wave length. This implies that photons of squared four-momentum Q^2 can only "see" the partons of transverse momentum $k_T^2 \leq Q^2$, i.e., of a size $r_T^2 \geq 1/Q^2$. So the parton distribution function introduced above is really a function of two variables,

$$f_i(x, Q^2) = \int_{\Lambda_{QCD}^2}^{Q^2} dk_T^2 \, f_i(x, k_T^2) \qquad (9.9)$$

obtained by integrating over the transverse momentum distribution of the partons up to the maximum value accessible to photons with resolution Q^2. The lower limit of k_T is determined by the basic scale Λ_{QCD} of QCD.

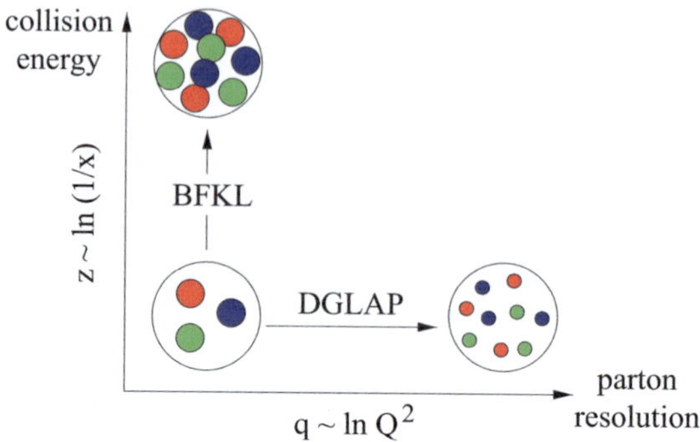

Fig. 9.7 Parton evolution

When such a beam of partons hits a target, the interaction can cause each parton
to split into further partons. This branching is clearly reminiscent of the resonance
evolution patterns discussed above, and the formal basis is in fact very similar. We
had found there that the splitting of a hadronic resonance or fireball into further such
objects is determined by the number of partitions $p(n)$ of an integer n, leading to

$$p(n) \sim \exp\{a\sqrt{n}\}, \tag{9.10}$$

with constant a. Here one finds that the number of partons grows as

$$f_i(y, q) \sim \exp\left\{2\sqrt{\alpha_s(q)yq}\right\}, \tag{9.11}$$

where $y = \ln(1/x)$ and $q = \ln(Q^2/Q_0^2)$ measure the incident parton energy and
the inverse transverse parton size, respectively. In addition, the coupling parameter
$\alpha_s(q)$ specifies the interaction strength. The parton distribution function (9.11) is the
solution of the so-called DGLAP equation[1]

$$\frac{\partial^2 f_i(y, q)}{\partial y \partial q} = \alpha_s(q)f_i(y, q), \tag{9.12}$$

which specifies the evolution of the partons in the collision. Closer inspection of
the solution (9.11) shows that decreasing either the average transverse partonic size
$(1/Q^2)$ or increasing the collision energy $(1/x)$ increases the corresponding number
of partons. Increasing q at fixed y leads to more partons of smaller and smaller size
(DGLAP evolution), while increasing y at fixed q leads to more partons of the same
average size (BFKL evolution[2]); the different cases are illustrated in Fig. 9.7.

---

[1]Dokshitzer-Gribov-Lipatov-Altarelli-Parisi [3–5].
[2]Balitsky-Fadin-Kuraev-Lipatov [6–8].

Returning now to nucleus-nucleus collisions, we have a "beam" of partons of the different species colliding with another such beam. In the interactions, branching further increases the number of partons, particularly of gluons, and the Lorentz contraction of the colliding nuclei will cause collisions of partons from different nucleon-nucleon interactions to overlap. It is this dense partonic system which is expected to equilibrate and thus produce a thermal partonic medium, the quark-gluon plasma. Starting from the non-equilibrium configuration of the two colliding nuclei, the evolution of the collision is thus assumed to have the form illustrated in Fig. 9.8. After the collision, there is a short pre-equilibrium stage, in which the primary partons of the colliding nuclei interact, multiply and eventually thermalize to form a quark-gluon plasma. This then expands, cools and hadronizes to form the observed hadronic secondaries.

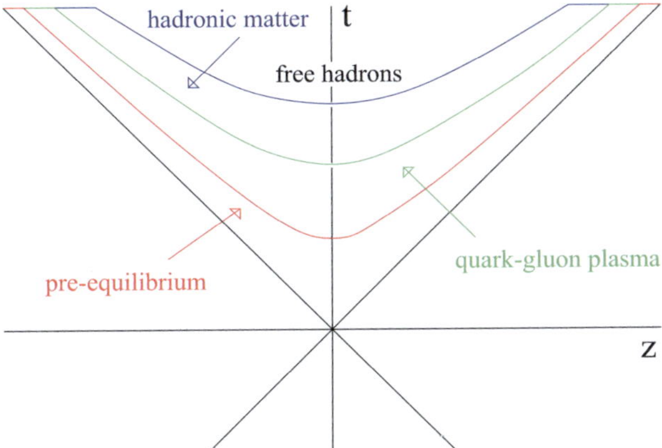

Fig. 9.8 Expected evolution of a nuclear collision

Clearly an essential assumption for quark-gluon plasma formation in such a scenario is that the initially incident collinear parton beams in the collision multiply and eventually form a locally thermalized medium, i.e., create bubbles of quarks and gluons in some kind of thermal equilibrium state. One such possibility is provided by partonic interactions on a perturbative scale: numerous well-defined partons interact sufficiently often to form kinetically an equilibrated system, where the interactions include elastic ($2 \rightarrow 2$) scattering as well as branching ($1 \rightarrow 2$) and fusion ($2 \rightarrow 1$). This is the "classical" point of view, which initiated the various programs for high energy nuclear collision experiments. The crucial question here is if the "droplets" of quarks and gluons formed in nuclear collisions live long enough to thermalize - or, in other words, what is the thermalization time, compared to the expansion time of the formed medium? This topic has been addressed for many years, both conceptually [9–11] and through numerical codes [12–15]. The main results are:

- In the initial stages, gluons dominate; they multiply and thermalize very rapidly, arriving at an isotropic momentum distribution after times of the order of 0.3 fm.
- The evolution of the quark distributions takes longer, since the relevant interaction cross sections are smaller by a factor 2–3, due to color coefficients.

More recent work [16] has confirmed the crucial role of the gluons and concluded that in the early stage, the primary, relatively hard gluons produce many soft gluons, which quickly thermalize, and the resulting heat bath then draws energy from the remaining hard gluons, leading eventually to full thermalization.

As mentioned before, the work just discussed assumes that the thermalizing medium consists of independent, perturbatively treated partons. This assumption, as we shall see, is not tenable at sufficiently high energies and/or large nuclei; let us now turn to this problem.

9.4   Parton Percolation and Saturation

The distribution of the partonic constituents in the initial state of a high energy nuclear collision is given by a superposition of the corresponding distributions inside the nucleons of the colliding nuclei. Here it should be kept in mind that for nucleons inside nuclei the distributions will be modified, compared to those in single nucleons, because already the presence of other nucleons affects the partonic structure of a given nucleon ("shadowing" or "antishadowing"). Nevertheless, the general picture in the transverse plane is similar to that for colliding hadrons, except that for an $A - A$ interaction at the same collision energy, the parton density is now much higher (Fig. 9.6). The density of partons increases further with $\sqrt{s}$, and at some critical point, parton percolation must occur [17–19] and a "global" color connection set in (see Fig. 9.9). In the resulting "condensate", the partons are expected to lose their independent existence and well-defined origin. In other words, at this point the parton model as such presumably should break down, become ill-defined. The evolution schemes mentioned above, DGLAP and BFKL, apply only in a relatively dilute medium, in which the concept of a parton remains meaningful.

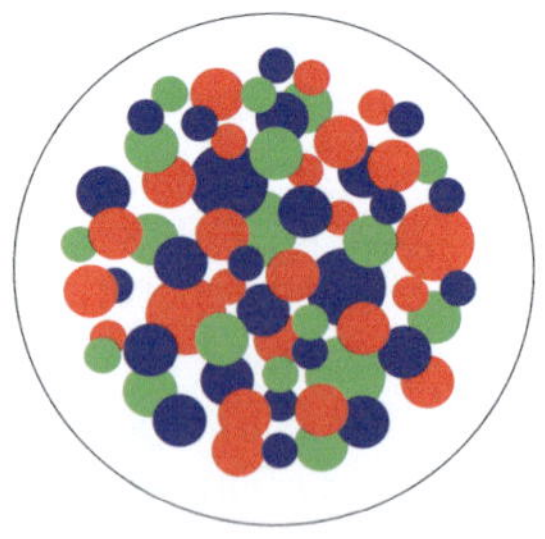

Fig. 9.9 Parton distribution in the transverse plane of an incident nucleus

The geometric considerations leading to parton percolation are in fact signals of a more fundamental problem. It can be shown [20–22] that the total cross-section for hadron-hadron scattering cannot increase arbitrarily fast with collision energy $\sqrt{s}$. General physics principles, i.e., probability conservation (unitarity), causality and relativistic invariance, lead to the Froissart bound

$$\sigma(s) \leq (\ln s)^2. \tag{9.13}$$

This limit had originally been conjectured by Heisenberg [20] on phenomenological grounds and was subsequently proved in quantum field theory by Froissart and Martin [21, 22]. The BFKL evolution, however, leads with parton distributions of the form (9.11) to a number of partons growing at fixed transverse area $1/Q^2$ as $f_i \sim \exp A \sqrt{y}$. Since $y \sim \ln s$, the total area of interacting partons increases for large s as

$$\exp\{A \sqrt{\ln s}\} > \ln^2 s. \tag{9.14}$$

The number of partons can therefore not continue to grow according to the BFKL evolution in the limit of small x – this would violate the Froissart bound.

An end to the conventional parton considerations thus seems quite reasonable on very basic grounds. This does not tell us, however, how to describe the new medium formed of dense, overlapping partons. This topic, which we introduced through parton percolation, was initially addressed by allowing a non-linear parton fusion correction to the perturbative DGLAP/BFKL treatment [23, 24]. So the final state now presents a competition between the two opposing processes, and as a result, Eq. (9.12) becomes correspondingly modified and acquires a non-linear term [23, 24]; for gluons, this has the form

$$\frac{\partial^2 f_i(z, q)}{\partial z \partial q} = \alpha_s(q) f_i(z, q) \left[1 - \frac{\alpha_s(q)}{R^2 Q^2} f_i(z, q) \right], \tag{9.15}$$

where $\sim f_i^2$ describes the fusion of two into one. Here $1/Q^2 \sim r^2$ specifies the transverse parton size, and R^2 that of the nucleus; hence the square bracket vanishes when

$$\frac{f_i(z, q)}{R^2} \simeq \frac{1}{\alpha_s(q) r^2}, \tag{9.16}$$

i.e., when the sum of all parton areas is just that of the nucleus, taking into account interaction effects through $\alpha_s(q)$. At this point, the parton density stops increasing – the medium is saturated. This is essentially a percolation condition, see Sect. 2.2. Whichever approach we take, percolation or the balance of splitting and fusion, there will be a region in the y, q plane in which the concept of individual partons is no longer meaningful. We illustrate this in Fig. 9.10.

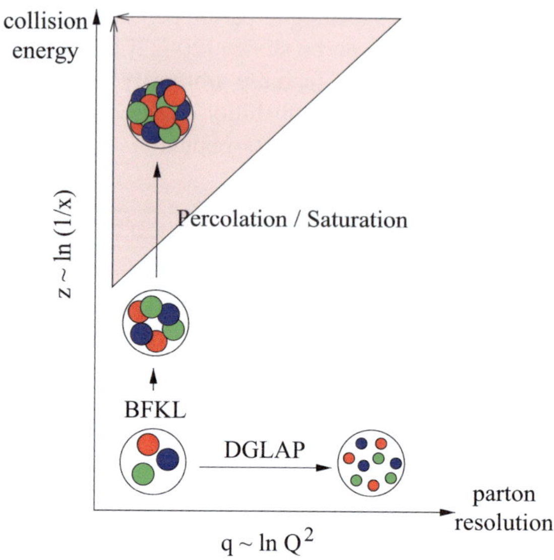

Fig. 9.10 Parton saturation

It might be of interest to consider this saturation limit in the context of a simple percolation picture. Let us attribute an average transverse size a to partons, determined by their average transverse momenta. In Chap. 2 we had found that in two dimensions, percolation sets in (see Eq. 2.33) at a parton density $n \simeq 1.13/a$. In a slab of 1 fm thickness, we thus have an initial entropy density

$$s_0 = 1.13/a. \tag{9.17}$$

The most recent lattice studies indicate that the deconfinement transition occurs at the pseudo-critical temperature of 155 Mev, where the entropy density is found to be $2.5 \pm 0.3\,\mathrm{fm}^{-3}$ [25]. This implies that at the saturation/percolation threshold, the partons have an intrinsic radius of about 0.38 fm, corresponding to an intrinsic transverse momentum of a little more than half a GeV.

In recent years, such partonic connectivity or saturation, and the properties of a connected pre-thermal primary state ("color glass condensate") have attracted much attention and led to a much better understanding of the initial state of high energy collisions [26]. This is by now a topic on its own, and a detailed treatment is obviously beyond the scope of this book; moreover, there are quite a number of excellent surveys [27–29]. We want to present here only the basic ideas and their implications on the fundamental problem of how to produce a thermalized quark-gluon plasma in nuclear collisions.

9.5 Color Glass Condensate and Glasma

Looking at an incident hadron or nucleus, we see a dense beams of partons. Keeping our resolution constant, but increasing the collision energy, means increasing the parton density, eventually leading to a percolation or saturation limit, in which the many colored partons combine to form some new non-perturbative medium. The system beyond this point contains many overlapping partons; we can picture its formation as a plate, onto which more and more drops of liquid are deposited, until the drops finally condense to a large connected domain of liquid. The new medium formed by the fusion of the colored drops is thus a kind of "colored condensate".

In the center of mass system, the partons of the incident hadrons or nuclei are contained in a Lorentz-contracted sphere of longitudinal thickness $D = (2m_h/\sqrt{s})2R_h$, which in the high energy limit becomes effectively a thin pancake. The passage time τ_0 of the two collision partners thus becomes arbitrarily short; if we assume $R_h \sim 1/m_h$, we have

$$\tau_0 \sim \frac{1}{\sqrt{s}}.$$

(9.18)

The soft gluon has a momentum k, and through the uncertainty relation this leads to a formation time scale

$$\tau_g \sim \frac{1}{k}$$

(9.19)

for the gluon evolution. Equivalently, we have in the target rest frame a finite passage time given by the target size, while the formation time of the gluon $\sqrt{s}/k \sim 1/x$ (and thus its coherence length) diverges in the high energy limit. During the interaction time (9.18), the gluon thus appears frozen ("ordered"), while on the much larger formation time scale (9.19), it evolves (becomes "disordered"). Media endowed with such different order states at different time scales are generally referred to as glasses, appearing as solid on a short and as liquid on a long time scale. This has led to the name color *glass* condensate for the saturated small x parton medium.

The gluons in the saturated medium no longer appear as independent partons; instead, they act coherently and therefore form classical fields. The sources of these fields are the hard initial partons, which through splitting lead to the abundant soft gluons. Because of their large momenta $k_q \sim \sqrt{s}$, their time scale is also short and so they can be considered as "static" in time. Since the incident hadrons are essentially contracted to discs orthogonal to the collision axis, the color fields are in the corresponding (two-dimensional) transverse plane (see Fig. 9.11 (left)).

In the collision itself, the two color field discs combine, and as the two hadrons separate, additional longitudinal color fields are formed, so that we now have a medium consisting of three dimensional classical color fields. This state, immediately after the collision, has been referred to as a "glasma". It expands in an anisotropic (mainly longitudinal) fashion (see Fig. 9.11 (right)), and when it has become sufficiently dilute, independent quarks and gluons are possible: the quark-gluon plasma appears.

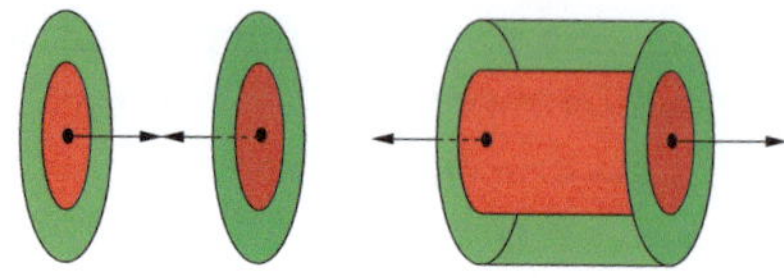

Fig. 9.11   Color glass condensate (left) and glasma (right)

The scenario proposed by the color glass condensate approach thus addresses hadronic or nuclear collisions in a regime in which the conventional parton model leads to saturation, so that an independent parton description must break down. The incident hadrons are now no longer considered as "beams of (independent) partons", but rather as two-dimensional discs of frozen classical color fields. In the collision and the subsequent expansion, the combined new color fields become three-dimensional and expand, eventually leading to quark-gluon plasma formation.

In summary: the parton model itself leads to two distinct regimes of nuclear collisions. For sufficiently low parton density and transverse size, in the pre-saturation regime, the conventional parton picture is applicable and parton thermalization can lead to QGP formation. Beyond the saturation point, be it through sufficiently high collision energy or large A, a thermalization of independent partons does not make sense. Here a possible new approach is given by the color glass condensate and its evolution to a glasma, which eventually also leads to a QGP.

References

 1. V.N. Gribov, Sov. Phys. JETP **29**, 483 (1969)
 2. J.D. Bjorken, Phys. Rev. D **27**, 140 (1983)
 3. V.N. Gribov, L.N. Lipatov, Sov. J. Nucl. Phys. **15**, 438 (1972)
 4. G. Altarelli, G. Parisi, Nucl. Phys. B **126**, 298 (1977)
 5. Y.L. Dokshitzer, Sov. Phys. JETP **46**, 641 (1977)
 6. L.N. Lipatov, Sov. J. Nucl. Phys. **23**, 338 (1976)
 7. E.A. Kuraev, L.N. Lipatov, V.S. Fadin, Sov. Phys. JETP **45**, 199 (1977)
 8. Y.Y. Balitzky, L.N. Lipatov, Sov. J. Nucl. Phys. **28**, 822 (1978)
 9. R. Anishetty, P. Köhler, L. McLerran, Phys. Rev. D **22**, 2793 (1980)
10. R.C. Hwa, K. Kajantie, Phys. Rev. Lett. **56**, 696 (1986)
11. J.P. Blaizot, A.H. Mueller, Nucl. Phys. B **289**, 847 (1987)
12. D. Boal, Phys. Rev. C **33**, 2206 (1986)
13. K. Geiger, B. Müller, Nucl. Phys. B **369**, 600 (1992)
14. K. Geiger, Phys. Rev. D **47**, 133 (1993)
15. T.S. Biró et al., Phys. Rev. C **48**, 1275 (1993)

16. R. Baier et al., Phys. Lett. B **502**, 51 (2001)
17. N. Armesto et al., Phys. Rev. Lett. **77**, 3736 (1996)
18. M. Nardi, H. Satz, Phys. Lett. B **442**, 14 (1998)
19. H. Satz, Nucl. Phys. A **661**, 104c (1999)
20. W. Heisenberg, Z. Phys. **133**, 65 (1952)
21. M. Froissart, Phys. Rev. **123**, 1053 (1961)
22. A. Martin, Nuovo Cim. **42**, 930 (1966)
23. L.V. Gribov, E.M. Levin, M.G. Ryskin, Nucl. Phys. B **188** (1981)
24. L.V. Gribov, E.M. Levin, M.G. Ryskin, Phys. Rep. **100**, 1 (1983)
25. A. Bazavov et al. (HotQCD), Phys. Rev. D **90**, 094503 (2014)
26. L. McLerran, R. Venugopalan, Phys. Rev. D **49** 2233 and 3352 (1994) and D **50**, 2225 (1994)
27. For surveys, see e.g. L. McLerran, Nucl. Phys. A **702**, 49 (2002)
28. For surveys, see e.g. H. Weigert, Prog. Part. Nucl. Phys. **55** 461 (2005)
29. For surveys, see e.g. D. Banerjee, J.K. Nayak, R. Venugopalan, in *The Physics of the Quark-Gluon Plasma*. Springer Lecture Notes in Physics, vol. 785 (Heidelberg, 2010)

Chapter 10
Probing the Quark-Gluon Plasma

Look, it cannot be seen – it is beyond form.
Listen, it cannot be heard – it is beyond sound.
Grasp, it cannot be held – it is beyond touch.

Lao Tzu, *Tao Te Ching 14,* ~ 600 B. C.

Abstract Assuming that high energy collisions of two heavy nuclei results in the formation of a quark-gluon plasma, we want to consider in this section some possible probes which can provide information about its structure and its thermodynamic properties. In particular, we address hadronic and electromagnetic radiation, quarkonium dissociation and jet quenching.

10.1 Tools to Probe

Let us begin in a very general way. If we want to probe the internal structure of a system of overall linear size L, we need a probe of intrinsic linear dimension (e.g., wave length) $\lambda < L$, so that it can see if there is a substructure. If the medium is made up of many building blocks of size $a < L$, we need in fact a probe with $\lambda < a$, so that it can resolve the elements of the substructure. What does that mean in QCD? The basic scale of strong interactions is $\Lambda_{\rm QCD} \simeq 1$ fm, the size of a hadron, so that any probe should have a wavelength not larger than this scale. If we want to reach sub-hadronic structures, λ must be correspondingly smaller, and if we want to probe the hot quark-gluon plasma, temperature becomes the relevant scale, so that $\lambda < 1/T$.

This gives us some hints of possible probes. If we want to test the source size for hadron emission in nuclear collisions, with heavy and hence large nuclei, hadrons themselves are a reasonable tool: they measure things in hadronic scales. As we shall see, they are also very suitable to study the hadronisation transition; this will be dealt with in detail in Chaps. 11 and 12.

© Springer International Publishing AG 2018
H. Satz, *Extreme States of Matter in Strong Interaction Physics,*
Lecture Notes in Physics 945, https://doi.org/10.1007/978-3-319-71894-1_10

If we want to go below this scale, into the genuine plasma regime, the wavelength has to become shorter, to obtain the counterpart of X-rays for the matter to be studied. This leaves us with three possibilities.

To test the substructure of hadrons, we need (real or virtual) photons of a wavelength much shorter than Λ_{QCD}, i.e., we have to resort to deep inelastic lepton-hadron scattering. To probe a hot quark-gluon plasma, we need wavelengths $\lambda < 1/T$, so that we have to consider high p_T photons or high mass dilepton pairs (Drell-Yan production).

The hard scattering of hadrons at large momentum transfer is generally due to gluon exchange, so that such processes involve quark-gluon interactions within the hadron. As a result, an energetic parton is emitted in the transverse direction, eventually hadronizing to form a jet. In nuclear collisions, the parton has to traverse the hot quark-gluon plasma they form, so that here the in-medium modification of jet production provides a probe.

The heavy quark resonances J/ψ and Υ have binding energies much larger than Λ_{QCD} and radii much smaller than $1/\Lambda_{QCD}$; they therefore provide a quite novel probe, surpassing the usual technique of "looking" with photons or gluons. The large quark mass brings a perturbative calculation of quarkonium production in hadronic interactions within reach, and the modification of quarkonium formation in the presence of a hot quark-gluon plasma can in principle be studied in statistical QCD.

The specific question we want to address here is how to determine the state and the properties of a macroscopic (i.e., super-hadronic) volume of hot QCD matter in equilibrium and of zero overall baryon number density, by studying specific observable processes. We will assume that we are provided with such a system with the help of (for simplicity central, i.e., head-on) nucleus-nucleus collisions, and that we can vary its energy density by varying the collision energy. What physical phenomena will allow us to "measure" the temperature or other thermodynamic properties of the medium thus produced (see Fig. 10.1), and what changes will we observe when the energy density is varied?

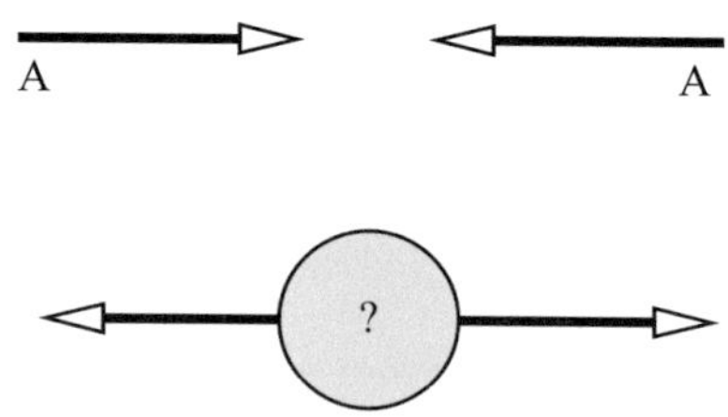

Fig. 10.1 Schematic view of a nuclear collision

This problem is evidently a simplified version of that presented in any real experimental study of strongly interacting media in nuclear collisions. In that case, one has to consider in addition a variety of specific collision features, such as the structure of the initial state, its evolution in time, the onset of equilibration, inhomogeneities, and much more. These are features which arise when we try to "make matter" in a high energy heavy ion collision. We shall return to some of these aspects, but in general, we shall assume that the matter to be studied is already "made".

As we have just seen, there are several "tools" we can use to study the unknown medium, which we assume to be situated in a physical vacuum environment:

- hadronic radiation,
- electromagnetic radiation,
- dissociation of quarkonium states,
- energy loss of passing hard jets.

As mentioned, we assume the system under consideration to be in overall thermal equilibrium and to evolve adiabatically in time.

10.2 Hadronic Radiation and Source Size

Any medium is, by definition, hotter than its environment (the vacuum) and hence emits radiation. An outside observer will detect the emission of thermal hadrons, predominantly light mesons; these, however, cannot exist in the interior of a hot QGP and hence must be formed through hadronization at the cooler surface, or more generally, when the medium has cooled down to the hadronization temperature. Such radiation will therefore carry information only about the hadronization stage of the medium; but in doing so, it can tell us something about two features which are absolutely crucial to the entire concept of producing strongly interacting matter through nuclear collisions.

- The formation of hadrons should occur at the hadronization temperature determined in statistical QCD studies, i.e., at about 160 MeV for systems of negligible overall baryon density. The hadronization temperature plays a crucial role for both the relative abundances of the hadron species ("hadrochemistry") and for their momenta (see caveat below).
- If an increase of collision energy results in an increase of the initial energy density of the thermal medium, i.e., if it produces a hotter QGP, the medium must expand more in order to cool down to the hadronization point. The source size for hadron emission must therefore increase with collision energy.

The first point will be dealt with in detail in the next chapter, where we will see that hadron abundances in high energy collisions indeed lead to a universal hadronization temperature of the expected value; this is evidently one of the most important results of the entire experimental study so far. The second raises the question of determining the source size for hadron emission, and that is what we want to address here.

Before turning to this, a caveat seems appropriate. We are here considering global equilibrium aspects only. If the medium evolves from hot QGP to the hadronic stage through hydrodynamic expansion, the momenta of the emitted hadrons will reflect a collective overall flow in addition to the thermal hadronization momentum, and for non-central collisions there will furthermore be directed flow, depending on the orientation of the pressure gradient. Such flow components in turn do depend on the *initial* energy density of the medium, i.e., of the hot QGP. Hadron momentum spectra could thus provide indirect information also about the pre-hadronic stages.

The determination of source sizes through radiation measurements is a classical problem of astronomy, where it is needed to measure the size of stars. The key to its solution was found in 1956 by the astronomers R. Hanbury-Brown and R. Q. Twiss [1]; the application of the resulting "HBT" technique in particle or nuclear physics is based on an interference of two identical particles [2], emitted from an extended source and measured simultaneously in two separate detectors. The geometry is illustrated in Fig. 10.2. The particles are emitted at points x_1 and x_2, with momenta k_1 and k_2, and measured at detectors situated at points A and B.

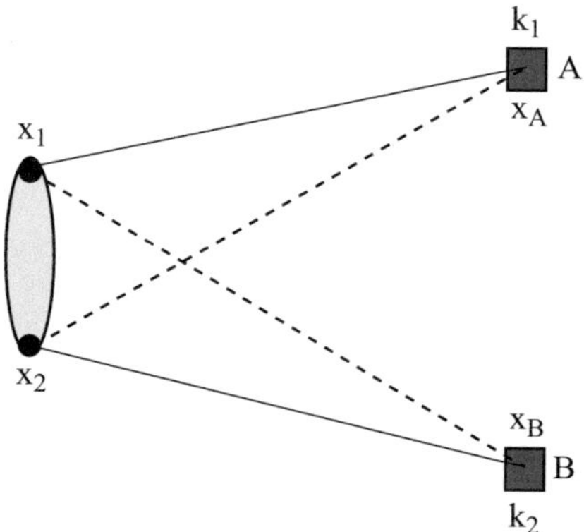

Fig. 10.2 HBT geometry lay-out

Since the particles are by assumption identical, A cannot specify whether its particle came from x_1 or x_2, and the same for B. Hence the corresponding amplitude must be symmetrized (bosons) or antisymmetrized (fermions), depending on the statistics of the observed particles. Moreover, the emission is taken to be incoherent, i.e., with random relative phases. The wave function for the process is then given by

$$A(k_1, k_2) = \frac{1}{\sqrt{2}} \left[\exp\{ik_1(x_A - x_1) + ik_2(x_B - x_2)\} \right.$$

$$\left. \pm \exp\{ik_1(x_A - x_2) + ik_2(x_B - x_1)\} \right]; \tag{10.1}$$

this leads to the detection rate

$$P(k_1, k_2) = |A(k_1, k_2)|^2 = 1 \pm \cos[(k_1 - k_2)(x_1 - x_2)]. \tag{10.2}$$

The simultaneous measurement of two identical bosons or fermions (most studies use bosons, since pions are the most abundantly produced hadrons) of similar momenta is thus enhanced due to interference effects and, for known $\Delta k = (k_1 - k_2)$, allows the determination of the source size $R = (x_1 - x_2)$. Obviously, this is only a very schematic description of the actual HBT analysis; for more details, see [3].

The HBT technique was in fact employed in nuclear collision studies from the very start of the experiments [3]. Correlations were measured for various collision energies in the three space directions, generally taken along the beam axis ("longitudinal"), orthogonal to the beam axis and along the momentum of the two bosons ("out"), and orthogonal to both of these ("side"). The resulting source volume is formed by multiplying the three distinct source dimensions, and it did not seem to show the expected well-defined growth, particularly in the high energy region. This "HBT puzzle" was finally resolved by the recent advent of data at very high energy [4], taken in the first nuclear collision studies at the Large Hadron Collider (LHC) at CERN. In Fig. 10.3, the source volume is shown for central heavy nuclear collisions (Pb-Pb and Au-Au) at various energies. As seen there, the new LHC point now does establish the expected linear increase with the average hadron multiplicity (see Eq. (9.7)) and hence with the conjectured initial energy density. One essential feature for the use of nuclear collisions for QGP formation is thus now established.

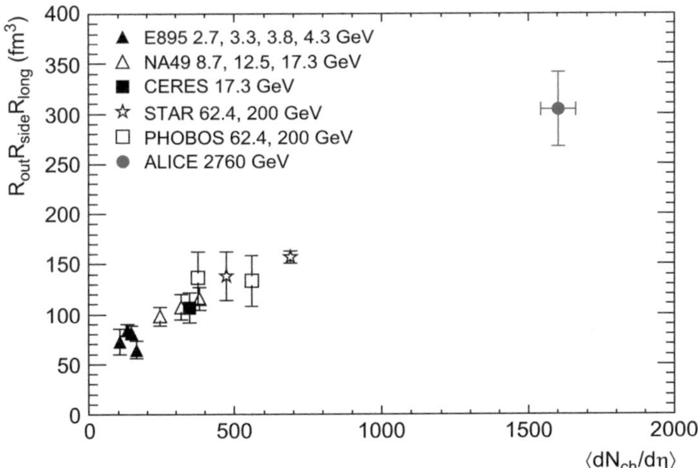

Fig. 10.3 HBT results for source sizes $V = R_{out}R_{side}R_{long}$ as a function of the average multiplicity at the given energy [4]

10.3   Electromagnetic Radiation

The hot medium emits electromagnetic radiation into the vacuum, i.e., it radiates photons and dileptons (e^+e^- or $\mu^+\mu^-$ pairs). These are formed either by the interaction of quarks and/or gluons, or by quark-antiquark annihilation (see Fig. 10.4).

Since the photons and leptons interact only electromagnetically, they will, once they are formed, leave the medium without any further strong interaction effects. Hence their spectra provide information about the state of the medium at the space-time point where they were formed. Photons and dileptons thus provide in principle an excellent probe of the hot QGP [5–8].

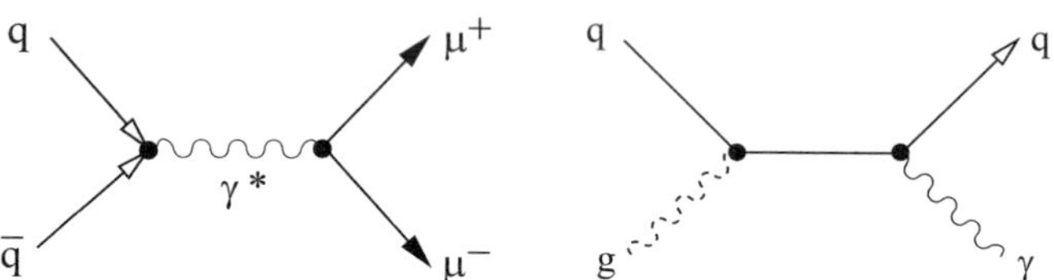

Fig. 10.4 Dilepton production through $q\bar{q}$ annihilation (left) and photon production through gluonic Compton scattering (right)

The application of these probes is, however, not so simple, since there are competing production processes, and the evolution stage of the specific production process enters as a crucial element. Before any thermal medium is formed, at the very early stages of the collision, hard primary partonic interactions can lead to Drell-Yan lepton pair production or prompt hard photon emission. These are also formed by quark-quark annihilation or quark-gluon interaction, i.e., by processes of the same nature as those in the thermal medium. In contrast, however, the parton distributions are now determined by the initial state of the collision partners, not by a produced thermal system. Furthermore, the produced hot medium evolves in time, from QGP to hadronic matter, and electromagnetic signals can be formed also at later evolution stages. The task of making electromagnetic radiation a viable tool for evolving media is therefore the identification of the hot "thermal" radiation emitted by the QGP in a definite stage, after the hard primary and before the soft hadronic emission. Let us consider this in a little more detail.

To calculate the mass spectrum of dileptons emitted from a hot QGP, the cross-section $\sigma(q\bar{q} \rightarrow \mu^+\mu^-)$ for quark-antiquark annihilation (see Fig. 10.4) has to be convoluted with thermal quark and antiquark momentum distributions

$$f(k_q/T) \sim \exp\{-|k_q|/T\}, \tag{10.3}$$

where k_q is the three-momentum of the (massless) quark and T the temperature of the medium. We thus obtain for the production rate

$$\frac{dN}{dM} \sim \int d^3k_q f(k_q) d^3k_{\bar{q}} f(k_{\bar{q}}) \sigma(q\bar{q} \rightarrow \mu^+\mu^-), \tag{10.4}$$

where M is the invariant mass of the dilepton. The convolution leads to the schematic result

$$\frac{dN}{dM} \sim \exp\{-M/T\}, \tag{10.5}$$

so that a measurement of a thermal dilepton spectrum provides the temperature of the medium. As already indicated, the medium undergoes an evolution and cools down; the observed dileptons originate from all stages, so that a temperature measurement is not straight-forward and will in general depend on the evolution pattern. In addition, there is the mentioned competing production of non-thermal dileptons through hard primary Drell-Yan production.

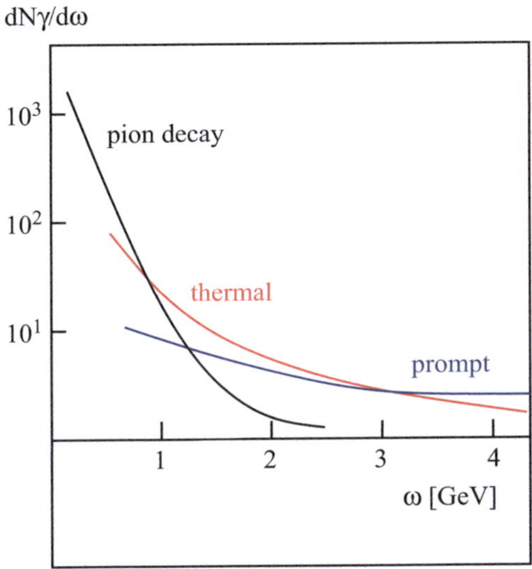

Fig. 10.5 Sources of photon emission in nuclear collisions

For photon production, the situation is similar. Here the dominant partonic interaction is a gluonic Compton effect, as illustrated in Fig. 10.4 (right). The thermal rate is now given by a convolution of a thermal quark with thermal gluon distribution, integrating over the perturbative Compton cross section $\sigma(qg \to q\gamma)$. The result is

$$\frac{dN}{d\omega} \sim \exp\{-\omega/T\},\tag{10.6}$$

where ω denotes the absolute value of the momentum of the emitted photon. Here again the basic problem of electromagnetic probes arises: the thermal photons originate from all evolution stages and are in competition with those from non-thermal sources, in particular from early "prompt" hard photons and from late soft photons originating in pion decay ($\pi^0 \to e^+ e^- \gamma$, $\pi^0 \to \gamma\gamma$). The situation is schematically illustrated in Fig. 10.5; the crucial question thus is if it is possible to locate a "window" for the observation of thermal photons from the QGP.

A recent analysis of data from the PHENIX experiment at the Relativistic Heavy Ion Collider (RHIC), Brookhaven National Laboratory, has given indications that this may be feasible. In Au-Au collisions at $\sqrt{s} = 200$ GeV, they find an "anomalous" excess of photons between Dalitz decay and prompt photons [9]; the resulting spectrum leads to a temperature $T = 221 \pm 19(\text{stat.}) \pm 19(\text{syst.})\,\text{MeV}$, above the expected deconfinement value. First results from the CERN LHC confirm the possibility of such a temperature determination and lead to a continuing increase: for central Pb-Pb collisions at $\sqrt{s} = 5\,\text{GeV}$ [10], one finds $T = 297 \pm 12(\text{stat.}) \pm 41(\text{syst.})$.

10.4   Quarkonium Dissociation

The quark-gluon plasma consists by definition of deconfined and hence colored gluons, quarks and anti-quarks. One of the essential features of any plasma is charge screening, which for electromagnetic interactions reduces the long-range Coulomb potential in vacuum to a much shorter range screened in-medium form,

$$\frac{e^2}{r} \rightarrow \frac{e^2}{r}\,\exp\{-\mu r\}, \tag{10.7}$$

where μ is the screening mass specifying the Debye or screening radius $r_D = 1/\mu$. In a plasma of color-charged constituents, one expects a similar behavior, and this is indeed observed in lattice studies, as discussed in Chap. 8: in the QGP just above T_c, μ increases much more than linearly (see Fig. 8.4), and hence r_D decreases correspondingly. Asymptotically, perturbation theory suggests $\mu \simeq g(T)T$, with $g(T)$ for the temperature-dependent strong coupling. The range of strong interactions thus shows a striking in-medium decrease with increasing temperature.

Quarkonia are a special kind of hadrons, bound states of a heavy quark and its antiquark. They are thus pairs of charm ($m_c \simeq 1.3\,\text{GeV}$) or bottom ($m_b \simeq 4.7\,\text{GeV}$) quarks. The large quark mass makes possible a spectroscopy based on non-relativistic potential theory [12]. Hence the Schrödinger equation

$$\left\{2m_c - \frac{1}{m_c}\nabla^2 + V(r)\right\}\,\Phi_i(r) = M_i\Phi_i(r), \tag{10.8}$$

using the "Cornell" form for the confining potential [13, 14]

$$V(r) = \sigma\,r - \frac{\alpha}{r}, \tag{10.9}$$

in terms of the string tension $\sigma \simeq 0.2\,\text{GeV}^2$ and the gauge coupling $\alpha \simeq \pi/12$, determines the masses M_i and the radii r_i of the different charmonium and bottomonium states. The results are summarized in Table 10.1 and are seen to give a good account of quarkonium spectroscopy, with an uncertainty of less than 1% in the mass determination $\Delta M = M_{\text{exp}} - M_{\text{th}}$ for all (spin-averaged) states.

state	J/ψ	χ_c	ψ'	Υ	χ_b	Υ'	χ_b'	Υ''
exp. mass [GeV]	3.07	3.53	3.68	9.46	9.99	10.02	10.26	10.36
ΔM [GeV]	0.02	-0.03	0.03	0.06	-0.06	-0.06	-0.08	-0.07
ΔE [GeV]	0.64	0.20	0.05	1.10	0.67	0.54	0.31	0.20
radius [fm]	0.25	0.36	0.45	0.14	0.22	0.28	0.34	0.39

Table 10.1 Quarkonium spectroscopy in non-relativistic potential theory [11]

For the ground state J/ψ, the binding energy $\Delta E = 2M_D - M_{J/\psi}$ is about $0.6\,\text{GeV}$; here M_D denotes the mass of the lowest open charm (light-heavy $c\bar{q}$) meson. Thus ΔE is for the J/ψ much larger than the typical hadronic scale $\Lambda \sim 0.2\,\text{GeV}$, and with a radius $r_{J/\psi}$ of about $0.25\,\text{fm}$, the J/ψ is also much smaller than the typical hadron, with $r_h \simeq 1$ fm. For the bottomonium ground state Υ, both ΔE and r move even further away from the hadronic scales. The fate of heavy quark bound states in a quark-gluon plasma depends on the size of the color screening radius r_D in comparison to the quarkonium radius r_Q: if $r_D \gg r_Q$, the medium does not really affect the heavy quark binding. Once $r_D \ll r_Q$, however, the two heavy quarks cannot "see" each other any more and hence the bound state will melt [15]. It is therefore expected that quarkonia will survive in a quark-gluon plasma through some range of temperatures above T_c, and then dissociate once T becomes large enough.

The higher excited quarkonium states are less tightly bound and hence larger, although their binding energies are in general still greater and their radii smaller than those of the usual light quark hadrons. Take the charmonium spectrum as example: the radius of the $J/\psi(1S)$ is about $0.25\,\text{fm}$, that of the $\chi_c(1P)$ about $0.36\,\text{fm}$, and that of the $\psi'(2S)$ $0.45\,\text{fm}$. Since melting sets in when the screening radius reaches the binding radius, we expect that the different charmonium states have different "melting temperatures" in a quark-gluon plasma. To illustrate how these temperatures can be determined, we consider the Schrödinger equation (10.8) with a screened potential. The simplest form for such a potential is given by [16]

$$V(r, T) = \sigma r \left\{ \frac{1 - e^{-\mu r}}{\mu r} \right\} - \frac{\alpha}{r} e^{-\mu r} = \frac{\sigma}{\mu} \{ 1 - e^{-\mu r} \} - \frac{\alpha}{r} e^{-\mu r}, \qquad (10.10)$$

replacing the vacuum form (10.9). Here $\mu(T)$ now denotes the color screening mass; for $\mu = 0$, the vacuum form is recovered, and for large r, the potential reaches a finite value, $V(r, T) \to \sigma/\mu$, which corresponds to two separated quarks in a deconfined medium, surrounded by gluon polarization clouds. For each quarkonium state i the resulting Schrödinger equation provides the dissociation value μ_i; for $\mu < \mu_i$, there exists a bound state, while for $\mu \geq \mu_i$, it has become dissociated. Given the temperature dependence of the screening mass, the dissociation mass μ_i then determines the corresponding dissociation temperature T_i. Hence a spectral analysis based on in-medium quarkonium dissociation should provide a **QGP thermometer** [16, 17]; see Fig. 10.6.

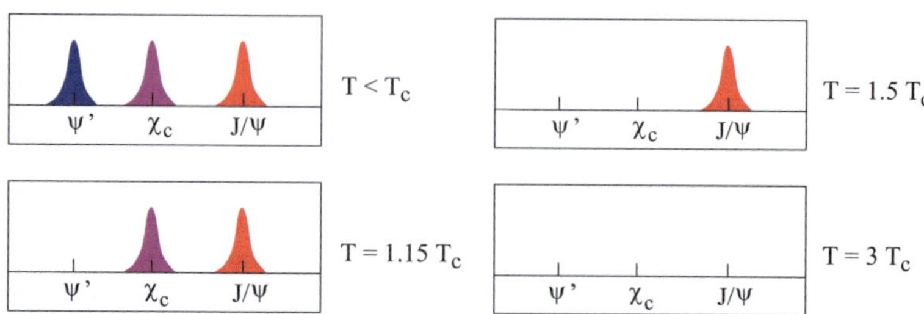

Fig. 10.6 Charmonia as thermometer

The calculation of the dissociation points for the different states, just illustrated on the basis of a simplistic screened potential, has been pursued both in more realistic potential theory and through direct lattice studies [18, 19]. In both approaches, there are presently still ambiguities; we give an indicative summary in Table 10.2, but emphasize that here work is still in progress.

state	$J/\psi(1S)$	$\chi_c(1P)$	ψ'	Υ	χ_b	Υ'	χ_b'	Υ''
T_d/T_c	2.0	1.2	1.1	> 4.0	1.8	1.6	1.2	1.2

Table 10.2 Quarkonium dissociation temperatures T_d in units of the deconfinement temperature T_c [18, 19]

The dissociation of quarkonium states in a deconfined medium, as compared to their survival in hadronic matter, can also be considered on a more dynamical level, using the J/ψ as example. The J/ψ is a hadron with characteristic short-distance features; in particular, rather hard gluons are necessary to resolve or dissociate it, making such a dissociation accessible to perturbative calculations. J/ψ collisions with ordinary hadrons made up of the usual u, d and s quarks thus probe the local partonic structure of these 'light' hadrons, not their global hadronic aspects, such as mass or size. It is for this reason that J/ψ's can be used as a confinement/deconfinement probe.

This can be illustrated by a simple example. Consider an ideal pion gas as a confined medium. The momentum spectrum of pions has the Boltzmann form $f(p) \sim \exp-(|p|/T)$, giving the pions an average momentum $\langle|p|\rangle = 3\,T$. With the pionic gluon distribution function $xg(x) \sim (1-x)^3$, where $x = k/p$ denotes the fraction of the pion momentum carried by a gluon, the average momenta of gluons confined to pions becomes

$$\langle|k|\rangle_{\text{conf}} \simeq 0.6\,T. \qquad (10.11)$$

On the other hand, an ideal QGP as prototype of a deconfined medium gives the gluons themselves the Boltzmann distribution $f(k) \sim \exp-(|k|/T)$ and hence average momenta

$$\langle|k|\rangle_{\text{deconf}} = 3\,T. \qquad (10.12)$$

Deconfinement thus results in a hardening of the gluon momentum distribution. More generally speaking, the onset of deconfinement will lead to parton distribution functions which are different from those in vacuum, as determined by deep inelastic scattering experiments. Since hard gluons are needed to resolve and dissociate J/ψ's, one can use J/ψ's to probe the in-medium gluon hardness and hence the confinement status of the medium.

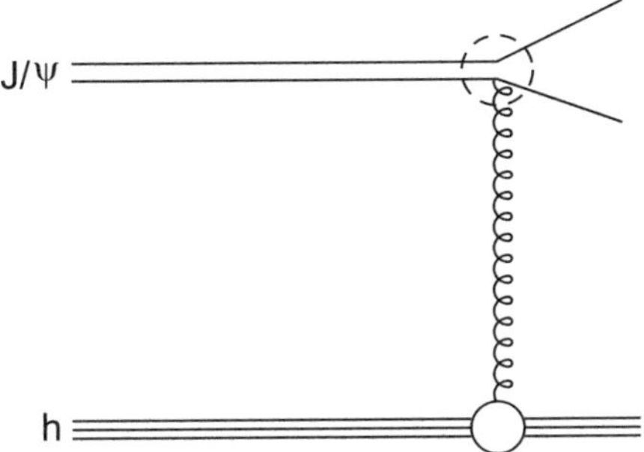

Fig. 10.7 J/ψ dissociation by hadron interaction

These qualitative considerations can be put on a solid theoretical basis provided by short-distance QCD [20–23]. In Fig. 10.7 we show the relevant diagram for the calculation of the inelastic J/ψ-hadron cross section, as obtained in the operator product expansion framework (essentially a multipole expansion for the charmonium quark-antiquark system). The upper part of the figure shows J/ψ dissociation by gluon interaction; the cross section for this process,

$$\sigma_{g-J/\psi} \sim (k - \Delta E_\psi)^{3/2} k^{-5}, \tag{10.13}$$

constitutes the QCD analogue of the photo-effect. Convoluting the J/ψ gluon-dissociation with the gluon distribution in an incident meson, $xg(x) \simeq 0.5(1-x)^3$, we obtain

$$\sigma_{h-J/\psi} \simeq \sigma_{\text{geom}} (1 - \lambda_0/\lambda)^{5.5} \tag{10.14}$$

for the inelastic J/ψ-hadron cross section, with $\lambda \simeq (s - M_\psi^2)/M_\psi$ and $\lambda_0 \simeq (M_h + \Delta E_\psi)$; again s denotes the squared J/ψ-hadron collision energy. In Eq. (10.14), $\sigma_{\text{geom}} \simeq \pi r_\psi^2 \simeq 2 - 3\,\text{mb}$ is the geometric cross section attained at high collision energies with the mentioned gluon distribution. In the threshold region and for relatively low collision energies, $\sigma_{h-J/\psi}$ is very strongly damped because of the suppression $(1-x)^3$ of hard gluons in mesons, which leads to the factor $(1-\lambda_0/\lambda)^{5.5}$ in Eq. (10.14). In Fig. 10.8, we compare the cross sections for J/ψ dissociation by gluons ("gluo-effect") and by mesons, as given by Eqs. (10.13) and (10.14). Gluon dissociation shows the typical photo-effect form, vanishing until the gluon momentum k passes the binding energy ΔE_ψ; it peaks just a little later and then vanishes again when sufficiently hard gluons just pass through the much larger charmonium bound states. In contrast, the J/ψ-meson cross section remains negligibly small until rather high hadron momenta (3–4 GeV). In a thermal medium, such momenta correspond to temperatures of more than 1 GeV. Hence confined media in the temperature range of a few hundred MeV are essentially transparent to J/ψ's, while deconfined media of the same temperatures very effectively dissociate them and thus are J/ψ-opaque. Similar considerations can of course be applied to the higher excited charmonium or the bottomonium states.

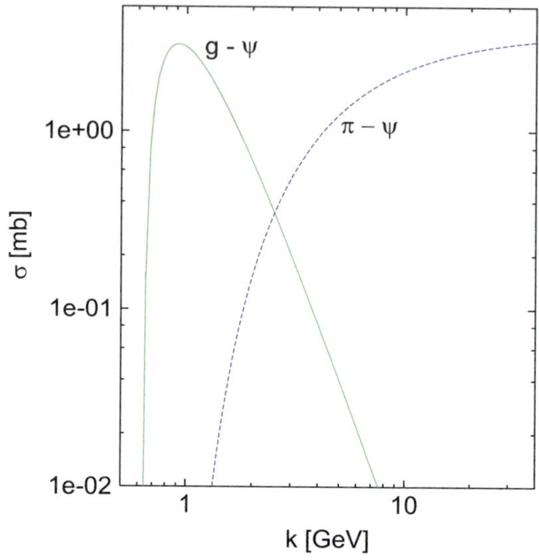

Fig. 10.8 J/ψ dissociation by gluons and by pions; k denotes the momentum of the projectile incident on a stationary J/ψ

To obtain an observable probe, we thus have to study quarkonium production in nuclear collisions. The initial step, the production of heavy quark pairs in nucleon-nucleon collisions, is a hard process which (with some grains of salt) is theoretically calculable [24, 25]. At high energies, it is dominantly given by gluon fusion, leading to the production of a colored $c\bar{c}$ or $b\bar{b}$ pair. A certain (very small) fraction f_i of these pairs subsequently neutralises its color in the field of the interaction and binds to form quarkonium state i (see Fig. 10.9).

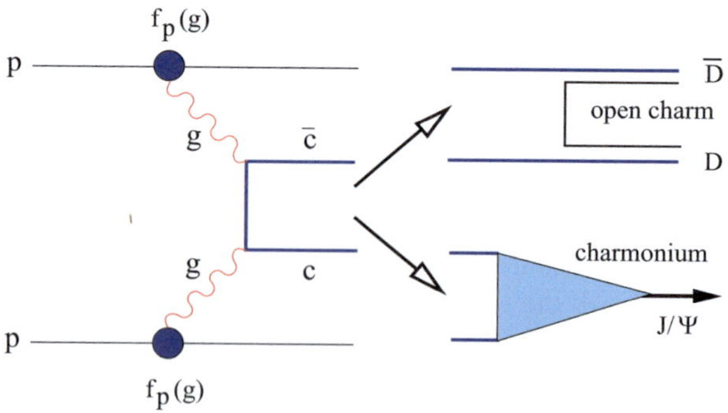

Fig. 10.9 Charm quark production through gluon fusion in *pp* collisions

This color neutralisation is no longer perturbatively calculable, at least not for the production of quarkonia at low transverse momentum. An operationally viable road to prediction is to measure these fractions at some collision energy; since they are found to be energy-independent, this "color evaporation model" [26–29] has in fact predicted the production rates very well at all energies so far.

The extension of such a model to nucleus-nucleus collisions encounters a number of obstacles. The gluon distribution functions $f(g)$ needed to calculate the $Q\bar{Q}$ production are modified in nuclei ("shadowing", "energy loss"). This initial state effect can be quantified by studying the process in $p - A$ collisions and thus determining the effect of cold nuclear matter on the production process. The same is true for possible final state modifications, due to absorption of the produced quarkonia in the passing nuclear media of target and projectile. If and when these "normal" effects are taken into account, one can study the remaining "anomalous" suppression. As we shall see, the only model-independent way of achieving this is by considering the ratio of quarkonium production to open charm (D meson) production [30, 31]. We shall return to this point shortly.

From what was said above, the survival probability for a given quarkonium state depends on its size and binding energy. Hence the excited states will be dissolved at a lower initial energy density than the more tightly-bound ground states. In actual production, however, one encounters "feed-down": only a fraction (about 60%) of the observed J/ψ is a directly produced $(1S)$ state, the remainder is due to the decay of excited states, with about 30% from $\chi_c(1P)$ and 10% from $\psi'(2S)$ decay [32–34]. A similar decay pattern arises for Υ production. The decay processes occur far outside the produced medium, so that the medium affects the excited states. As a result, the formation of a hot deconfined medium in nuclear collisions will produce a sequential quarkonium suppression pattern, as illustrated in Fig. 10.10 for J/ψ production. Increasing the energy density of the QGP above deconfinement first leads to ψ' dissociation, removing those J/ψ's which otherwise would have come from ψ' decays. Next the χ_c melts, and only for a sufficiently hot medium also the direct J/ψ's disappear. For the bottomonium states, a similar pattern will hold; it is included in Fig. 10.10, also with vacuum feed-down fractions [35].

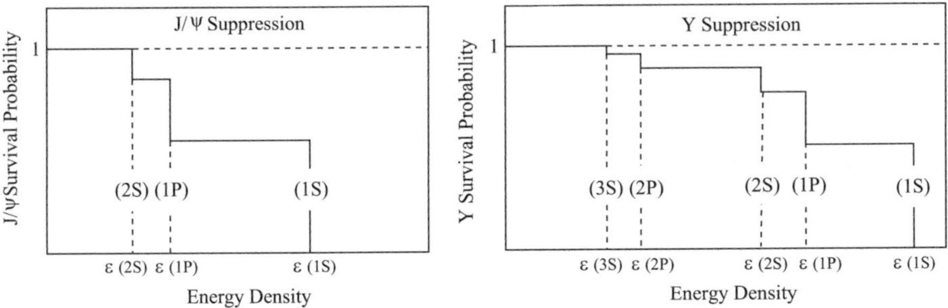

Fig. 10.10 Sequential quarkonium suppression

It is evident that a corresponding spectral analysis of nucleus-nucleus data is quite complex. Nevertheless, it has already produced quite promising results, indicating an anomalous J/ψ suppression of about 40–50% for central collisions in the energy range $\sqrt{s} = 17 - 200\,\mathrm{GeV}$ [36–38]. Experiments at the CERN-SPS measured $Pb - Pb$ and $In - In$ collisions at a c.m.s. energy $\sqrt{s} \simeq 17\,\mathrm{GeV}$; subsequently, the collision energy was increased to $\sqrt{s} = 200\,\mathrm{GeV}$ at the BNL-RHIC. Once cold nuclear matter effects were taken into account, this energy increase did not significantly increase the J/ψ suppression [39]. The present status up to $\sqrt{s} = 200\,\mathrm{GeV}$ is shown in Fig. 10.11, combining the results from SPS and RHIC experiments. It shows the J/ψ production rate, normalized to the scaled result for cold nuclear matter effects, as obtained from $p - A$ collisions at CERN and $d - A$ collisions at BNL. The observed behavior is in accord with the suppression of the excited states χ_c and ψ' and the survival of the directly produced $J/\psi(1S)$ [40].

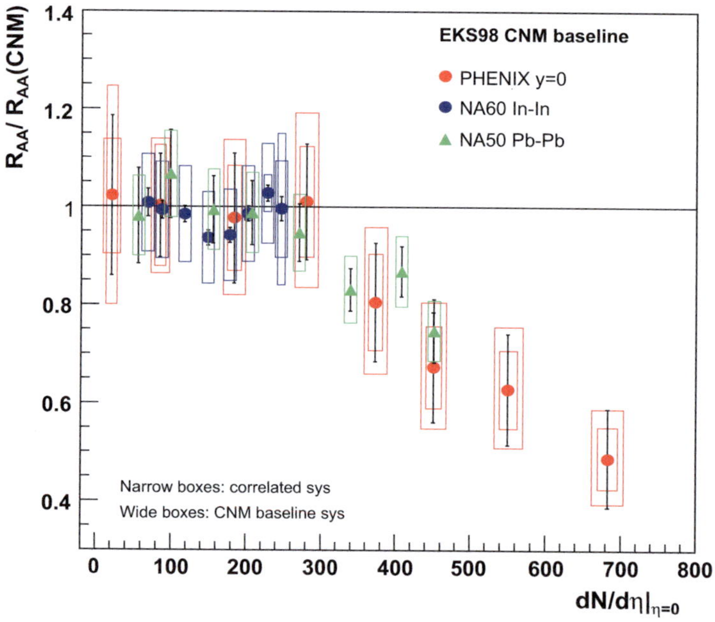

Fig. 10.11 J/ψ suppression in nucleus-nucleus collisions [39]

Before we turn to more recent results from the LHC, we have to address a crucial point. The $c\bar{c}$ production process in proton-proton interactions (see Fig. 10.9) leads, independent of the collision energy, to 90% open charm (mainly D production) and 10% hidden charm in the form of charmonia. The charmonium suppression formalism just discussed states that the presence of a medium modifies, more specifically decreases, this 10% fraction going into hidden charm. The formalism does not address the overall production rate of $c\bar{c}$ pairs. It is indeed quite possible that the overall number of such pairs produced in AA collisions differs from that in *pp* interactions, and this would in general change the number of charmonia produced in *AA* relative to scaled *pp* interactions. Nevertheless, the ratio 10:90 for hidden to open charm could still persist, and this would imply that the presence of the medium in *AA* interactions has no effect on charmonium production. In other words, the crucial issue is whether the ratio of charmonium to open charm production in *AA* collisions is changed in comparison to the 10:90 found for *pp*. A comparison of charmonium production in *AA* to that in *pp* collisions is clearly not indicative.

The correct analysis is illustrated in Fig. 10.12 [41–43]. The left box shows J/ψ and open charm production data in $Au - Au$ collisions at RHIC for large p_T, and it is seen that they agree: here the medium has no effect on charmonium production. In contrast, the right box shows the behavior at low p_T, which constitutes the dominant part of the overall production of both hidden and open charm. Here one finds a J/ψ suppression of up to 80% for the most central interactions. This is the predicted J/ψ suppression through color screening.

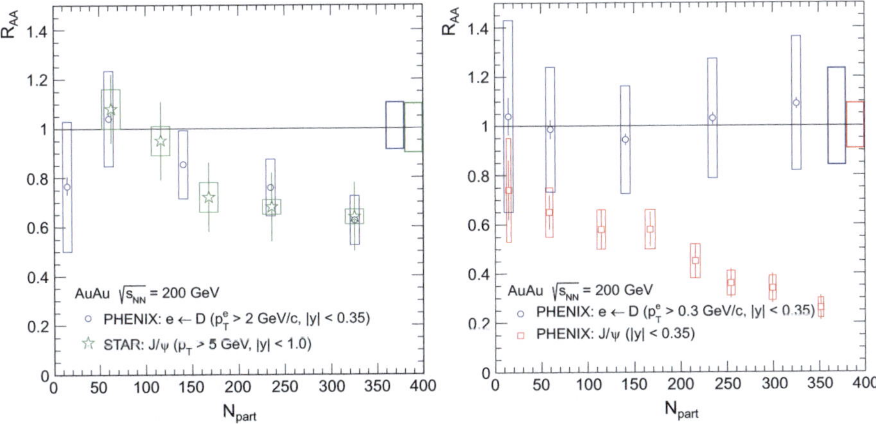

Fig. 10.12 Data from PHENIX & STAR at RHIC: J/ψ vs. open charm production at high & low transverse momenta [41–43]

The difficult part in such studies is the measurement of the open charm production, generally of *D* mesons, which are not easy to identify, particularly at low transverse momenta. For this reason, the extensive data on J/ψ production at the LHC is presently still inconclusive. One does find a somewhat weaker suppression in comparison to proton-proton rates, but as long as the reference data of open charm production is not available, this does not allow any conclusions. At high p_T there exist data on both hidden and open charm production, and here the rates agree, as they did at RHIC (left box of Fig. 10.12).

At the LHC, however, it is within reach to also determine the corresponding behavior for the bottomonium states. First results have just appeared [44], and they indicate the behavior expected for sequential quarkonium suppression, as seen in Fig. 10.13.

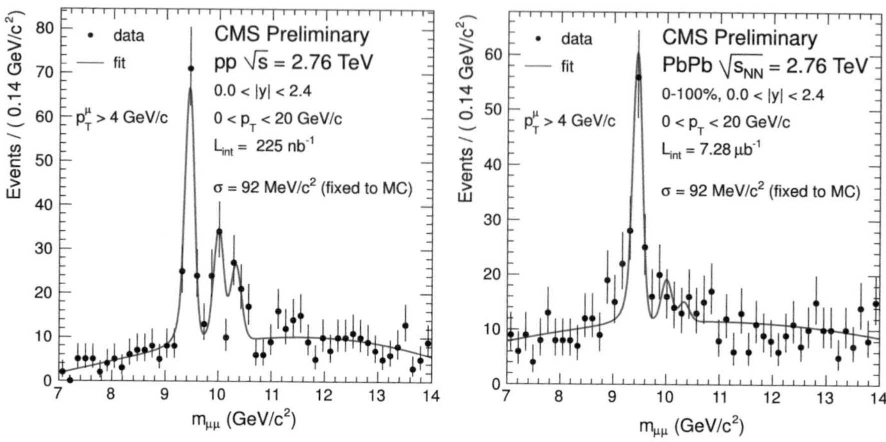

Fig. 10.13 Bottomonium production in *pp* (left) and *Pb − Pb* (right) collisions, as measured by the CMS collaboration at CERN–LHC [44]

- The production of the exited bottomonium states $\Upsilon(2S)$ and $\Upsilon(3S)$ in *Pb − Pb* collisions shows a suppression of about 70%, compared to that in *p−p* collisions.
- The $\Upsilon(1S)$ production in *Pb − Pb* collisions itself is reduced by about 40%, compared to the scaled results from *pp* interactions.

As indicated in Fig. 10.10, about 50% of the $\Upsilon(1S)$ rate measured in *p−p* collisions is due to feed-down from the higher excited states [45]. If the suppression of the $\Upsilon(2S)$ and $\Upsilon(3S)$ is indicative for the fate of all excited states (i.e., also for the $\chi_b(1P)$ and $\chi_b(2P)$), then the suppression of the feed-down sources in *Pb − Pb* collisions accounts for the observed reduction of $\Upsilon(1S)$ production. In other words, it would indicate that the direct $\Upsilon(1S)$ production is essentially not modified by the presence of the medium produced in nuclear collisions at the LHC. It would require still higher temperatures to achieve that (see Table 10.2). Once lattice studies give reliable quarkonium dissociation thresholds, such a program would allow a direct comparison between QCD calculations and nuclear collision data.

It would, of course, also be ideal to compare hidden to open bottom production; however, in the comparison of the different bottomonium states, one expects initial state effects (shadowing) and final state damping (parton energy loss) to cancel out when comparing closely adjacent resonances.

We close this subsection with an interesting twist to the story. The mentioned hard production of $c\bar{c}$ pairs can, in nuclear collisions of very high energy, lead to a medium containing an overabundance of charm quarks, i.e., more than would be found in a QGP at full thermal equilibrium. If charmonium production occurs only through the primary hard process, with a given fraction f_i forming charmonium state i, this does not affect the scenario just discussed. If, however, in the cooler plasma at the hadronization point there exists a possible "statistical" J/ψ formation process, with rates determined solely by the numbers of c and $\bar{c}$ present, then the suppression of the charmonia initially formed in individual nucleon-nucleon collisions becomes irrelevant. The large total number of available charm quarks, from unbound pair production and from charmonium dissociation, will in this case by regeneration at the hadronization point lead to more charmonia than obtained in primary production [46–49]. Since the $c\bar{c}$ production rate at a given collision energy increases with centrality, such a mechanism would also lead to an increase of J/ψ production with centrality, in contrast to what is presently observed, up to RHIC energy (see Figs. 10.11 and 10.12).

The modification due to such an effect is illustrated schematically in Fig. 10.14. It shows the J/ψ survival probability, as obtained from the ratio of J/ψ production relative to that of open charm. Sequential suppression removes charmonia from the total $c\bar{c}$ production, while statistical regeneration adds further charmonium states, and in the ratio to open charm, initial state effects such as shadowing or parton energy loss will effectively cancel out. Also here, for regeneration just as for color screening, the crucial predicted modification is the change of hidden to open charm production [30, 31], and this is presently not measured at the LHC.

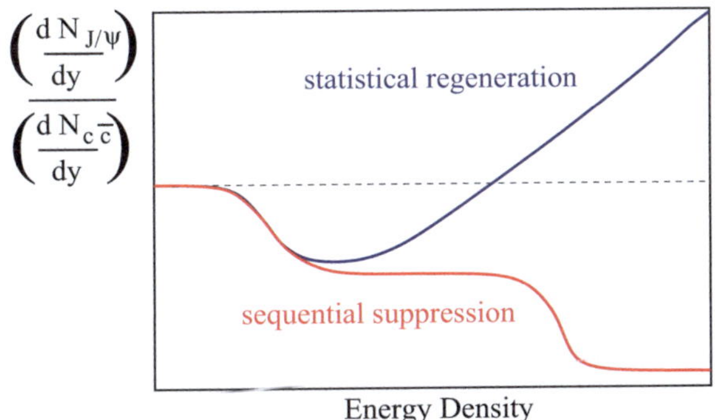

Fig. 10.14 Sequential J/ψ suppression vs. statistical regeneration

If observed, statistical regeneration would eliminate the possibility of a direct comparison of nuclear collision data with QCD predictions for charmonium dissociation at very high energies, where one finds enhanced $c\bar{c}$ production. An alternative would be given by detailed studies at lower energies, where the number of available charm quarks is still sufficiently small; this could be done at SPS and RHIC energies. For a comparison to QCD results at th extreme LHC energies, one would then have to turn to bottomonium states, where formation through recombination appears unlikely. On the other hand, the observation of statistical J/ψ formation would provide clear evidence for full thermalization, including even the heavier charm quarks. Incidentally, in Fig. 10.14 we have shown the sequential suppression rate remaining non-zero even above the threshold for direct J/ψ dissociation. The reason is that at the edge of the interaction region, production can still take place without being affected by the hot medium in the core region. We shall return to this "corona effect" in Sect. 10.6.

10.5   Jet Quenching

Another possible hard probe is provided by the study of jets in nucleus-nucleus collisions, consisting of one or more hadrons with very high momentum transverse to the collision axis. These are formed initially by a very energetic parton, quark or gluon, produced in the early hard collision stages and emitted in the transverse direction. Given a QGP formation time of about 1 fm, the nascent jet will pass through several fermi of hot deconfined matter before it escapes the interaction region and eventually hadronizes. How much energy it has lost when it finally emerges will tell us something about the density of the medium [50–53]. In particular, the density in a quark-gluon plasma is by an order of magnitude or more higher than that of a confined hadronic medium, and so the energy loss of a fast passing color charge is expected to be correspondingly higher as well. Let us consider this in more detail.

An electric charge, passing through matter containing other bound or unbound charges, loses energy by scattering. For charges of low incident energy E, the energy loss is largely due to ionization of the target matter. For sufficiently high energies, the incident charge scatters directly on the charges in matter and as a result radiates photons of average energy $\omega \sim E$. Per unit length of matter, the 'radiative' energy loss due to successive scatterings is thus proportional to the incident energy,

$$-\frac{dE}{dz} \sim E. \tag{10.15}$$

This probabilistic picture of independent successive scatterings breaks down at very high incident energies [54–56]. The squared amplitude for n scatterings now no longer factorizes into n interactions; instead, there is destructive interference, which for a regular medium (crystal) leads to a complete cancellation of all photon emission except for the first and last of the n photons. This Landau-Pomeranchuk-Migdal (LPM) effect greatly reduces the radiative energy loss.

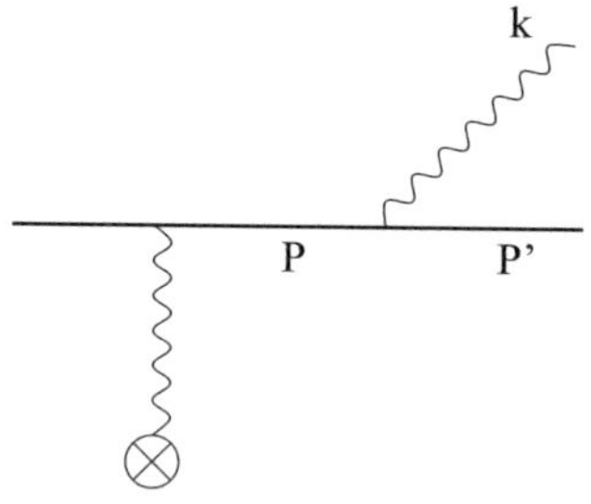

Fig. 10.15 Gluon emission after scattering

The physics of the LPM effect is clearly relevant in calculating the energy loss for fast colour charges in QCD media. These media are not regular crystals, so that the cancellation becomes only partial. Let us consider the effect here in a heuristic fashion; for details of the actual calculations, see [57–62]. The time t_c needed for the emission of a gluon after the scattering of a quark (see Fig. 10.15) is given by

$$t_c = \frac{1}{\sqrt{P^2}} \frac{E}{\sqrt{P^2}} = \frac{E}{2P'k}, \tag{10.16}$$

in the rest frame of the scattering center, where P^2 measures how far the intermediate quark state is off-shell; on-shell quarks and gluons are assumed to be massless, and $E/\sqrt{P^2}$ is the γ-factor between the lab frame and the proper frame of the intermediate quark. For gluons with $k_L \gg k_T$, we thus get

$$t_c \simeq \frac{\omega}{k_T^2}. \tag{10.17}$$

If the passing color charge can interact with several scattering centers during the formation time of a gluon, the corresponding amplitudes interfere destructively, so that in effect after the passage of n centers over the coherence length z_c, only one gluon is emitted, in contrast to the emission of n gluons in the incoherent regime. Nevertheless, in both cases each scattering leads to a k_T-kick of the charge, so that after a random walk past n centers, $k_T^2 \sim n$. Hence

$$k_T^2 \simeq \mu^2 \frac{z_c}{\lambda}, \tag{10.18}$$

where λ is the mean free path of the charge in the medium, so that $z_c/\lambda > 1$ counts the number of scatterings. At each scattering, the transverse kick received is measured by the mass of the gluon exchanged between the charge and the scattering center, i.e., by the screening mass μ of the medium. From Eq. 10.17 we have

$$z_c \simeq \frac{\omega}{k_T^2}, \tag{10.19}$$

so that the formation length in a medium characterized by μ and λ becomes

$$z_c \simeq \sqrt{\frac{\lambda}{\mu^2}\omega}. \tag{10.20}$$

For Eq. (10.20) to be valid, the mean free path has to be larger than the interaction range of the centers, i.e., $\lambda > \mu^{-1}$.

The energy loss of the passing color charge is now determined by the relative scales of the process. If $\lambda > z_c$, we have incoherence, while for $\lambda < z_c$ there is coherent scattering with destructive interference. In both cases, we have assumed that the thickness L of the medium is larger than all other scales. When the coherence length reaches the size of the system, $z_c = L$, effectively only one gluon can be emitted. This defines a critical thickness $L_c(E) = (E\lambda/\mu^2)^{1/2}$ at fixed incident energy E, or equivalently a critical energy $E_c = \mu^2 L^2/\lambda$ for fixed thickness L; for $L > L_c$, there is bulk LPM-behavior, below L_c there are finite-size corrections.

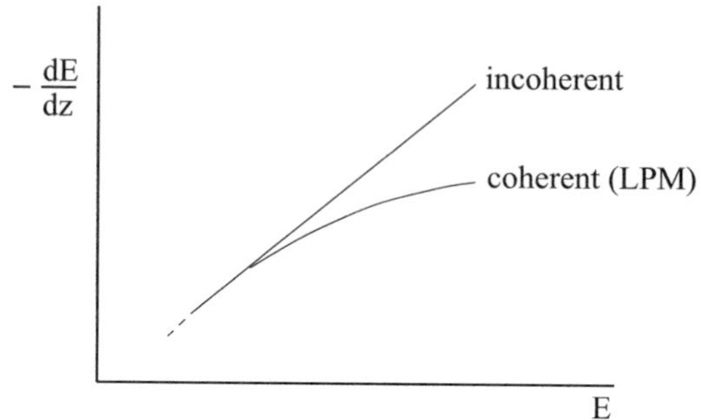

Fig. 10.16 Energy loss in incoherent and coherent interactions

We are thus left with three regimes for radiative energy loss. In case of incoherence, $z_c < \mu^{-1}$, there is the classical radiative loss

$$-\frac{dE}{dz} \simeq \frac{3\alpha_s}{\pi}\frac{E}{\lambda}, \tag{10.21}$$

where α_s is the strong coupling. In the coherent region, $\lambda > z_c$, the energy loss is given by the LPM bulk expression when $L > L_c$ [57–60],

$$-\frac{dE}{dz} \simeq \frac{3\alpha_s}{\pi}\sqrt{\frac{\mu^2 E}{\lambda}}. \tag{10.22}$$

The resulting reduction in the radiative energy loss dE/dz is illustrated in Fig. 10.16. Note that in earlier estimates the energy loss due to interactions of the gluon cloud accompanying the passing color charge had been neglected [63, 64]; this led to a considerably smaller energy loss, proportional to $\ln E$ instead of $\sqrt{E}$. Finally, in a medium of thickness $L < L_c$, there is less scattering and hence still less energy loss. Equation (10.22) can be rewritten as

$$-\frac{dE}{dz} \simeq \frac{3\alpha_s}{\pi}\frac{\mu^2}{\lambda}L_c(E), \tag{10.23}$$

and for $L < L_c$, this leads to

$$-\frac{dE}{dz} \simeq \frac{3\alpha_s}{\pi}\frac{\mu^2}{\lambda}L \tag{10.24}$$

as the energy loss in finite size media with $L \leq L_c$. The resulting variation of the radiative energy loss with the thickness of the medium is shown in Fig. 10.17, with saturated (i.e., bulk) LPM behavior setting in for $L \geq L_c$.

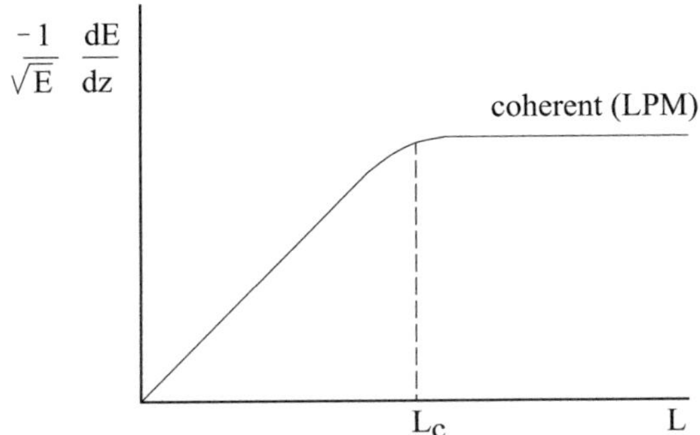

Fig. 10.17 Energy loss in coherent interactions as function of the thickness L of the medium

Equation (10.24) has been used to compare the energy loss in a deconfined medium of temperature $T = 0.25\,\mathrm{GeV}$ to that in cold nuclear matter of standard density [65]. For the traversal of a medium of 10 fm thickness, estimates give for the total energy loss

$$\Delta E = \int_{0\,\mathrm{fm}}^{10\,\mathrm{fm}} dz \frac{dE}{dz} \tag{10.25}$$

in a quark-gluon plasma

$$-\Delta E_{qgp} \simeq 30 \ \mathrm{GeV}, \tag{10.26}$$

corresponding to an average loss of $3\,\mathrm{GeV/fm}$. In contrast, cold nuclear matter leads to

$$-\Delta E_{cnm} \simeq 2 \ \mathrm{GeV} \tag{10.27}$$

and hence an average loss of $0.2\,\mathrm{GeV/fm}$. A deconfined medium thus leads to a very much higher rate of jet quenching than confined hadronic matter, as had in fact been suggested quite some time ago [50–53].

The advent of the RHIC at Brookhaven National Laboratory has made jet production in nuclear collisions experimentally feasible, and today a wealth of data exists. Clear evidence for jet quenching is found in an analysis of the azimuthal correlation structure in jet production. In $p - p$ collisions, a jet is determined by a pion (or other hadron) of very large transverse momentum, and this is generally found to be balanced by another high transverse momentum particle in the opposite direction. If in $A - A$ collisions the primary hard partonic process occurs near the edge of the interaction region, the balancing "away-side" jet has to traverse most of the newly formed hot and dense medium and is expected to be quenched. This in turn should lead to enhanced soft hadron production as momentum balance. Both cases are schematically illustrated in Fig. 10.18.

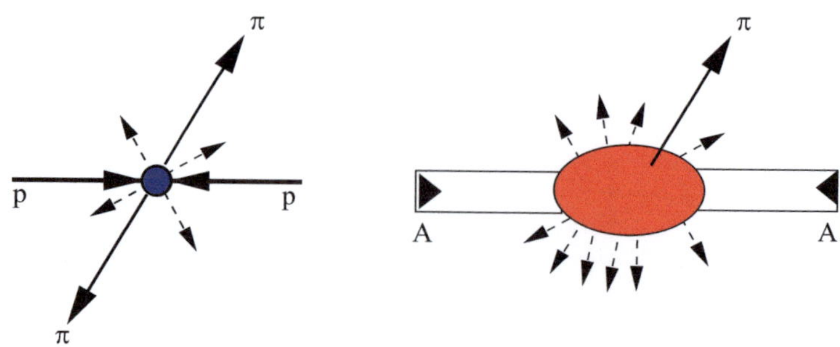

Fig. 10.18 Jet production in $p - p$ (left) and $A - A$ (right) collisions

The azimuthal distribution of jet production was studied at RHIC in $p - p$, $d - Au$ and $Au - Au$ collisions at $\sqrt{s} = 200$ GeV. Defining the azimuthal angle of the hardest hadron as zero ("near-side"), one clearly observes in both $p - p$ and $d - Au$ collisions the balancing "away-side" jet at $180°$. In $d - Au$ collisions, the jet does encounter a medium, but since this is normal (confined) nuclear matter, it is not significantly affected. On the other hand, in $Au - Au$ collisions the jet on the opposite side is strongly suppressed [66, 67]. The result is illustrated in Fig. 10.19, showing the azimuthal distribution of hadrons of transverse momenta above 5 GeV.

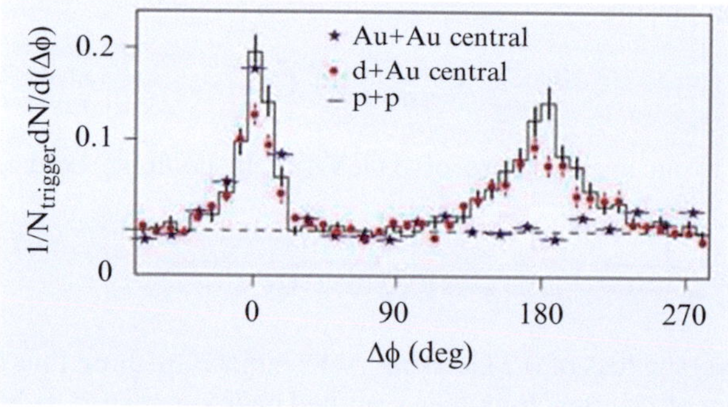

Fig. 10.19 Azimuthal distribution of hadrons with transverse momenta above 5 GeV in $p - p$, $d - Au$ and $A - A$ collisions [66, 67]

Corroborating evidence for jet quenching is obtained by considering directly the transverse momentum spectra of the hard hadrons in $p - p$ and $A - A$ collisions. Since jet production is a hard process, it is expected to scale with the number of primary collisions, as already found for the production of heavy quark pairs (previous subsection). A check for this is provided by hard photon spectra. Since the produced photons escape from the interaction region without any (strong) in-medium interactions, the spectra in central $A - Au$ interactions should coincide with those from the corresponding $p - p$ reaction, multiplied by the effective number of collisions. In Fig. 10.20, the ratio

$$R_{AA} = \frac{(dN/dk_T)_{AA}}{N_{\text{coll}}(dN/dk_T)_{pp}} \tag{10.28}$$

is shown for photon production in central $Au - Au$ vs. $p - p$ collisions and seen to be compatible with unity for $k_T \geq 3$–$5\,\text{GeV}$ [68]. Also shown in that Figure is the corresponding ratio for π^0 production; it is seen to be reduced by almost a factor of five, indicating strong jet quenching. This factor is in agreement with total quenching in the central interaction region and survival only in the surface shell [79]. The amount of quenching remains the same also for other hadrons, indicating that it is indeed the initially produced hard parton that is being quenched on its passage through the medium.

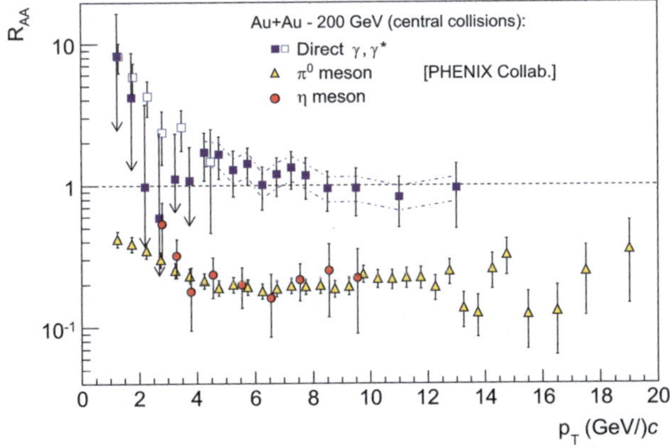

Fig. 10.20 Transverse momentum distributions of photons and hadrons in $Au - Au$ collisions at $\sqrt{s} = 200\,\text{GeV}$, normalized to the $p - p$ distributions scaled by the number of collisions [68]

Numerous other features of jet production were studied at RHIC and described in terms of QCD-based models; for more details, see e.g. [69]. They all indicate that high energy nuclear collisions indeed produce a very dense and very strongly interacting medium. Hopefully the advent of further data from the LHC will eventually also lead to a quantitative comparison with ab initio QCD calculations.

10.6   The Corona Effect

We have seen that two prominent signatures of QGP production in nucleus-nucleus collisions are jet quenching and quarkonium dissociation. For an infinite interaction volume, the suppression would in both cases be complete. In actual experiments, however, there are collisions at the rim of the interaction region (see Fig. 10.21), and these will be little or not at all affected by the central core of the medium. This so-called corona effect [70–78] will assure that even at the highest collision energies jets and quarkonia are still being produced. The azimuthal jet distributions mentioned in the previous section provide an example of such a situation: the away-side jet is effectively suppressed, if the near-side jet is produced at the surface of the interaction region. Evidently the role of the corona, relative to the overall interaction area, increases with peripherality; the most peripheral collisions are effectively just nucleon-nucleon interactions.

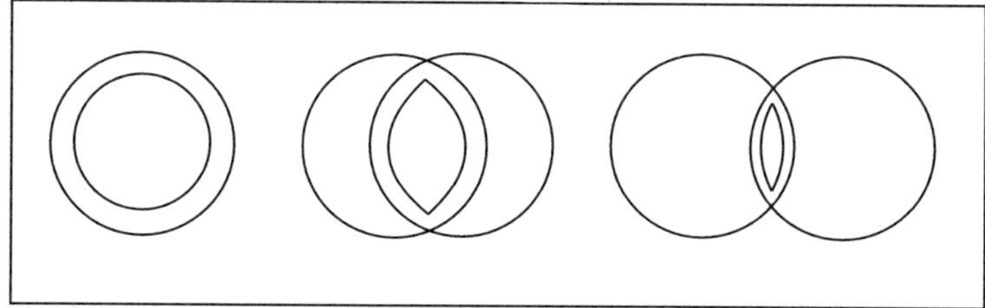

Fig. 10.21   Corona effect for decreasing centrality

The jet and quarkonium production rates as a function of the centrality of the collision should thus be determined by the variation with impact parameter of the ratio surface to volume. Moreover, since the primary production processes of jets and heavy quark pairs are both hard partonic interactions, they will depend on the number of collisions and thus lead to a very similar centrality dependence.

Geometric considerations lead to $R_{S/V} \simeq 0.17$ [79] for central collisions, and this is found to agree well with the jet quenching observed for central $Au - Au$ collisions at RHIC (see Fig. 10.20). If the origin of the surviving jets is indeed the corona, this fraction should remain the same also at higher energies, modulo slight possible modifications of the corona dimension.

In the case of J/ψ suppression, the feed-down from B production has to be taken into account in addition. Even if direct primary charmonium production is completely suppressed, the decay products from $B \rightarrow J/\psi + x$ will still be present, though conceivably not in scaled $p - p$ rates, but reduced due to nuclear modifications of parton distribution functions. Simple scaling of $p - p$ rates would lead to about 15–20% J/ψ production due to B decay; this applies at low transverse momenta, since the importance of B feed-down increases with p_T. The feed-down must be added to the corona production, which leads to about 30–40% for the remaining central J/ψ production rate, assuming total suppression in the core of the interaction region.

References

1. R. Hanbury-Brown, R.Q. Twiss, Phil. Mag. **45**, 663 (1954)
2. G. Goldhaber, S. Goldhaber, W. Lee, A. Pais, Phys. Rev. **120**, 300 (1960)
3. M.A. Lisa, S. Pratt, R. Soltz, U. Wiedemann, Ann. Rev. Nucl. Part. Sci. **55**, 357 (2005)
4. The Alice Collaboration (2011). arXiv:1012.4035v2
5. E.V. Shuryak, Phys. Rep. **61**, 71 (1980)
6. K. Kajantie, H.I. Miettinen, Z. Phys. C **9**, 341 (1981)
7. J. Kapusta, Phys. Lett. **136B**, 201 (1984)
8. L. McLerran, T. Toimela, Phys. Rev. D **31**, 545 (1985)
9. A. Adare et al. (PHENIX), Phys. Rev. Lett. **104**, 132301 (2010)
10. ALICE Coll., arXiv:1509.07324
11. See e.g., S. Digal et al., Eur. Phys. J. **43**, 71 (2005)
12. S. Jacobs, M.G. Olsson, C. Suchyta, Phys. Rev. **33**, 3338 (1986)
13. E. Eichten et al., Phys. Rev. D **17**, 3090 (1978)
14. E. Eichten et al., Phys. Rev. D **21**, 203 (1980)
15. T. Matsui, H. Satz, Phys. Lett. B **178**, 416 (1986)
16. F. Karsch, M.T. Mehr, H. Satz, Z. Phys. C **37**, 617 (1988)
17. F. Karsch, H. Satz, Z. Phys. C **51**, 209 (1991)
18. For more detailed discussion and references, see e.g. H. Satz, J. Phys. G **32**, R25 (2006)
19. For more detailed discussion and references, see e.g. L. Kluberg, H. Satz, arXiv:0901.3831
20. M.E. Peskin, Nucl. Phys. B **156**, 365 (1979)
21. G. Bhanot, M.E. Peskin, Nucl. Phys. B **156**, 391 (1979)
22. A. Kaidalov, in *QCD and High Energy Hadronic Interactions*, ed. by J. Tran Thanh Van, Editions Frontieres, Gif-sur-Yvette (1993)
23. D. Kharzeev, H. Satz, Phys. Lett. B **334**, 155 (1994)
24. R. Gavai et al., Int. J. Mod. Phys. **10**, 3043 (1995)
25. For a more recent summary see N. Brambilla et al., *Heavy Quarkonium Physics*, CERN-2005-005
26. M.B. Einhorn, S.D. Ellis, Phys. Rev. D **12**, 2007 (1975)
27. H. Fritzsch, Phys. Lett. B **67**, 217 (1977)
28. M. Glück, J.F. Owens, E. Reya, Phys. Rev. D **17**, 2324 (1978)
29. J. Babcock, D. Sivers, S. Wolfram, Phys. Rev. D **18**, 162 (1978)
30. H. Satz, K. Sridhar, Phys. Rev. D **50**, 3557 (1994)
31. H. Satz, Adv. High Energy Phys. **2013**, 242918 (2013)
32. L. Antoniazzi et al. (E705 Collaboration), Phys. Rev. Lett. **70**, 383 (1983)
33. For recent results and further references, see P. Faccioli et al., JHEP **0810**, 004 (2008)
34. A. Adare et al. (PHENIX Collaboration), Phys. Rev. D **85**, 092004 (2011)
35. N. Brambilla et al. (QWG), *Heavy Quarkonium Physics*, CERN -2005-005
36. B. Alessandro et al. (NA50), Eur. Phys. J. C **39**, 335 (2005)
37. R. Arnaldi et al. (NA60), Nucl. Phys. A **830**, 345c (2009)
38. L.A. Linden-Levy, Nucl. Phys. A **830**, 353c (2009)
39. N. Brambilla et al. (QWG), *Heavy Quarkonium: Progress, Puzzles and Opportunities*, 2010, p. 128; arXiv:1010.5827
40. D. Kharzeev, F. Karsch, H. Satz, Phys. Lett. B **637**, 75 (2006)
41. Phenix Coll., Phys. Rev. C **84**, 044905 (2011)
42. Phenix Coll., Phys. Rev. Lett. C **98**, 232301 (2007)
43. Star Coll., Phys. Lett. B **722**, 55 (2013)
44. The CMS Collaboration, Phys. Rev. Lett. **107**, 052302 (2011)
45. T. Affolder et al. (CDF Collaboration), Phys. Rev. Lett. **84**, 2094 (2000)
46. P. Braun-Munzinger, J. Stachel, Nucl. Phys. A **690**, 119 (2001)
47. R.L. Thews et al., Phys. Rev. C **63**, 054905 (2001)
48. L. Grandchamp, R. Rapp, Nucl. Phys. A **709**, 415 (2002)

49. R.L. Thews, M.L. Mangano, Phys. Rev. C **73**, 014904 (2006)
50. J.D. Bjorken, Fermilab-Pub-82/59-THY (1982) and Erratum
51. M. Gyulassy, X.-N. Wang, Nucl. Phys. B **420**, 583 (1994)
52. R. Baier et al., Phys. Lett. B **345** (1995)
53. B. G. Zakharov, JETP Lett. **63**, 952 (1996)
54. L.P. Landau, I.Y. Pomeranchuk, Doklad. Akad. Nauk SSSR **92**, 535, 735 (1953)
55. A.B. Migdal, Phys. Rev. **103**, 1811 (1956)
56. E.L. Feinberg, I.Y. Pomeranchuk, Suppl. Nuovo Cim. III, Ser. X, **4**, 652 (1956)
57. R. Baier et al., Phys. Lett. B **345**, 277 (1995)
58. R. Baier et al., Nucl. Phys. B **483**, 291 (1997)
59. R. Baier et al., Nucl. Phys. B **484**, 265 (1997)
60. R. Baier et al., Nucl. Phys. B **531**, 403 (1998)
61. B.G. Zakharov, JETP Lett. **63**, 952 (1996)
62. B.G. Zahkarow JETP Lett. **65**, 615 (1997)
63. M. Gyulassy, X.-N. Wang, Nucl. Phys. B **420**, 583 (1994)
64. M. Gyulassy, M. Plümer, X.-N. Wang, Phys. Rev. D **51**, 3436 (1995)
65. D. Schiff, Acta Phys. Polon. B **30**, 3621 (1999)
66. C. Adler et al. (STAR Collaboration), Phys. Rev. Lett. **90**, 082302 (2003)
67. A. Adare et al. (PHENIX Collaboration), Phys. Rev. C **78**, 014901 (2008)
68. D. d'Enterria, J. Phys. G **34**, S53 (2007)
69. D. d'Enterria, B. Betz, in *The Physics of the Quark-Gluon Plasma*, ed. by S. Sarkar, H. Satz, B. Sinha. Springer Lecture Notes in Physics, vol. 785 (Springer Verlag, Berlin/Heidelberg, 2010)
70. R. Stock, Phys. Rept. **135**, 259 (1986)
71. F. Becattini et al., Phys. Rev. C **69**, 024905 (2004)
72. C. Hohne, F. Puhlhofer, R. Stock, Phys. Lett. B **640**, 96 (2006)
73. K. Werner, Phys. Rev. Lett. **98**, 152301 (2007)
74. F. Becattini, J. Manninen, J. Phys. G **35**, 104013 (2008)
75. F. Becattini, J. Manninen, Phys. Lett. B **673**, 19 (2009)
76. A. Andronic et al., Nucl. Phys. A **789**, 334 (2007)
77. J. Aichelin, K. Werner, Phys. Rev. C **82**, 034906 (2010)
78. S. Digal, H. Satz, R. Vogt, Phys. Rev. **C85**, 034906 (2011)
79. B. Müller, Phys. Rev. C **67**, 061901 (2003)

Chapter 11
The Fireball Paradigm

> *Es wurde schon vor längerer Zeit geschlossen, daß bei einem
> sehr energiereichen Stoß eines Kernteilchens auf ein anderes
> viele Mesonen mit einem Schlag erzeugt werden können*
>
> Werner Heisenberg, *Zeitschrift für Physik 126 (1949) 569*
>
> *(It was concluded quite some time ago that in an energetic
> collision of one nuclear particle with another, many mesons
> could be created with one bang)*
>
> (Werner Heisenberg)

Abstract In this chapter we show how multihadron production can be related to thermodynamical considerations. Following a general introduction to the topic, we discuss the statistical hadronization model, in which each species is produced according to its phase space weight, and show that this leads to a universal hadronization temperature found in e^+e^- annihilation as well as in hadron-hadron and nucleus-nucleus collisions. In the final part we address deviations from a universal statistical hadronization description.

11.1 Statistical Multihadron Production

Multiparticle production in high energy collisions of strongly interacting particles has fascinated physicists for well over half a century. As predicted by Heisenberg [1], the little bang of such collisions produces with increasing energy an ever growing number of mesons and baryons of different quantum states, and from the beginning, the large numbers were a challenge to describe these reactions by collective or statistical approaches. It was tempting to go even further, to imagine that what they produced were really droplets of strongly interacting matter, thus providing a means to access in the laboratory the thermodynamics of strong interaction physics.

The main features observed in high energy collisions are the multiplicity, i.e., the number of produced particles as a function of the collision energy, the momentum

© Springer International Publishing AG 2018

H. Satz, *Extreme States of Matter in Strong Interaction Physics*,
Lecture Notes in Physics 945, https://doi.org/10.1007/978-3-319-71894-1_11

spectra of the particles, their correlations, and the relative abundances of the different species. These then are also the basic quantities which any theoretical framework has to provide.

The first statistical treatment was formulated by Fermi [2]. He assumed that the collision deposits a great amount of energy in a small spatial region around the colliding particles, and that the energy of this fireball is then distributed among the various observable degrees of freedom, the emitted mesons and nucleons, according to statistical laws. The derivation of the model is summarized in "Appendix 1: Scattering Matrix and Phase Space", where we show that the description of the production process is determined by the grand canonical phase space volume $Q(E, V)$ of a gas of non-interacting hadrons; here E denotes the collision energy in the center of mass of the colliding particles, and V the interaction volume. This phase space volume was to be calculated with whatever constraints are imposed by conservation laws (charge, baryon number, etc.).

The main features obtained from Fermi's model are:

- a multiplicity $n(E) \sim E^{3/4}$ growing as a power of E,
- isotropic production of secondary particles; and
- average secondary momenta $\langle |p| \rangle \sim E^{1/4}$ also increasing as a power of E.

Modified versions of the model [3, 4] lead to slightly changed powers, but the basic features remain. The available collision energy is equidistributed among the isotropically emitted secondaries; any increase of E goes partially into making more secondaries and partially into making each constituent more energetic.

Experimental data showed that with increasing collision energy, this picture became untenable for two main reasons:

- in a nucleon-nucleon collision, the incident nucleons always retained a considerable fraction of the collision energy ("leading particle effect");
- the secondaries were not emitted isotropically; their average transverse momenta (orthogonal to the collision axis) reached a constant value, independent of the incident energy, while the average longitudinal momenta increased with E.

Clearly this meant that not all information about the initial state was lost in the collision; the reaction retained a memory both of the conserved quantum numbers of the incident particles and of the collision axis. Fermi had already suggested that the spatial volume, as seen in the overall center-of-mass system, should be Lorentz-contracted along the collision axis. However, as long as there is no interrelation between the momenta and the coordinates of the secondaries, this does not produce anisotropic particle production.

A collective scenario for producing such an anisotropy was the hydrodynamical model proposed by Landau [5] soon afterwards. He assumed that the medium created in the collision was a fluid contained in the interaction region and which subsequently expanded according to hydrodynamics. For a fluid, the increase of pressure due to the Lorentz-contraction along the collision axis leads to a stronger flow in that direction. The resulting secondaries thus have larger longitudinal than transverse momenta; however, the leading particle effect, being of a more microscopic nature, does not arise in such a description, if it is based on one compressed fluid.

Looked at from another point of view, the combination of the anisotropic secondary momentum distributions and the leading particle effect seemed to indicate that at high energy the incident nucleons could not fully stop each other. Instead, they seemed to "pass through" one another, losing only part of their energy in the process. This "transparency" was subsequently explained by Gribov [6] as a consequence of hadronic size and the finite speed of information transmission.

It thus became evident that high energy collisions could not be understood in terms of the formation of one fireball, in the sense of a single isotropic energy deposit into a small spatial volume. The collision instead seems more like the passage of an energetic charge through a medium, leaving behind a condensation trail of smaller fireballs superimposed along the collision axis. Each of these bubbles could now in principle have the phase space structure envisioned by Fermi, and if one attributes the conserved baryon numbers to the fastest bubbles in each direction, the scenario would also provide the leading particle effect.

While this does bring in the desired longitudinal momentum growth, the energy of each bubble could also still increase, and this in turn results in an energy dependence (albeit weaker) also for the average transverse momenta. The basic puzzle of the field or, looked at in a more positive way, the most important hint provided by nature, was the constancy of the average transverse momenta p_T of the secondaries. Making this even more tantalizing was the observation that while different species of secondaries led to different (energy-independent) transverse momentum distributions, the transverse energies (also called *transverse masses*) m_t^i appeared to follow one universal pattern, with

$$\frac{dN_i}{dm_t^i} \sim \exp\{-\lambda m_t^i\}, \quad m_t^i = \sqrt{(p_t^i)^2 + m_i^2} \tag{11.1}$$

describing the functional form of the distribution of all species i of different masses m_i in terms of one universal parameter λ.

The first explanation of this universality was proposed by Hagedorn [7, 8], based on the resonance structure governing the interaction of the different hadron species (see Chap. 3). Experiment had shown that multiparticle production with increasing energy did not simply produce a shower of many pions, kaons and nucleons. Instead, it led to the production of more and different excited resonant states which decayed strongly into less excited states and finally into the ground state hadrons. Starting in the 1960s, an ever growing number of such hadronic resonances were discovered, and today the standard compilation [9] lists hundreds of them. The main theoretical approaches to resonance composition laws were studied in Chap. 3, based on partitioning and self-similarity. The number of states $\rho(m)$ of a resonance of mass m, its "degeneracy", is then given as the number of different composition patterns. This led to an exponential growth of $\rho(m)$,

$$\rho(m) \sim m^a \exp\{bm\}; \tag{11.2}$$

while the power a depends on the details of the partition problem, the coefficient b was expressed in terms of fundamental features of strong interaction physics, such as hadronic size, the range of the strong force, or the Regge resonance pattern.

At this stage then, experiment had shown several basic deviations from Fermi's original fireball picture. The assumption of a completely random production process failed: the system retained some information of the initial state, secondary particle distributions were different in directions along and orthogonal to the collisions axis, and there were leading particles carrying a baryon number. Moreover, the emitted pions, kaons and nucleons, had gone through some intermediate interactive stage, with resonance formation and decay as the dominant process. The first of these features, anisotropy and leading particles, could be accounted for through a superposition of fireballs, using initial state dynamical information as input.

To solve the resonance problem, Hagedorn invoked a result first obtained by Beth and Uhlenbeck [10] and subsequently generalized by Dashen, Ma and Bernstein [11]. They had argued that if the interaction of a gas of constituents is indeed dominated by resonance formation, then one can replace the interacting system of elementary particles by a non-interacting system of all possible resonances. The relevant phase space for the states of multiparticle production would thus be that of an ideal resonance gas, with an exponentially growing resonance mass spectrum, and contained in an interaction volume V_0.

We have shown in Chap. 3 that the grand-canonical partition function $Z(T, V_0)$ of such a resonance gas diverges for

$$T > T_H = \frac{1}{b}, \tag{11.3}$$

so T_H becomes an upper bound on the temperature. Increasing the energy of such a system does not increase the momentum of the secondaries and hence its temperature; instead, it leads to more species of more massive hadrons. As a result, the momentum spectra now have the form (11.1); the experimentally observed universal transverse mass pattern with a parameter λ is thus accounted for as the universal limiting temperature of an ideal resonance gas. This indeed agreed quite well with the observed transverse energy spectra. The energy-dependent and unbounded longitudinal momenta, on the other hand, arise from the superposition of an energy-dependent number of such Hagedorn-type fireballs.

In the previous chapters, we have seen that today the Hagedorn temperature is essentially interpreted as the critical point associated to the confinement/deconfinement transition, when approached from the confined hadronic side. In a parton model picture, the initial state of any hadronic collisions consists of beams of partons moving and colliding in the longitudinal (beam) direction. Their interaction leads to momentum loss and eventually to a hadronization of the partons. This occurs in the form of the Bjorken "inside-outside" cascade [12], with the slowest parton clusters hadronizing first. Conceptually, this provides the basis for the superposition of fireballs moving at different rapidities, as a phenomenological picture of high energy multihadron production.

11.2   The Abundance of the Species

For an ideal resonance gas, the universal temperature T_H determines not only the momentum spectra, but also the relative abundances of the different species. At a fixed temperature, a heavy meson is less likely to be present than a lighter one. In other words, the ratio of species i to j is predicted as the ratio of the corresponding phase space weights,

$$\frac{N_i}{N_j} \simeq \left(\frac{m_i}{m_j}\right)^2 \frac{K_2(m_i/T_H)}{K_2(m_j/T_H)} \simeq \left(\frac{m_i}{m_j}\right)^{3/2} \exp\{-(m_i - m_j)/T_H\}. \tag{11.4}$$

Hence, if high energy collision results are specified simply by the pure phase space of a resonance gas, they should also lead to corresponding production ratios. Let us consider this in more detail.

The statistical hadronization model assumes that hadronization in high energy collisions is a universal process proceeding through the formation of multiple colorless massive clusters, or fireballs, of finite spacial extension and fixed temperature T_H. These clusters are assumed to decay into hadrons according to a purely statistical law: every multi-hadron state of the fireball phase space defined by its mass, volume and charges is equally probable. The mass distribution and the distribution of charges (electric, baryonic and strange) among the clusters and their (fluctuating) number are, however, in principle determined in the prior dynamical stage of the process, which determines how fireballs are emitted along the collision axis.

Hence one would seem to need this dynamical information in order to make definite quantitative predictions to be compared with data. Nevertheless, for Lorentz-invariant quantities such as multiplicities, one can introduce a simplifying assumption and thereby obtain a simple analytical expression in terms of the temperature. The key point is to assume that the distribution of masses and charges among clusters is again purely statistical [13–15], so that, as far as the calculation of multiplicities is concerned, the set of many clusters becomes equivalent, on the average, to one large cluster (an *equivalent global cluster*) whose volume is the sum of the individual proper cluster volumes and whose charge is the sum of cluster charges (and thus the conserved charge of the initial colliding system). In such a framework, the global cluster can be hadronized on the basis of statistical equilibrium.

We shall consider this here for a system without conserved charges; the exact conservation of general "charges" (baryon number, strangeness, electric charge, isospin, etc.) has been solved [16–18] and will be summarized in "Appendix 2: Exact Charge Conservation". For the application of this more general formalism to species abundance data, we refer to [19, 20]. The primary multiplicity of a hadron species i due to fireball decay in the Boltzmann limit of phase space is then given by

$$\langle n_i \rangle = d_i \gamma_s^{\nu_i} \frac{V T m_i^2}{2\pi^2} K_2(m_i/T_H), \tag{11.5}$$

where d_i specifies the degeneracy of species i and V the overall production volume (the sum of all fireballs). In abundance ratios, V cancels out, so it is a parameter needed only for absolute multiplicities.[1] We denote $\langle n_i \rangle$ as *primary* multiplicity, since all heavier resonances will still decay into lighter ones, so that the actually observed multiplicities are obtained from (11.5) by

$$\langle n_i \rangle = \langle n_i \rangle^{\text{primary}} + \sum_k \langle n_k \rangle \text{Br}(k \to i), \tag{11.6}$$

summing over the various branching ratios Br as measured. The final pion multiplicity, for example, is in fact several times larger than the primary one.

Special note should be given to the parameter γ_s, which was introduced somewhat ad hoc [21] in order to achieve a general reduction in the production of hadrons containing strange quarks. One had observed that in *elementary* hadron collisions and in $e^+ e^-$ annihilation, fewer kaons, ϕ's, Λ's , Ξ's, Ω's, etc., were produced than predicted by phase space alone. On the other hand, it was found that one reduction parameter γ_s per strange quark contained in the hadron could account quite universally for the entire observed strangeness suppression, with $\gamma_s \simeq 0.5\text{--}0.6$. To illustrate, the multiplicity of kaons according to Eq. (11.5) is multiplied by a factor γ_s, that of ϕ's by γ_s^2, for Ω's by γ_s^3, and similarly for further strange hadrons and antihadrons; in each case this reduces the corresponding production rate in comparison to that of hadrons of a comparable mass but containing only u, d quarks. The origin of the factor γ_s has been discussed extensively, though without definite conclusion. One possibility might be a reduction of strange quark formation in the hadronization process, due to the larger strange quark mass. And in fact, $\exp\{-m_s/T_H\} \simeq 0.6$, using $m_s \simeq 80\,\text{MeV}$ and $T_H \simeq 160\,\text{MeV}$. On the other hand, in nucleus-nucleus collisions the suppression is essentially removed, as we shall see shortly. This suggests local charge conservation as origin [22]. In elementary collisions, with a comparably small number of hadrons containing strange quarks, the formation of a strange quark has to be compensated locally, and this leads to a reduced production rate [23]. In nuclear collisions, the much larger amount of strangeness allows compensation by strange quarks from different parent collisions and thus weakens or removes the local conservation constraints. Very recently, it was shown that strangeness suppression indeed follows a universal pattern in all hadronic collisions, from pp to AA interactions. This will be addressed in Sect. 11.3.

The other parameter in Eq. (11.5), apart from the temperature, is the degeneracy d_i. We have noted above and discussed in detail in Chap. 3 how theoretical modelling based on self-similarity and/or partitioning led to a degeneracy $\exp\{bm\}$ increasing exponentially in mass (see Eq. (11.1)). Such an increase in turn leads to a critical point $T_H = 1/b$, limiting hadron thermodynamics to $T \leq T_H$. A more empirical alternative is the *physical* resonance gas, in which the resonance spectrum is taken

---

[1]This is no longer correct in the presence of exactly conserved charges, see "Appendix 2: Exact Charge Conservation".

to consist of the actually measured and tabulated resonances, with only their spin and isospin degeneracy taken into account. This approach is obviously limited in resonance mass, since little is known about states above 2.5–3.0 GeV. With an upper limit in resonance mass, the partition function is analytic and hence there is no critical point; the energy density and all higher derivatives remain finite for all values of the temperature, also those above a Hagedorn T_H. As far as the measurable abundances are concerned, the missing high mass states do not play a significant role: including all excited states up to 2.0 GeV covers almost all of the feed-down sources for pions, for example, since the higher mass states are strongly suppressed, for $m \gg T_H$. Studies comparing species abundances with cuts at 1.5 or 2.0 GeV thus show very little difference. The measured low mass hadron states follow a Regge pattern and thus produce the correct beginning of the limiting temperature pattern. The contribution of the "missing" high mass states is essentially limited to the energy density or the speed of sound in a very narrow region around T_H, as shown in Chap. 3.

If the basic assumption of statistical hadronization – the equivalence of interacting hadron gas and ideal resonance gas – is indeed correct, the abundances of the species can thus be used to determine the hadronization temperature. Surprisingly, this approach has turned out to be correct far beyond all expectations. Over the past years, the resulting predictions were tested in a variety of collision configurations, from $e^+ e^-$ annihilation [13, 24–26] over $p - p/p - \bar{p}$ [19, 27] to nucleus-nucleus collisions [28–33]. With some caveats to be elaborated, they were found to provide a most remarkable account of what is observed, both of species abundances and, where applicable, of transverse momentum spectra [13]. Moreover, the temperature obtained for high energy experiments turned out to be quite universal, always lying around 160–180 MeV, i.e., in a range which partitioning arguments as well as studies of critical phenomena in QCD had pre- and postdicted.

To illustrate, we show in Table 11.1 the production rates measured [34–40] for 18 different hadron species in proton-proton collisions at a center-of-mass energy of 200 GeV^2; they are shown together with the fit rates obtained in the statistical model for a temperature $T_H = 170.1 \pm 3.5$ MeV [27]. The fit was made with exact baryon number and strangeness conservation, implemented (see "Appendix 2: Exact Charge Conservation") in a description of the form of Eqs. (11.5) and (11.6). The overall χ^2 per degree of freedom is 1.11, showing that the fit is indeed quite good. We should, however, remember that the statistical hadronization model is an average description of hadron emission, and sufficiently precise data must eventually show deviations.

---

^{2}The data are taken in a unit interval at midrapidity. In principle, statistical hadronization applies to full (4π) production rates; at high energies, however, the rapidity distributions are sufficiently flat to allow midrapidity studies.

Particle	Measurement (E)	Relative error	Model (M)	(M − E)/E (%)
	pp collisions at 200 GeV			
π^+	1.44 ± 0.11	0.076	1.403	−2.62
π^-	1.42 ± 0.11	0.077	1.384	−2.59
K^+	0.150 ± 0.013	0.087	0.152	1.48
K^-	0.145 ± 0.013	0.090	0.146	0.68
p	0.138 ± 0.012	0.087	0.149	7.42
$\bar{p}$	0.113 ± 0.010	0.088	0.1120	5.56
ϕ	0.0180 ± 0.0029	0.16	0.01130	−59.3
Λ	0.0436 ± 0.0041	0.094	0.04348	−0.28
$\bar{\Lambda}$	0.0398 ± 0.0038	0.095	0.03686	−7.96
Ξ^-	0.0026 ± 0.00092	0.35	0.003070	15.3
$\bar{\Xi}^+$	0.0029 ± 0.00104	0.36	0.002728	−6.29
$\Omega + \bar{\Omega}$	0.00034 ± 0.00019	0.56	0.0005712	40.5
K^0_S	0.134 ± 0.011	0.082	0.147	9.5
ρ^0	0.259 ± 0.039	0.15	0.1861	−28.1
$(K^{*0} + \bar{K}^{*0})/2$	0.0508 ± 0.0063	0.12	0.05151	1.4
$\Sigma^{*+} + \Sigma^{*-}$	0.0107 ± 0.00146	0.14	0.01028	−3.9
$\bar{\Sigma}^{*+} + \bar{\Sigma}^{*-}$	0.0089 ± 0.00126	0.14	0.008650	−2.8
$\Lambda(1520) + \bar{\Lambda}(1520)$	0.0069 ± 0.0011	0.16	0.00561	−18.7

Table 11.1 Measured and fitted hadron abundances in *pp* collisions at 200 GeV [27]; the data are from the STAR experiment at BNL-RHIC [34–40]

Next we compare in Table 11.2 the results of the mentioned *pp* data to those from other initial state configurations (e^+e^- and $A - A$) at high collision energies. To start from a comparable basis, we have restricted the data set in all three reaction channels as far as possible to the same 12 long-lived hadron species; the rates for short-lived and hence broader resonances are in general more difficult to measure. We see that all channels indeed appear to converge to a hadronization temperature value of about 160–170 MeV.

Collision	pp	e^+e^-	$Au - Au$
CMS Energy [GeV]	200	91.25	200
Temperature [MeV]	169.8 ± 4.2	164.7 ± 0.9	168.5 ± 4.0
Average relative deviation data vs. fit [%]	12.5	9.4	11.7
Strangeness Suppression γ_s	0.57 ± 0.03	0.66 ± 0.01	0.93 ± 0.04

Table 11.2 Fit results for a set of 12 long-lived particles in high energy *pp*, $Au - Au$ and e^+e^- collisions [27]

Combining the results from e^+e^- annihilation, elementary hadron-hadron interactions and nucleus-nucleus collisions at different high energies leads to one of the truly striking observations in high energy strong interaction physics: the existence of a universal hadronization temperature T_H. An overall view of this result is given in Fig. 11.1.

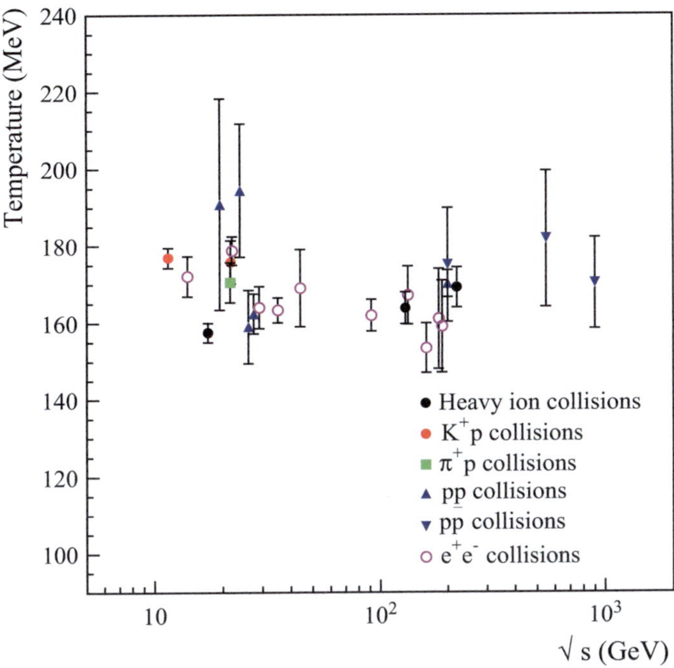

Fig. 11.1 Hadronization temperatures obtained for various initial collision configurations at different (high) energies

On the theoretical side, this suggests that there must be an underlying universal production mechanism, i.e., that the hadronization temperature is indeed determined by or closely related to the confinement/deconfinement transition of strongly interacting matter.[3] On the experimental side, it leads to the remarkable prediction that the relative hadron abundances produced in high energy collisions become with increasing energy independent of the collision energy. One could thus predict such ratios with considerable confidence for the experiments at the CERN-LHC.

Nevertheless, one important feature distinguishes elementary from nuclear interactions. It is seen in Table 11.2 that the strangeness suppression observed in elementary reactions has essentially disappeared in nucleus-nucleus collisions. We had mentioned this already above – the suppression factor γ_s is now compatible with unity. The fully thermal behavior of strange hadrons in the hot medium produced in central nuclear collisions is indeed a first indication for collective features present

---

[3]We note here that over the years, both the deconfinement temperature T_c obtained in lattice studies for $\mu \simeq 0$ and the hadronization temperature determined in the analysis of species abundances have shown some slight fluctuations. The ideal resonance gas is, of course, a priori a model. Nevertheless, all versions have remained in the range between 150 and 170 MeV, and this range is in accord with the lattice results.

in such interactions. For its further understanding, it will be very helpful to see if *pp* collisions at the very much higher energies of the CERN-LHC also provide, as predicted [41], a γ_s approaching unity.

A further point to note is that while in elementary collisions we had to implement an exact conservation of baryon number and strangeness, in nucleus-nucleus collisions both "charges" are present in sufficiently large numbers to allow a grand-canonical treatment. This means that the rate for a species of strangeness S_i, baryon number B_i and charge Q_i, again in the Boltzmann limit, is now given by

$$\langle n_i \rangle = d_i \gamma_s^{v_i} \frac{V T m_i^2}{\pi^2} K_2(m_i/T_H) \; \sinh\{(S_i\mu_S + B_i\mu_B + Q_i\mu_Q)/T\}; \qquad (11.7)$$

note that $\langle n_i \rangle$ specifies the *net* overall number of species i, e.g., the number of baryons minus that of antibaryons, etc. The chemical potentials μ_S, μ_B and μ_Q are then fixed such that the overall strangeness is zero and the overall baryon number and charge are those of the incident nuclei. At mid-rapidity, for sufficiently large energies all three tend to zero, since the nuclear remnants are concentrated in the fragmentation regions at large rapidity.

11.3   Universal Strangeness Suppression

In contrast to the hadronization temperature T_H, which is found to reach a constant value of some 160 MeV in all high energy collisions from e^+e^- annihilation to *pp*, *pA* and *AA* collisions, the strangeness suppression factor γ_s differs by almost a factor two between *pp* and *AA* collisions at the same collision energy of 200 GeV, see Table 11.2. The overall energy dependence of γ_s for the different collision configurations is shown in Fig. 11.2.

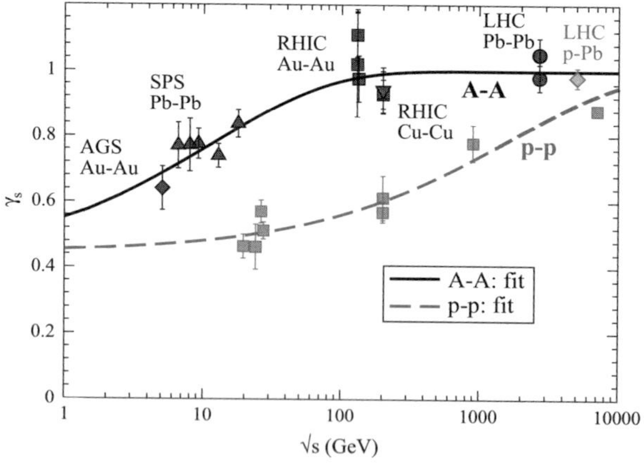

Fig. 11.2 Strangeness suppression in *pp* and *AA* collisions at cms energy $\sqrt{s}$ [42–48]

The results shown in Fig. 11.2 suggest an obvious question: is there a unified description for the strangeness suppression observed in *AA* and as well as in *pp* collisions? It is indeed found [49] that if we simply replace the collision energy $\sqrt{s}$ by more natural thermal variables, then γ_s in fact becomes one universal function.

We recall that the evolution of a high energy hadronic collision is schematically taken to proceed as shown in Fig. 11.3. At proper time τ_0, a thermal medium is reached, which at proper time τ_h hadronizes, i.e., turns into an ideal resonance gas.

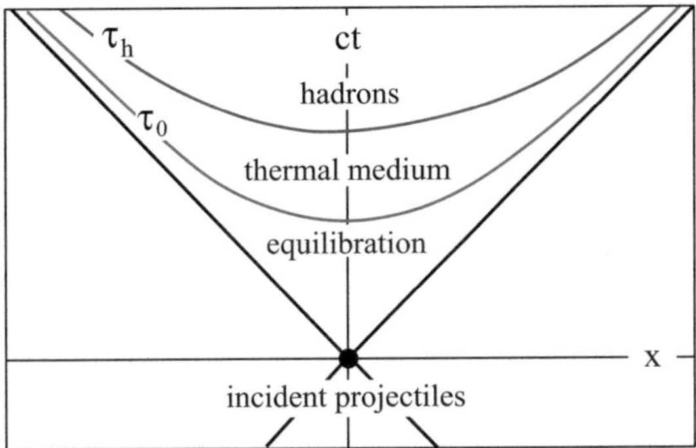

Fig. 11.3 High energy collision in terms of longitudinal space x and time t in the center of mass

The initial entropy density s_0 at the time of thermalisation is for one-dimensional hydrodynamic expansion given by the Bjorken relation

$$s_0 \tau_0 \simeq \frac{1.5 A^x}{\pi R_x^2} \left(\frac{dN}{dy} \right)^x_{y=0}, \text{ with } x \sim pp, pA, AA, \tag{11.8}$$

where $(dN/dy)^x(s)$ denotes the multiplicity per unit central rapidity in the corresponding reaction x at collision energy $\sqrt{s}$, while R_x gives the radius of the associated transverse area. This multiplicity has been measured at different collision energies for *pp* as well as for *AA* collisions, with the result shown in Fig. 11.4.

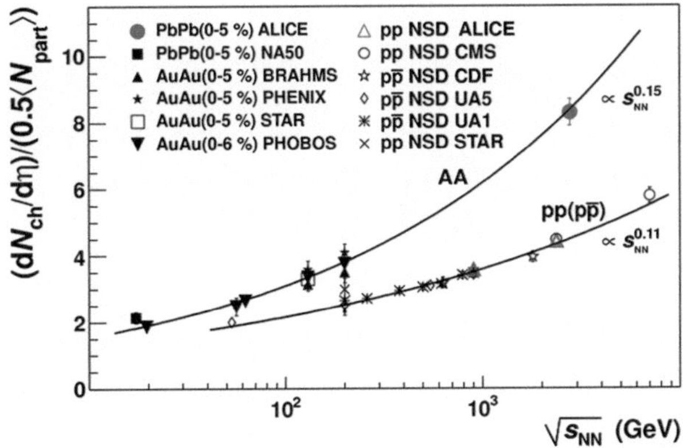

Fig. 11.4 Multiplicities in *pp* and *AA* collisions as function of the cms energie $\sqrt{s}$ [50]

The fit curves shown in this figure are given by

$$\left(\frac{dN}{dy}\right)_{y=0}^{AA} = a_A(\sqrt{s})^{0.3} + b_A \tag{11.9}$$

and

$$\left(\frac{dN}{dy}\right)_{y=0}^{pp} = a_p(\sqrt{s})^{0.22} + b_p \tag{11.10}$$

with $a_A = 0.7613$, $b_A = 0.0534$; $a_p = 0.797$; $b_p = 0.04123$ [50]. We can use these results to obtain the initial entropy density $s_0(s)$ in Eq. (11.8) as function of the collision energy. Next we then use that to convert the strangeness suppression factor γ_s into a function of the initial entropy density. The result is shown in Fig. 11.5.

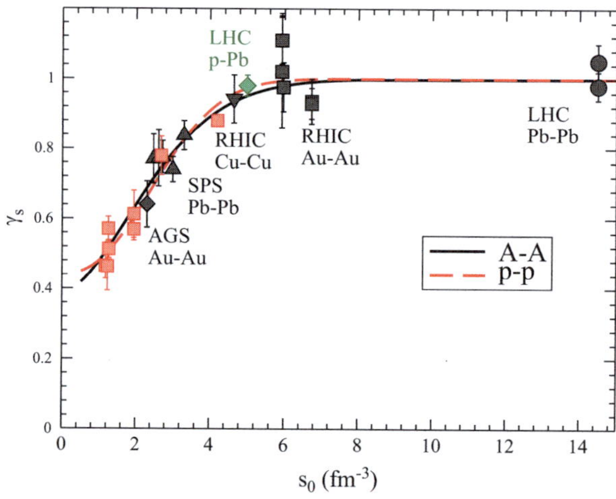

Fig. 11.5 Strangeness suppression γ_s as function of the initial entropy density s_0 [49]

We see that we now indeed obtain a universal form of strangeness suppression, with which pp and AA curves as well as data coincide. The crucial variable for the variation of the strangeness suppression factor is thus the initial entropy density.

Since finite temperature lattice QCD calculations [51] provide us with the dependence of s_0 on the initial temperature T, we can use this result to express γ_s as a function of T, see Fig. 11.6.

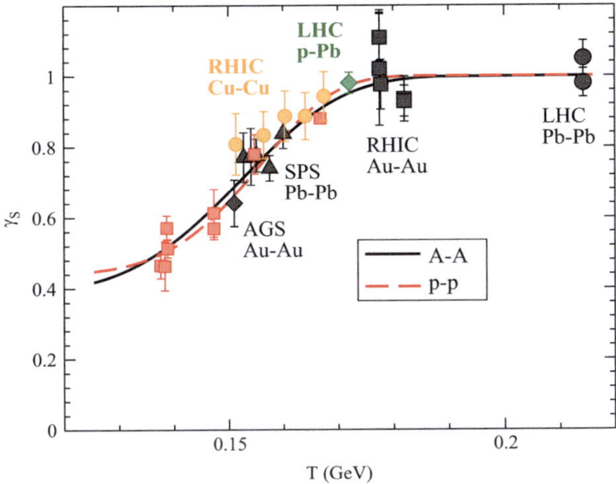

Fig. 11.6 Strangeness suppression γ_s as function of the initial temperature T of the thermal medium [49]

Here we note in particular that if the initial temperature of the thermal medium produced in the collision lies below the deconfinement temperature, i.e., if the medium is in a hadronic state, there is strangeness suppression, with $\gamma_s \simeq 0.5 - 0.6$. For media of initial temperatures above the transition region, such a suppression is no longer present; it disappears in a very narrow temperature range around $T_c \simeq 155$ MeV, for pp as well as for nuclear collisions. In other words, strangeness suppression disappears with deconfinement. The same relation to the confinement/deconfinement point is of course also present in Fig. 11.5: here the transition occurs in a narrow band around the corresponding entropy density at the transition point, $s_0^c \simeq 2.5$ fm^{-3}.

It is thus evident that as a function of thermal variables such as the initial entropy density or the initial temperature, the strangeness suppression factor follows a universal pattern for pp, pA and AA collisions at different collision energies. Moreover, with color deconfinement, strangeness suppression ends and the species abundances are in accord with equilibrium thermodynamics.

11.4   The Hadronic Resonance Gas and Its Limits

We have seen that the abundance of the species produced in high energy collisions is reproduced very well by the statistical hadronisation model, based on an ideal resonance gas, and that the resulting temperature, for small or vanishing baryon number, is in accord with the critical temperature obtained in statistical QCD. One may therefore ask if also other thermodynamic features of such a hadronic resonance system agree with the behavior found in lattice studies of the confined medium. The starting point for such a model of strongly interacting matter below T_c is the free energy density

$$f(T, \mu_S, \mu_B, \mu_Q) = \frac{T^4}{\pi^2} \sum_i d_i \left(\frac{m_i}{T^2}\right) K_2(m_i/T) \cosh\{(B_i\mu_B + S_i\mu_S + Q_i\mu_q)/T\},$$

$$(11.11)$$

in the Boltzmann limit and for $\gamma_s = 1$; as above, the sum is over all physical resonances with their observed degeneracy d_i. All other thermodynamic variables are obtained as derivatives of this function with respect to T, μ_B, μ_S and μ_Q, where μ_B, μ_S and μ_Q denote the chemical potentials for baryon number, strangeness and charge. In the following, we restrict ourselves to only the baryon number dependence and define $\mu_B = \mu$. We want to show here that the agreement between the resulting hadronic resonance gas thermodynamics and statistical QCD should break down once the system gets close enough to the critical point in T and μ. There are several aspects to this issue.

First of all, we recall in Fig. 11.7 the phase diagram expected in QCD with $(2 + 1)$ quark flavors for vanishing ($m_q = 0$) and non-vanishing light quark mass ($m_Q \neq 0$). In particular, the continuous second order transition for $m_q = 0$ becomes a rapid cross-over only, which turns the tricritical endpoint P of the first order transition into a critical endpoint E. We nevertheless still hope to find reflections of critical behavior for sufficiently small light quark masses, which, as we shall see, is not so easy. It requires the study of higher order moments of the free energy density (11.11).

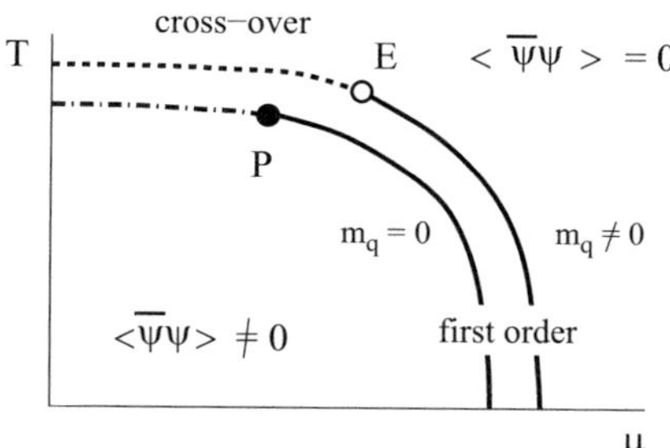

Fig. 11.7 Phase diagram for QCD with $m_q \neq 0$ compared to the case $m_q = 0$

As we had already noted at the end of Chap. 3, critical behavior is obtained in an ideal resonance gas only if the degeneracy grows exponentially in mass. Since the physical resonance gas is based on the measured species and their spin-isospin degeneracy, it will lead to thermodynamic observables which exist for all temperatures and, with the form (11.11), also for all chemical potential values. In other words, it does not provide critical behavior, even if the mass range is extended beyond the presently used range. This has lead to considering an exponential extension of the actual physical resonance spectrum [52]. However, the critical exponents for such a Hagedorn spectrum have been determined [53], and they do not agree with the critical exponents of the $O(4)$ universality class, proposed for the chiral limit of two-flavor QCD.

Neglecting strangeness and charge, we obtain from Eq. (11.11) for the moments

$$\chi_B^n = \frac{\partial^n (f(T,\mu)/T^4)}{\partial(\mu/T)^n} \tag{11.12}$$

a form which, keeping in mind that only baryons with $B = 1$ enter, leads to relations

$$\frac{\chi_B^{(3)}}{\chi_B^{(1)}} = \frac{\chi_B^{(4)}}{\chi_B^{(2)}} = \frac{\chi_B^{(5)}}{\chi_B^{(3)}} = \ldots = 1 \tag{11.13}$$

for ratios two units apart. For those separated by one unit, one has

$$\frac{\chi_B^{(2)}}{\chi_B^{(1)}} = \coth(\mu/T), \quad \frac{\chi_B^{(3)}}{\chi_B^{(2)}} = \tanh(\mu/T), \quad \frac{\chi_B^{(4)}}{\chi_B^{(3)}} = \coth(\mu/T), \tag{11.14}$$

and so on. If the confinement transition has led to a hadronic resonance gas in equilibrium, all memories of previous stages are lost, and so an agreement with relations (11.13) and (11.14) would indeed confirm the production of *hadronic matter* of a freeze-out temperature $T_f = T_H$ equal to the transition value.

Lattice studies have so far shown that such an ideal resonance gas behavior is also what QCD predicts below the deconfinement point. As an illustration, we show in the left box of Fig. 11.8 results for the second baryon number cumulant χ_2^B in $(2+1)$ flavor QCD, compared to the corresponding hadron resonance gas prediction. Up to the transition temperature, the agreement is excellent. Experimentally, the

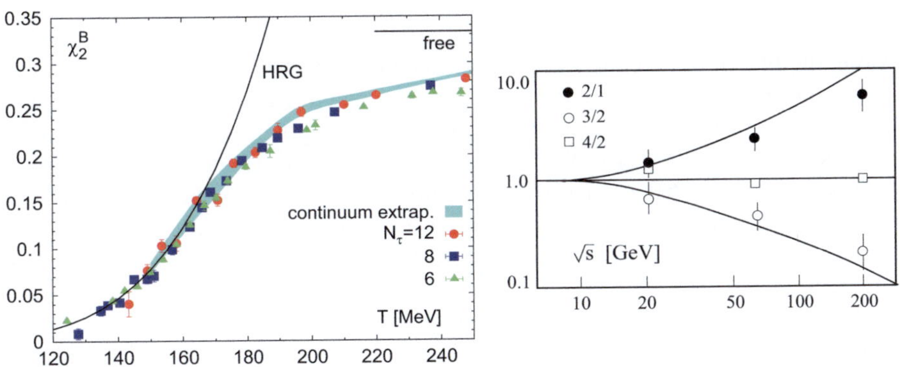

Fig. 11.8 Left: lattice results $(2+1$ quark flavors, $\mu = 0)$ for the second order baryon number cumulant, compared to the hadron resonance gas form. Right: STAR data [54] on baryon cumulant ratios, Eqs. (11.13) and (11.14), as function of collision energy $\sqrt{s}$, vs. hadron resonance gas predictions [55] (solid lines)

cumulants can be measured in terms of higher moments of the baryon number dispersion, $N_B - \langle N_B \rangle$, and such data has been obtained by the STAR collaboration at RHIC [54]. Species abundances determine T_f and $\mu(T_f)$ at collision energies of 19.6, 62.4 and 200 GeV. The baryon number fluctuations allow a determination of the cumulant ratios, and the results for three ratios are compared in the right box of Fig. 11.8 to the resonance gas predictions (11.13) and (11.14). The agreement is also seen to be quite good.

We thus find that high energy heavy ion collisions produce a medium which can be considered as hadronic matter in equilibrium, formed at the pseudo-critical hadronization temperature predicted by lattice QCD. And cumulant studies show that also correlations in this medium appear to follow the pattern of an ideal resonance gas, which for the lower orders is again in accord with QCD. On the other hand, it is evident that the hadronic resonance gas as conventionally formulated does not give any critical behavior in T, and if extended in the Hagedorn form, it gives a different universality class than chiral QCD. Moreover, by simply introducing the baryon number dependence through grand-canonical fugacities for each species, one cannot arrive at an interrelated $T - \mu$ singular functional form, as obtained in statistical QCD. In other words, the present hadronic resonance gas formulation does not lead to a $T - \mu$ phase boundary, and in the actual critical region, it can therefore not correctly describe chiral QCD. And as far as the data is concerned, we would like to see more: we want to find in the collision experiments some sign of the actual transition, of something like critical behavior. Sufficiently close to a continuous transition, correlations appear at all scales, the correlation length diverges, and in QCD this must produce strong deviations from the ideal hadron gas behavior just studied. What is "close enough", and what observables should one look at?

11.5   Critical Behavior: Fluctuations and Correlations

Along the line of the second-order transition and at the tricritical point, thermodynamic observables exhibit continuous critical behavior; this implies in particular that higher order derivatives of the thermodynamic potential diverge in a well-defined functional form specified in terms of the critical exponents of the relevant symmetry group. These derivatives, in turn, express calculable fluctuations of physical observables. For the idealized case considered here, the chiral limit of QCD, we thus obtain predictions for the fluctuations of baryon number, charge and strangeness.

There are two regions of critical behavior of particular interest for heavy ion studies. If the pseudocritical cross-over line for physical values of m_q and small μ is sufficiently close to that of the second order transition in the chiral limit, we may there expect remnants of O(4) criticality, and we can look for them at top RHIC energy and at the LHC. At larger μ and finite m_q, we expect to encounter critical behavior near the end point E of the first order regime, and the main aim of present beam energy scans at RHIC as well as that of the NA61 experiment at the CERN-

SPS is the search for this end point through its effect on the freeze-out line. Here the exponents are those of the Z_2 group. In both cases, near $\mu = 0$ and near the critical endpoint at finite μ, we are thus looking for *remnants of criticality* (see Fig. 11.9). The pseudo-critical line near $\mu = 0$ approaches the critical line only in the chiral limit $m_q \to 0$, and the freeze-out line at finite μ is off the line of first order transition and the critical endpoint by the presence of non-resonant interactions.

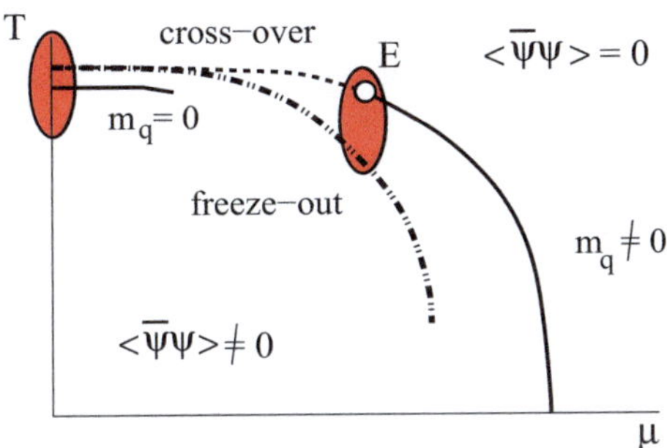

Fig. 11.9 Phase space regions proposed to look for remnant critical behavior

What can we calculate and what can we look for? Since the overall critical structure of QCD as function of T, μ and m_q is somewhat involved, let us first look at a simpler case for illustration, the Ising model of spins $s_i = \pm 1$ on a three-dimensional spatial lattice of N^3 sites, introduced in some detail in Chap. 2. The basic quantity to study is the density $f = F/N^3$ of the Gibbs free energy $F = -T \log Z(T, H)$. Consider the first four moments of f, obtained as derivatives with respect to T for $H = 0$. Both $f(T)$ and $\epsilon(T) = f^{(2)} \sim (\partial F/\partial T)$ are still continuous, but the higher derivatives diverge for $t = |T - T_c| = 0$. In particular, the specific heat has the form

$$c_v(T, H = 0) \sim \left(\frac{\partial \epsilon(T, H)}{\partial T} \right)_{H=0} \sim \left(\frac{\partial^2 f(T, H)}{\partial T^2} \right)_{H=0} \sim |t|^{-\alpha}, \qquad (11.15)$$

In Fig. 11.10, we illustrate the resulting patterns schematically. Included are here also the next two derivative orders, and in all cases, we also show the expected pattern if the symmetry is slightly broken by the presence of a small external field H. We note in particular the non-monotonic behavior of all derivatives higher than the energy density; this behavior signals criticality and clearly deviates from any resonance gas pattern.

The other variable H leads to a similar pattern. The first derivative with respect to H gives at $H = 0$ the spontaneous magnetisation,

$$m(t, H = 0) \sim \left(\frac{\partial f(T, H)}{\partial H} \right)_{H=0} \sim |t|^{\beta}, \qquad (11.16)$$

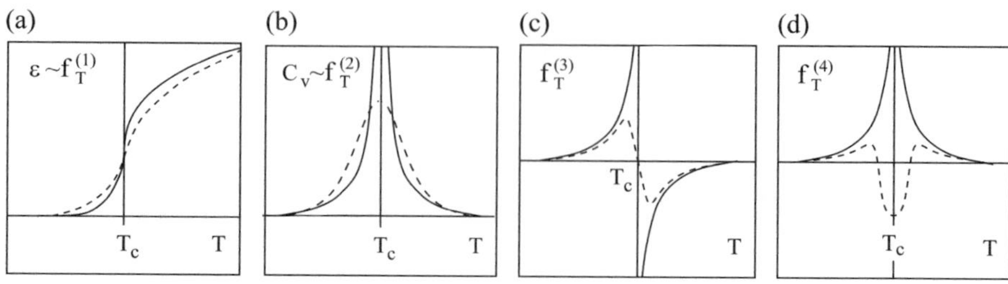

Fig. 11.10 3d Ising model response functions up to fourth order, for $H = 0$, solid line, and for small finite external field H, dashed line

for $t < 0$; for $t \geq 0$, $m(t, H = 0) = 0$: the magnetisation is the order parameter and thus vanishes above the critical point for $H = 0$. For finite H, it remains finite even there, vanishing as

$$m(t = 0, H) \sim \left(\frac{\partial f(T, H)}{\partial H} \right)_{t=0} \sim h^{1/\delta}, \tag{11.17}$$

along the critical isotherm, with $h = H/T_c$. The second derivative

$$\chi_T(t, h = 0) \sim \left(\frac{\partial m(t, h)}{\partial h} \right)_{h=0} \sim \left(\frac{\partial^2 f(t, h)}{\partial h^2} \right)_{h=0} \sim |t|^{-\gamma}, \tag{11.18}$$

gives the isothermal susceptibility, the rate at which the magnetisation vanishes at the Curie point; it also diverges there. In summary: critical behavior means that higher order derivatives of the free energy diverge in a functional form specified by critical exponents (we here had α, β, γ and δ) which are fixed once the symmetry group is given.

The singular behavior of the response functions can in turn be related to that of fluctuations and correlations. The specific heat determines the energy fluctuation over the lattice,

$$c_V(T) \sim \langle (\sum_{i,j} s_i s_j)^2 \rangle - \langle \sum_{i,j} s_i s_j \rangle^2 \tag{11.19}$$

while the susceptibility measures the fluctuation of the spin,

$$\chi_T(t, H = 0) \sim \langle (\sum_i s_i)^2 \rangle - \langle \sum_i s_i \rangle^2; \tag{11.20}$$

in the absence of spin-spin correlations, it vanishes. The divergence of the response functions at the critical point is thus connected to fluctuations diverging there, which in turn is a consequence of diverging correlations: at the critical point, the spin-spin correlation length ξ diverges as

$$\xi(t) \sim |t|^{-\nu}, \tag{11.21}$$

so that constituents of all scales become correlated (critical opalescence). Given sets of lattice configurations at various temperatures, we can thus, through the calculation of response functions, determine the onset and the analytical form of the critical behavior shown by the system. Using the form (11.21), one can also express the response functions in terms of the correlation length. For the 3d Ising model, the specific heat becomes

$$\chi_T \sim \xi^{\gamma/\nu} \sim \xi^2, \tag{11.22}$$

using $\gamma \simeq 1.2$, $\nu \simeq 0.6$ for the 3d Ising model; higher derivatives grow as higher powers of ξ.

In the case of QCD, we have in addition to the temperature the chemical potentials of the conserved quantum numbers as thermodynamic parameters, μ_B for the baryon number, μ_Q for the electric charge, and μ_S for the strangeness, respectively:

$$f_{\text{Ising}}(T,H) \ \rightarrow \ P_{\text{QCD}}(T,\mu_B,\mu_Q,\mu_S,m_q). \tag{11.23}$$

The one-dimensional temperature space T is thus generalized to a four-dimensional space T,μ_B,μ_Q,μ_S; variations of the chemical potentials do not affect the intrinsic symmetry of the system. The light quark mass m_q plays, as mentioned, the role of the external field: for $m_q \neq 0$, the chiral symmetry of the Lagrangian is broken explicitly. If we restrict ourselves again to the baryon number case, the critical point in T thus becomes a critical line in the $T - \mu$ plane, which renders the relation between variables and critical exponents somewhat more complex, see e.g. [56]. The principle remains the same, however: to find experimental evidence of criticality, we have to look for non-monotonic behavior of response functions measured in heavy ion collisions, to be compared to such behavior obtained in lattice studies. The finite interaction volume in actual collisions, together with time evolution effects, will limit the maximum size of correlations and thus mask a possible divergence. Since the higher derivatives depend on higher powers of ξ, they could provide a more sensitive tool.

The first task is thus to determine in lattice QCD some evidence for critical behavior in the hadronic state, behavior deviating from that of an ideal resonance gas. Here it has to be noted that the partition function in QCD depends on *squared* electric charges and baryochemical potential, since it is left invariant under a change of sign of these quantities. As a result, the non-monotonic behavior shown for the Ising model is shifted to higher order cumulants: the third order spin behavior is expected for the sixth order in QCD, and so on. So far, derivatives up to the second order were seen to agree with the non-critical resonance gas, see Fig. 11.8(left). First evidence for deviations has been found quite recently [41], studying the sixth order cumulant of the electric charge (χ_6^Q); the result is illustrated schematically in Fig. 11.11; it is the counterpart of the third moment in Fig. 11.10. We see that at the critical point, i.e., at the freeze-out value of the resonance gas, the sixth order cumulant vanishes for QCD, in contrast to the continued monotonic increase found for the hadronic resonance gas. Such differences are expected to continue for higher orders: the eighth cumulant should become negative in critical QCD, large and positive for the resonance gas. The onset of critical behavior, difficult if not impossible to detect in lower order cumulants, should thus become more and more evident with increasing order.

We should here emphasize that the forms shown in Fig. 11.11 are obtained in physical finite temperature lattice studies, i.e., for $2+1$ quark flavors of physical masses, and they give the behavior of the full cumulants, not just a singular part. They are therefore *bona fide* predictions: if these quantities become measurable in heavy ion collisions and there do not show the predicted form, the produced systems are not governed by equilibrium QCD thermodynamics. There can, of course, be various reasons for why this might be the case, but the result as such would be an observation decisive for our understanding of high energy nuclear collisions.

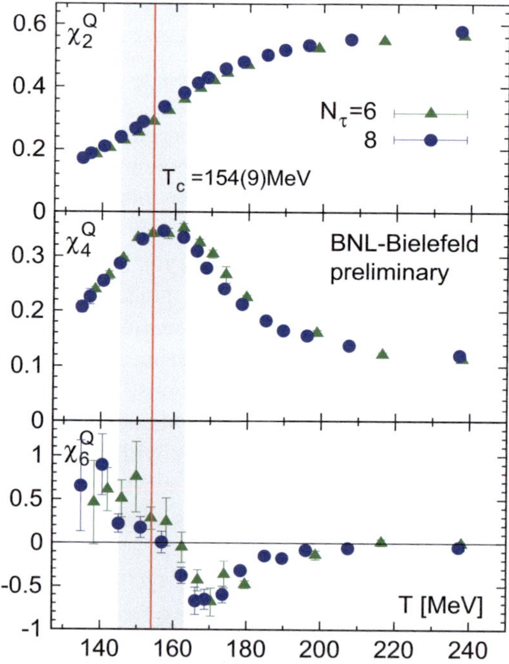

Fig. 11.11 Behavior of the second (χ_2^Q), fourth (χ_4^Q) and sixth (χ_6^Q) order cumulants of the electric charge in the critical region, from lattice QCD calculations [57]

The search for the critical endpoint will require a similar program, here focused on the baryon density, to be varied by varying the collision energy. The data obtained will again be analysed in terms of the ideal resonance gas, even though we now expect some non-resonant baryon-baryon interactions. And any onset of criticality should produce deviations from monotonic behavior under variations of the collision energy.

Predictions from lattice QCD here are not as readily obtained, since the conventional simulation scheme breaks down for finite μ; see the discussion in Sect. 7.5. Based on a Taylor expansion in μ, it was shown there that the baryon number susceptibility, which governs baryon number fluctuations, indicate an onset of non-monotonic variation around the critical temperature.

The remnant critical behavior of QCD thermodynamics thus is encoded in the higher cumulants of conserved quantum number distributions. Theoretically, these have been and are being studied in considerable detail, for baryon number, electric charge and strangeness. To what extent these are measurable in heavy ion collisions is another issue, beyond the scope of this report. The high statistics experiments at the LHC and the extended beam scan efforts at RHIC give rise to hope for precise enough data, e.g. of charge and baryon number distributions, to allow such studies.

11.6 Dynamical Effects

Finally, we return to the role of collision dynamics in the statistical hadronization picture. The basic statistical assumption had removed all such effects for the determination of species abundances, by invoking an equivalent global cluster. There are, however, cases where dynamics plays a crucial role, which has to be taken into account correctly before attempting a statistical description.

One case is given by e^+e^- annihilation. This process leads to the formation of an intermediate vector boson state (γ, Z_0), which then decays into a quark-antiquark pair (see Fig. 11.12); these primary quarks subsequently hadronize. The relative rates of the different initial quark states u, d, s, c, b are determined by electroweak theory as a function of the annihilation energy; at $\sqrt{s} = 91.25$ GeV, one obtains for the relative rates $R_{u+d} \simeq 40\%$, $R_s \simeq 20\%$, $R_c \simeq 20\%$, and $R_b \simeq 20\%$. This means in particular that 40% of the resulting events contain a charm or bottom quark pair, which is far more than would be statistically expected. In contrast to strange quark production, that of heavy flavors is thus enhanced in e^+e^- annihilation, and this enhancement has to be correctly included as a function of the annihilation energy. The relative rates of *open* heavy flavor hadrons can then be described by statistical considerations, and it turns out that this indeed works very well, with the same T_H as observed for light hadrons [25]. The fraction of primary charm and bottom quarks must, however, be fixed from electroweak dynamics. What thus appears to be universal is the hadronization of quarks, not their initial production rates. We return to the production of heavy flavor bound states (quarkonia) shortly.

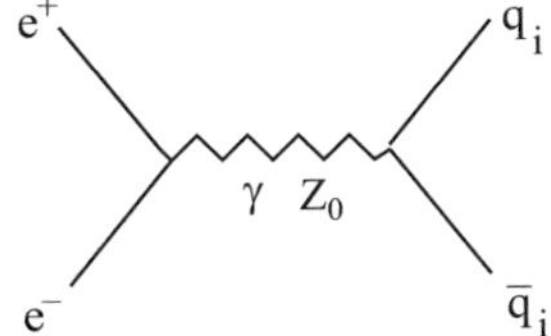

Fig. 11.12 $q\bar{q}$ production through e^+e^- annihilation, with $q_i \sim u, d, s, c, b$

Heavy flavor production in *pp* and nucleus-nucleus collisions shows a similar effect. It is a hard partonic interaction process (see Fig. 10.9 for charm production), for which the rate can be calculated in perturbative QCD; it grows power-like with energy and thus more rapidly than the logarithmic multiplicity dependence found for usual hadrons. In contrast to the relative abundances of light hadrons, which become energy independent for large $\sqrt{s}$, the relative rates of heavy to light flavor hadrons continue to grow with energy. To cite an example: the total $c\bar{c}$ cross-section grows from $\sqrt{s} = 20$ to $40\,\mathrm{GeV}$ by a factor 10, while the light hadron multiplicity increases by only 20%. Hadronic heavy flavor production thus also originates from a dynamical process which cannot be accounted for in any statistical scenario. Here again, however, we can proceed as above: the overall heavy flavor production is calculated in perturbative QCD, but given this rate, the relative channels into which it goes is described by the statistical model. The resulting predictions for the different open charm or bottom states also here agree very well with the measured rates, with the same universal T_H. This confirms the conclusion reached above: once the quarks are produced, their subsequent hadronization is universal, with one and the same temperature for all.

The final point in this section is the only exception to universal hadronization, the production of quarkonia. In elementary hadron collisions, such as in *pp*, gluon fusion leads to the perturbative production of a heavy quark pair; consider charm as example (see Fig. 10.9). In most cases, the c and the $\bar{c}$ separate and eventually hadronize; the resulting hadron species show the mentioned statistical relative abundances. In a small fraction of such events, however, the almost pointlike $c\bar{c}$ pair forms a bound charmonium state, such as J/ψ, ψ' or χ_c. Of the total open charm production in *pp* collisions, about 1% result in J/ψ formation; this holds true at all collision energies. Although this fraction is small, it is still considerably larger (by almost an order of magnitude) than what one would expect from a statistical hadronization of the available charm quarks. In e^+e^- annihilation, the situation appears similar. Here the bulk of charmonium production is due to the decay of open bottom mesons arising from the primary $b\bar{b}$ pair. All estimates of the remaining "prompt" production lead to rates far above statistical predictions [58].

On the other hand, quarkonium binding can be described very well by potential models [59, 60] and by non-relativistic QCD applied to heavy quarks [61, 62]. In particular, these studies predict that the relative rates of S-wave states, e.g., of J/ψ and ψ', are given by the ratio of their wave-functions at the origin [63],

$$\frac{\langle N(\psi')\rangle}{\langle N(J/\psi)\rangle} \simeq \frac{|\phi_{\psi'}(0)|^2}{|\phi_{J/\psi}(0)|^2} \simeq 0.25. \tag{11.24}$$

This is in good agreement with all corresponding *pp* production data over a wide range of energies [64]. In contrast again, statistical hadronization, Eq. (11.5), predicts a ratio which is a factor 5 lower.

Quarkonium binding thus seems to be per se a dynamical process, described by the very early and very short-range interaction between two heavy quarks. In elementary interactions, such as *pp* or e^+e^-, it appears as unrelated to the much later and longer range hadronization process, which leads to the statistical species abundances. This conclusion remains valid also if one allows the overall fraction of heavy quarks to be dynamically determined. Although the relative open charm/bottom states then are indeed statistically distributed, based on the universal hadronization temperature, the abundances of quarkonium states do not follow any statistical pattern.

An interesting proposal suggests that this situation changes in high energy nucleus-nucleus collisions [65–67]. As discussed in Chap. 9, the formation of a hot quark-gluon plasma in such interactions is expected to melt charmonia [68]. But once these primary charmonia have disappeared, one may ask if at hadronization there could not be a new, secondary statistical charmonium formation process, in which randomly meeting c and $\bar{c}$ quarks combine to a charmonium state. At lower energies, the charm production rate is not large enough to allow measurable charmonium rates through such statistical hadronization; in other words, charmonium formation through the "statistical" combination of a c from one primary nucleon-nucleon collision with a $\bar{c}$ from another is negligible. Due to the rapid growth of $c\bar{c}$ the production cross section with energy, the chance of a given charm quark to "meet" a charm antiquark from "different parents" increases, and for nuclear collisions at LHC energies, one would expect considerably more charm quarks than the statistically predicted fraction. The basis of this effect is that the hard charm production rate grows with the number $\sim N^2$ of nucleon-nucleon collisions, while that of light hadrons is essentially determined by the number $\sim 2N$ of participant nucleons; here N denotes the number of nucleons in each of the colliding nuclei. If this initial charm excess survives the subsequent evolution of the produced medium, the enhancement will cause the ratio between open charm hadrons and light hadrons grow from below to above the statistical value, as illustrated schematically in Fig. 11.13 for the ratio of D^+ to π^+ meson production.

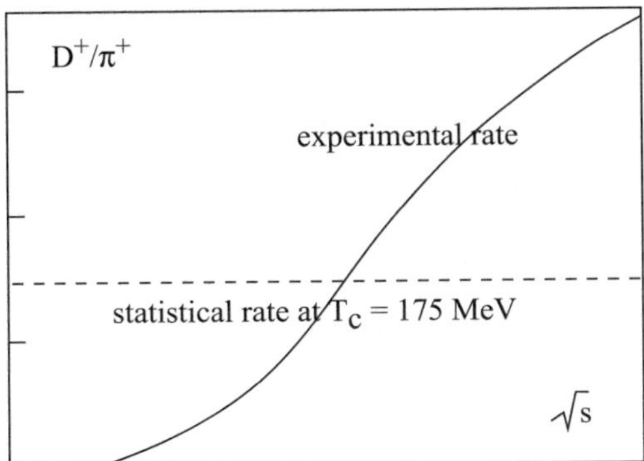

Fig. 11.13 The ratio of D^+ to π^+ in hadronic collisions, as function of the collision energy

At hadronization, the medium will then be over-saturated with charm quarks. If the only condition for charmonium formation is the fractional abundance of charm and anti-charm quarks relative to light quarks, simple quark combination will lead to more charmonium formation than present in a proton-proton collision. In other words, if for proton-proton collisions one finds a ratio $R_0(J/\psi/c\bar{c})$ for hidden to open charm production, then in nucleus-nucleus collisions, one should observe a ratio $R_A(J/\psi/c\bar{c}) \gg N_c R_0(J/\psi/c\bar{c})$, where N_c scales the pp value by the number of collisions in $A - A$. This enhancement, which has been estimated to be up to a factor 10, is due to charmonium formation by constituents from different nucleon-nucleon collisions.

If, on the other hand, charmonium formation retains the dynamical features observed in pp interactions, such as extremely short-distance binding only under suitable kinematic constraints, then statistical charmonium formation seems unlikely in the hadronizing medium, in which constituents are separated typically by about one fermi and move randomly relative to each other. The charmonium production experiments in nucleus-nucleus collisions at the LHC will hopefully clarify this point.

11.7   Conclusions

The success of the statistical hadronization model in reproducing the species abundances and in determining a universal hadronization temperature is undoubtedly one of the milestones in high energy multihadron production. On the other hand, it has also brought up an equally profound question. While nuclear collisions, with many superimposed nucleon-nucleon collisions, could perhaps be considered as a truly statistical system, e^+e^- interactions with as little as 2 or 3 hadrons per unit rapidity nevertheless fulfill the predictions just as well. How and why do these states reach something like "thermal equilibrium"? And what determines the universal temperature? Is there some common origin of all statistical multihadron production, which applies to e^+e^- as well as to hadronic and nuclear collisions? Heisenberg had noted that the many secondaries are created in "one bang", and Hagedorn thought that they do not somehow attain equilibrium, but that instead "they must be born in equilibrium". A first, rather vague answer to this basic question was given by the idea that any high energy collision in strong interaction physics leads to color charges traversing the vacuum, which thereby is disturbed and recovers from this such as to maximize entropy. The specific nature of the passing charges is irrelevant in this – it is the disturbed vacuum which leads to the statistical emission of hadrons. But this does not tell us what the emission temperature is. We shall return to a more quantitative version of such a scenario in the next chapter.

Appendix 1: Scattering Matrix and Phase Space

The probability for the production of N particles in the collision of two incident hadrons of four-momenta q_1 and q_2 is determined by

$$P_N(W) \sim \int \frac{d^3 p_1}{2p_{10}} \cdots \frac{d^3 p_N}{2p_{N0}} \delta^{(4)}\left(\sum p_i - q_1 - q_2\right) |\langle p_1 \ldots p_N | S | q_1 q_2 \rangle|^2, \quad (11.25)$$

where $W^2 = (q_1 + q_2)^2$ is the (squared) center-of-mass collision energy and $p_1, \ldots, p_N$ are the four-momenta of the final hadrons. For simplicity, we assume here for the time being only one species of identical scalar hadrons of mass m_0 and neglect any effects of quantum statistics. The unitary scattering matrix S maps the initial state onto the final state. The Fermi model [69] is obtained by assuming that the squared S-matrix element is proportional to the probability of finding the N final hadrons as free particles inside a spatial volume V,

$$|\langle p_1 \ldots p_N|S|q_1q_2\rangle|^2 \sim \int_V d^3x_1 \ldots d^3x_N \prod_1^N |\Phi_{\mathbf{p}_i}(\mathbf{x}_i)|^2. \tag{11.26}$$

With plane wave states

$$\Phi_{\mathbf{p}_i}(\mathbf{x}_i) = \sqrt{\frac{2p_{i0}}{(2\pi)^3}} \exp\{i\mathbf{p}_i\mathbf{x}_i\} \tag{11.27}$$

for the final hadrons we obtain the standard form for the relativistic N-particle phase space volume,

$$Q_N(W, V) = \left[\frac{V_0}{(2\pi)^2}\right]^N \int d^3p_1 \ldots d^3p_N \delta^{(3)}\left(\sum \mathbf{p}_i - \mathbf{P}\right)\delta\left(\sum p_{i0} - E\right), \tag{11.28}$$

with $W^2 = E^2 - \mathbf{P}^2$ for the invariant center of mass energy of the system. The delta functions project the N-particle state onto fixed total energy E and momentum P. The production probability can then be expressed in terms of the phase space volume $Q_N(W)$ as

$$P_N(W, V) = \frac{Q_N(W, V)}{Q(W, V)}, \tag{11.29}$$

where the normalization $Q(W)$ is given by the sum over all states at fixed energy and volume

$$Q(W, V) = \sum \frac{1}{N!} Q_N(W, V), \tag{11.30}$$

assuming, as mentioned, that all particles are identical.

The Laplace transforms of $Q_N(W)$ and $Q(W)$ are just the canonical and grand canonical partition functions of an ideal gas in the Boltzmann limit,

$$\mathcal{Z}_N(T, V) = \int d^3P dW\, Q_N(W, V) \exp\{-W/T\} \simeq [Vz(T)]^N \tag{11.31}$$

and

$$\mathcal{Z}(T, V) = \int d^3P dW\, Q(W, V) \exp\{-W/T\} = \sum_N \frac{1}{N!} \mathcal{Z}_N(T, V) \simeq \exp\{Vz(T)\}, \tag{11.32}$$

with

$$z(T) = \frac{m^2 T}{2\pi^2} K_2(m/T); \tag{11.33}$$

here $K_2(x)$ is the Hankel function of purely imaginary argument. For small x, $K_2(x) \simeq (2/x^2)$, so that for high temperatures

$$\ln \mathcal{Z}(T, V) \simeq \frac{V T^3}{\pi^2}. \tag{11.34}$$

This can be used to invert the Laplace transform (11.32), leading to [70]

$$\ln Q(W, V) \simeq \left[\frac{V W^3}{27\pi^2} \right]^{1/4}. \tag{11.35}$$

Hence the average multiplicity

$$\langle N \rangle = V \left(\frac{\partial \ln Q}{\partial V} \right)_W \simeq \frac{1}{4} \left[\frac{V W^3}{27\pi^2} \right]^{1/4} \tag{11.36}$$

increases as $W^{3/4}$, so that the average energy per secondary grows as $W^{1/4}$.

Appendix 2: Exact Charge Conservation

The phase space integral (11.28) specifies the totality of allowed states consisting of a fixed number N of constituents, subject to the conservation of overall energy and momentum and contained in a volume V. The corresponding partition function (11.30) then provides the grand canonical sum over all N-body states, with the temperature T determining the average overall energy; the effect of momentum conservation actually becomes asymptotically negligible [71]. Quantum-mechanically, we write the grand canonical partition function as

$$Z(T, V) = \mathrm{Tr}\left[\exp\{-\mathcal{H}/T\}\right] = \sum_N \mathrm{Tr}_N \left[\exp\{-\mathcal{H}/T\}\right], \tag{11.37}$$

where the trace runs over all possible N-particle states and $\mathcal{H}$ denotes the Hamiltonian. In the grand canonical framework, the conservation of a discrete charge Q (for simplicity, we restrict ourselves to only one additive conserved charge) is taken into account through fugacities $\exp\{-\mu \mathcal{Q}\}$, with $\mathcal{Q}$ for the charge operator and μ for the corresponding chemical potential,

$$Z(T, V, \mu) = \mathrm{Tr}\left[\exp\{-(\mathcal{H} - \mu\mathcal{Q})/T\}\right]. \tag{11.38}$$

For illustration, consider a system containing particles of charge zero and mass m_0, together with particles of charge ± 1 and a common mass m_1. The grand canonical partition function then becomes

$$\ln Z(T, V, \mu) = V\left[z_0 + z_1(e^{\mu/T} + e^{-\mu/T})\right] = V\left[z_0 + 2z_1 \cosh(\mu/T)\right],$$
(11.39)

and the average charge density $q = \langle Q \rangle / V$ is given by

$$\langle q \rangle = T\frac{\partial \ln Z(T, V, \mu)}{\partial \mu} = 2z_1 \sinh(\mu/T),$$
(11.40)

with

$$z_i = d_i \frac{m_i^2 T}{2\pi^2} K_2(m_i/T), \ \ i = 0, 1.$$
(11.41)

for the weight factor of a single particle of mass m_i and degeneracy d_i, in the Boltzmann limit. For a system of vanishing average charge density, Eq. (11.40) thus implies $\mu = 0$, and from Eq. (11.39) we see that the system contains an equal number of positive and negative particles, in addition to the neutral ones. We thus have

$$\ln Z(T, V, \mu = 0) = V\left[z_0 + 2z_1\right]$$
(11.42)

for the corresponding grand canonical partition function.

Such a grand canonical formulation provides a correct description of the system only if it contains sufficiently many charged particles, so that fluctuations can be neglected. If there are only very few such particles, charge conservation has to be implemented exactly, not on the average. One thus has to "project" out the section of phase space associated to a fixed quantum number. We define the partition function at fixed Q as

$$Z_Q(T, V) = \mathrm{Tr}_Q\left[\exp\{-(\mathcal{H}/T)\}\right] = \mathrm{Tr}\left[\exp\{-(\mathcal{H}/T)\}\mathcal{P}_Q\right],$$
(11.43)

where $\mathcal{P}_Q$ is the corresponding projection operator onto a fixed charge Q [72–75]. The partition function $Z_Q(T, V)$ is thus a sum over all possible $N = 2, 3, \ldots$ particle clusters of fixed total charge Q. Using it, the grand canonical partition function (11.30) can be "re-organized", i.e., rewritten as a sum over Q instead of as sum over N,

$$Z(T, V, \mu) = \sum_{Q=-\infty}^{Q=+\infty} Z_Q(T, V)\lambda^Q,$$
(11.44)

where

$$\lambda = \exp\{\mu/T\}$$
(11.45)

is the corresponding fugacity. Each term in the sum (11.44) is grand-canonical in terms of particle number, but canonical in charge. The series (11.44) can be inverted to obtain the partition function $Z_Q(T, V)$ at fixed charge Q, using the Cauchy formula. The result is

$$Z_Q(T, V) = \frac{1}{2\pi} \int_{-\pi}^{-\pi} d\phi\, e^{-iQ\phi} \tilde{Z}(T, V, \phi), \tag{11.46}$$

with the function $\tilde{Z}(T, V, \phi)$ obtained from the grand canonical partition function $Z(T, V, \mu)$ by a Wick rotation $\mu_s/T \to i\phi$, so that $\lambda \to \exp\{i\phi\}$.

For our system containing charges $0, \pm 1$, we thus obtain

$$Z_Q(T, V) = \frac{1}{2\pi} \int_{-\pi}^{\pi} d\phi\, e^{-iQ\phi} e^{V[z_0 + 2z_1 \cos\phi]}, \tag{11.47}$$

so that the "charge-canonical" partition function for exactly vanishing overall charge $Q = 0$ becomes

$$Z_0(T, V) = \frac{1}{2\pi} e^{Vz_0} \int_{-\pi}^{\pi} d\phi\, e^{2Vz_1 \cos\phi} = \frac{1}{\pi} e^{Vz_0} I_0(2Vz_1), \tag{11.48}$$

where $I_0(x)$ is the Bessel function of purely imaginary argument. From this we get

$$\ln Z_0(T, V) = V\left[z_0 + \frac{1}{V} \ln I_0(2Vz_1) \right]. \tag{11.49}$$

Comparing Eq. (11.45) to the grand canonical form (11.39), we see that the exact conservation of charge leads to the replacement

$$z_1 \;\to\; \frac{1}{V} \ln I_0(2Vz_1). \tag{11.50}$$

Since for large argument

$$I_0(x) = \frac{e^x}{\sqrt{2\pi x}} \left[1 + O(1/x)\right], \tag{11.51}$$

the two forms become identical in the large volume limit, as expected. For small V, corresponding to small particle numbers, the contribution of the charged particles is suppressed relative to that of the neutral particles, if the overall charge is fixed to zero. It can be shown, in fact, that the canonical (c) and grand canonical (gc) densities n_Q of particles of charge Q are related by [75]

$$n_Q^c \simeq n_Q^{gc} \left\{ \frac{I_Q(2Vz_Q)}{I_0(2Vz_Q)} \right\}. \tag{11.52}$$

This relation is approximate, since it neglects the possibility of multiply-charged states; the role of these is addressed in [75]. Since $I_1(x)/I_0(x) \to 1$ for $x \to \infty$, exact charge conservation leads to an effective charge suppression by a factor approaching unity in the large volume limit. Since the volume V is determined by the overall multiplicity, which in turn grows with the collision energy, the suppression of the form (11.52) disappears with increasing $\sqrt{s}$.

The formalism just sketched for the case of an Abelian (additive) charge $0, \pm 1$ can be extended to several Abelian as well as non-Abelian charges [72–75]; for details, we refer to the cited works. We note here only that while exact baryon number conservation does lead to the correct non-strange abundances in elementary hadron-hadron collisions and in e^+e^- annihilation, the strangeness reduction obtained through exact strangeness conservation is not sufficient to account for the reduced strange particle production observed in these reactions. This requires an additional suppression, as given, e.g., by the suppression factor γ_s applied per strange quark in the hadron in question, as discussed above.

References

1. W. Heisenberg, Z. Phys. **113**, 61 (1939)
2. E. Fermi, Progr. Theor. Phys. (Japan) **5**, 570 (1950)
3. P.P. Srivastava, G. Sudarshan, Phys. Rev. **110**, 765 (1958)
4. H. Satz, Il Nuovo Cim. **37**, 1407 (1965)
5. L.D. Landau, Izv. Akad. Nauk SSSr **17**, 51 (1953)
6. V.N. Gribov, Sov. Phys. JETP **29**, 483 (1969)
7. R. Hagedorn, Nuovo Cim. Suppl. **3**, 147 (1965)
8. R. Hagedorn, Nuovo Cim. A **56**, 1027 (1968)
9. W.-M. Yao et al., Rev. Part. Phys. J. Phys. G **33**, 1 (2006)
10. E. Beth, G.E. Uhlenbeck, Physica **4**, 915 (1937)
11. R. Dashen, S.-K. Ma, H.J. Bernstein, Phys. Rev. **187**, 345 (1969)
12. J.D. Bjorken, Lect. Notes Phys. (Springer) **56**, 93 (1976)
13. F. Becattini, G. Passaleva, Eur. Phys. J. C **23**, 551 (2002)
14. See also e.g., P. Braun-Munzinger, K. Redlich, J. Stachel, in *Quark-Gluon Plasma 3*, ed. by R.C. Hwa, X.-N. Wang (World Scientific, Singapore, 2003)
15. F. Becattini, R. Fries, arXiv:0907.1031 [nucl-th], and Landolt-Boernstein 1–23
16. F. Cerulus, Nuovo Cim. **19**, 528 (1961)
17. K. Redlich, L. Turko, Z. Phys. C **5**, 201 (1980)
18. For a general summary, see e.g., K. Redlich, F. Karsch, A. Tounsi
19. F. Becattini, U. Heinz, Z. Phys. C **76**, 268 (1997)
20. J. Cleymans, M. Marais, E. Suhonen, Phys. Rev. C **56**, 2747 (1997)
21. J. Letessier, J. Rafelski, A. Tounsi, Phys. Rev. C **64**, 406 (1994)
22. J.S. Hamieh, K. Redlich, A. Tounsi, Phys. Lett. B **486**, 61 (2000)
23. R. Hagedorn, Thermodynamics of Strong Interactions, CERN Yellow Report 71–12, 1971
24. F. Becattini, Z. Phys. C **69**, 485 (1996)
25. F. Becattini et al., Eur. Phys. J. C **56**, 493 (2008)
26. A. Andronic et al., Phys. Lett. B **675**, 312 (2009)
27. F. Becattini et al., Eur. Phys. J. C **66**, 377 (2010)
28. J. Cleymans et al., Phys. Lett. B **242**, 111 (1990)
29. J. Cleymans, H. Satz, Z. Phys. C **57**, 135 (1993)
30. K. Redlich et al., Nucl. Phys. A **566**, 391 (1994)

31. P. Braun-Munzinger et al., Phys. Lett. B **344**, 43 (1995)
32. F. Becattini, M. Gazdzicki, J. Sollfrank, Eur. Phys. J. C **5**, 143 (1998)
33. F. Becattini et al., Phys. Rev. C **64**, 024901 (2001)
34. B.I. Abelev et al. [STAR Collaboration], Phys. Rev. C **79**, 034909 (2009)
35. B.I. Abelev et al. [STAR Collaboration], Phys. Rev. C **75**, 064901 (2007)
36. J. Adams et al. [STAR Collaboration], Phys. Rev. Lett. **92**, 092301 (2004)
37. B.I. Abelev et al. [STAR Collaboration], Phys. Rev. C **79**, 034909 (2009)
38. J. Adams et al. [STAR Collaboration], Phys. Lett. B **612**, 181–189 (2005)
39. J. Adams et al. [STAR Collaboration], Phys. Rev. Lett. **98**, 062301 (2007)
40. B.I. Abelev et al. [STAR Collaboration], Phys. Rev. Lett. **97**, 132301 (2006)
41. P. Castorina, H. Satz, Eur. Phys. J. CA **52**, 200 (2016)
42. The γ_s values were obtained in the following data analyses, where also the original experimental references are given: F. Becattini, J. Manninen, M. Gazdzicki, Phys. Rev. C **73**, 044905 (2006)
43. F. Becattini et al., Eur. Phys. J. CC **66**, 377 (2010)
44. F. Becattini, J. Manninen, Phys. Rev. C **78**, 054901 (2008)
45. F. Becattini et al., Eur. Phys. J. CC **66**, 377 (2010)
46. M. Floris et al., Nucl. Phys. A **931**, 103 (2014)
47. F. Becattini, hep-ph/9701275
48. F. Becattini et al., Phys. Rev. Lett. **111**, 082302 (2013)
49. P. Castorina, S. Plumari, H. Satz, Int. J. Mod. Phys. E **26**, 1750081 (2017)
50. K. Aamodt et al. (ALICE Coll.), Phys. Rev. Lett. **105**, 252301 (2010)
51. A. Bazazov et al. (HotQCD), Phys. Rev. **D90**, 094503 (2014)
52. A. Majumdar, B. Müller, Phys. Rev. Lett. **105**, 252002 (2010)
53. H. Satz, Phys. Rev. D **19**, 1912 (1979)
54. M.M. Aggarwal et al. (STAR Collaboration), Phys. Rev. Lett. **105**, 22302 (2010)
55. F. Karsch, K. Redlich, Phys. Lett. B **695**, 136 (2011)
56. O. Kaczmarek et al., Phys. Rev. D **83**, 014504 (2011)
57. Christian Schmidt, Nucl. Phys. A904–905 **2013**, 865c (2013)
58. A. Andronic et al., Phys. Lett. B **678**, 350 (2009)
59. S. Jacobs, M.G. Olsson, C. Suchyta, Phys. Rev. **33**, 3338 (1986)
60. See e.g., L. Kluberg, H. Satz, arXiv:0901.3831 [hep-ph] and Landolt-Boernstein
61. See e.g., N. Brambilla et al., Heavy Quarkonium Physics, CERN Yellow Report CERN-2005-005
62. See e.g., N. Brambilla et al., Phys. Rev. D **78**, 014017 (2008)
63. R. Baier, R. Rückl, ZP C **19**, 251 (1983)
64. R.V. Gavai et al., Int. J. Mod. Phys. A **10**, 3043 (1995)
65. P. Braun-Munzinger, J. Stachel, Nucl. Phys. A **690**, 119 (2001)
66. R.L. Thews et al., Phys. Rev. C **63**, 054905 (2001)
67. L. Grandchamp, R. Rapp, Nucl. Phys. A **709**, 415 (2002)
68. T. Matsui, H. Satz, Phys. Lett. B **178**, 416 (1986)
69. E. Fermi, Progr. Theor. Phys. (Japan) **5**, 570 (1950)
70. H. Satz, Il Nuovo Cim. **37**, 1407 (1965)
71. R. Hagedorn, Thermodynamics of Strong Interactions, CERN Yellow Report 71–12, 1971
72. F. Cerulus, Nuovo Cim. **19**, 528 (1961)
73. K. Redlich, L. Turko, Z. Phys. C **5**, 201 (1980)
74. K. Redlich, F. Karsch, A. Tounsi, hep-ph/0302245; published in Physical and Mathematical Aspects of Symmetries, Paris (2002)
75. P. Braun-Munzinger, K. Redlich, J. Stachel, in *Quark-Gluon Plasma 3*, ed. by R.C. Hwa, X.-N. Wang (World Scientific, Singapore, 2003)

Chapter 12
The Event Horizon of Confinement

God does play dice, but He sometimes throws them   where they can't be seen.

Stephen Hawking

Abstract Color confinement constitutes for quarks and gluons something like an event horizon which they can never cross. Signals transmitted to the outside world from inside such a horizon cannot contain information and must thus be of thermal nature. In this chapter, we consider multihadron production in high energy collisions as the QCD counterpart of Hawking-Unruh radiation, encountered in black holes and for accelerated observers. This is shown to provide a common, "non-kinetic" origin for thermal multihadron production.

12.1   Black Holes and Event Horizons

The one feature which makes quantum chromodynamics basically different from all other atomistic theories is that its fundamental constituents, quarks and gluons, are colored, and by color confinement they are not allowed to exist as individual entities in the world we can observe. It is not that we cannot register their effects; but a single quark or gluon can never be observed as an isolated object, in contrast to a single proton or electron, for example.

When color confinement was first proposed, it seemed natural to recall the only other case where things remain in principle beyond our reach: black holes. A black hole is the final stage of a neutron star after gravitational collapse [1]. It has a mass M concentrated in such a small volume that the resulting gravitational field confines all matter and even photons to remain inside the "event horizon" R of the system: no causal connection with the outside world is possible.

Could it be that a hadron, containing colored constituents that cannot get out, is something like a black hole of strong interaction physics? In general relativity, forces are assumed to modify the underlying space-time manifold. The space-time metric of this manifold is given by

© Springer International Publishing AG 2018
H. Satz, *Extreme States of Matter in Strong Interaction Physics*,
Lecture Notes in Physics 945, https://doi.org/10.1007/978-3-319-71894-1_12

$$ds^2 = g_0 dt^2 - g_0^{-1} dr^2 - d^2\Omega, \tag{12.1}$$

with r and Ω specifying the spatial part, and t the time; for flat space, we have $g_0 = 1$. The event horizon of a (spherical) black hole is determined by the point at which this metric is so deformed that space and time interchange, i.e., the point at which $g_0 = 0$. For gravitation, the Einstein equations give

$$g_0 = \left(1 - \frac{2GM}{r}\right), \tag{12.2}$$

which leads to the Schwarzschild radius of a black hole,

$$R = 2GM, \tag{12.3}$$

where $G \simeq 6.7 \times 10^{-39}\,\mathrm{GeV}^{-2}$ is the gravitational constant and M the mass of the system.

It is therefore instructive to consider the Schwarzschild radius of a typical hadron, assuming a mass $m \sim 1\,\mathrm{GeV}$,

$$R_g^{\mathrm{had}} \simeq 1.3 \times 10^{-38}\,\mathrm{GeV}^{-1} \simeq 2.7 \times 10^{-39}\,\mathrm{fm}. \tag{12.4}$$

To become a gravitational black hole, the mass of the hadron would thus have to be compressed into a volume more than 10^{100} times smaller than its actual volume (with a radius of about 1 fm). On the other hand, if instead we increase the interaction strength from gravitation to strong interaction [2], we gain in the resulting "strong" Schwarzschild radius R_s^{had} a factor α_s/Gm^2, where α_s is the dimensionless strong coupling and Gm^2 the corresponding dimensionless gravitational coupling for the case in question. This leads to

$$R_s^{\mathrm{had}} \simeq \frac{2\alpha_s}{m} \tag{12.5}$$

which for the limiting value of the strong coupling [3], $\alpha_s \simeq 3$, gives $R_s^{\mathrm{had}} = 1.2\,\mathrm{fm}$. In other words, the confinement radius of a hadron is about the size of its "strong" Schwarzschild radius, so that we could picture quark confinement as the strong interaction version of the gravitational confinement in black holes [2, 4]. In this chapter, we want to show that the analogy between gravitational and color confinement can in fact lead to other interesting consequences.

At this point it is of interest to note that the event horizon can be specified also for more general black holes than the Schwarzschild type just considered. The only properties black holes are allowed to have, besides mass, are charge and spin. Both tend to reduce the event radius obtained through gravity alone. An overall charge results in Coulomb repulsion, an overall spin in centrifugal force. Both counteract the attraction by gravity and thus lower the event horizon. In particular, one finds [1]

$$R_{\mathrm{RN}} = GM(1 + \sqrt{1 - (Q^2/GM^2)}) \tag{12.6}$$

for the so-called Reissner-Nordström black hole having an overall charge Q. It is seen that for $Q^2 = GM^2$, the radius of the event horizon is reduced by a factor two, compared to an uncharged black hole.

To prove that QCD leads to a color event horizon implies proving color confinement and hence is a million dollar project [5]. In the case of gravitation, the solution of the Einstein equations provides the metric coefficient g_0 (Eq. 12.2). Another case is electrodynamics in a non-linear background field, where the motion of photons can be restricted to a compact region of space [6]. There it is shown that an effective Lagrangian $\mathcal{L}(F)$, depending on a background field $F = F_{\mu\nu}F^{\mu\nu}$, leads to a metric

$$g_0 = \mathcal{L}' - 4F\mathcal{L}'', \tag{12.7}$$

where the primes indicate the first and second derivatives of $\mathcal{L}$ with respect to F. Thus $g_0 = 0$ here defines the compact spatial region of the theory, i.e., the counterpart of a black hole. QCD is an inherently non-linear theory, with the physical vacuum playing the role of the medium [7]. Here one can also formulate an effective Lagrangian depending on a Yang–Mills background field [8–11],

$$\mathcal{L}_{QCD}(F) = \frac{1}{4}\epsilon(F)F_{\mu\nu}F^{\mu\nu}, \tag{12.8}$$

where $\epsilon(F)$ denotes the dielectric parameter of the medium; it is the F-dependence of $\epsilon(F)$ which makes the vacuum non-linear. The problem is that the solution of the metric relation $g_0(F) = 0$, with g_0 given by Eq. (12.7), requires a non-perturbative calculation of $\epsilon(F)$. So far, only a perturbative solution exists, which does suggest an event horizon [12, 13].

12.2   Accelerated Frames and Unruh Radiation

Starting from the basic concept of a black hole, studies in the course of the past decades have led to a unification of a number of seemingly different phenomena. First, Hawking had shown that black holes do send signals to the outside world: quantum fluctuations just outside the Schwarzschild surface (or, equivalently, tunnelling through that surface from inside the black hole) result in the emission of thermal radiation [14]. Unruh then found that, more generally, an observer in a reference frame undergoing constant acceleration would see the physical vacuum as a thermal medium of a temperature determined by the acceleration [15]. The equivalence principle between gravity and accelerated frames made the Hawking radiation a special case of that obtained by Unruh. More recently, it was furthermore noted that the Schwinger mechanism [16] for particle production in a strong electric field is in fact also very closely related to Unruh radiation [17–21]. It therefore seems quite possible that "thermal" multiparticle production in strong interaction

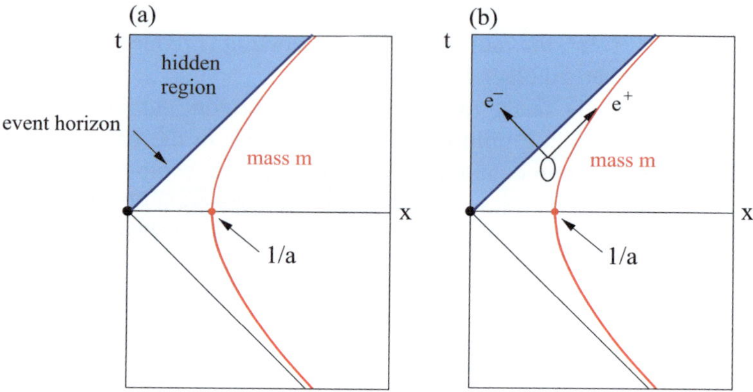

Fig. 12.1 Hyperbolic motion of a mass m in uniform acceleration a

physics is yet another instance of such radiation, with the acceleration and hence the temperature determined by the string tension confining color charges [13]. Let us look at these different aspects in more detail, beginning with the idea of Unruh radiation.

Assume that we have a mass m travelling through vacuum with a uniform acceleration a. The classical equation of motion then is

$$\frac{d}{dt} \frac{mv}{\sqrt{1-v^2}} = F \tag{12.9}$$

where $v = dx/dt$ is the velocity, $F = ma$ the force acting on m, and we have taken $c = 1$. Its solution has the hyperbolic form [22], with τ denoting the proper time,

$$x = \frac{1}{a} \cosh a\tau, \quad t = \frac{1}{a} \sinh a\tau, \tag{12.10}$$

which is illustrated in Fig. 12.1a for the boundary conditions $t = -\infty, x = +\infty$ and $t = \infty, x = \infty$. There thus exists a part of the world, a hidden region bounded by an "event horizon" $x = t$, which m can never reach as long as it maintains its acceleration. Similarly, no observer in the hidden region can ever communicate with m.

As the mass passes through the vacuum, a part of its acceleration energy can be used to excite on-shell one of the ever-present virtual vacuum fluctuations. In Fig. 12.1b, this is illustrated for an e^+e^- fluctuation: the e^+ is absorbed by a detector on m, while the e^- disappears beyond the event horizon. An equivalent way of describing this phenomenon is to note that an e^-, emitted from m, by quantum tunnelling through the event horizon reaches the hidden region. Such behavior is today called "quantum entanglement" (Einstein-Podolsky-Rosen effect [23, 24]): an observer on m measuring the e^+ has only incomplete information, since the

e^- has disappeared beyond the event horizon and is thus forever gone. The same holds for an observer in the hidden region, who can never access the e^+. Because of this principal lack of complete information, either observer can only see thermal radiation, unable to transmit an information-carrying signal across the horizon. As a consequence, the observer on m sees the physical vacuum as a thermal medium, a heat bath of electrons, of the Unruh temperature

$$T_U = \frac{\hbar a}{2\pi c}. \tag{12.11}$$

Correspondingly, an observer in the hidden region registers the passage of m by measuring radiation of the same temperature. We have here explicitly included c and $\hbar$, which otherwise are taken to be unity; they show that the phenomenon is a relativistic quantum effect, which would disappear for $c \to \infty$ or $\hbar \to 0$.

We can now use this formalism to derive the temperature of Hawking radiation. For acceleration due to gravity, the force is given by

$$F = ma = G\frac{Mm}{R^2}, \tag{12.12}$$

so that the acceleration becomes

$$a = \frac{GM}{R^2} = \frac{1}{4GM}; \tag{12.13}$$

for the last step, we have made use of the fact that the horizon of the black hole is defined by the vanishing of g_0, leading to the Schwarzschild radius (12.3). Inserting this into the Unruh formula (12.11) gives

$$T_{\mathrm{BH}}(M) = \frac{1}{8\pi GM}, \tag{12.14}$$

the temperature for Hawking radiation from a black hole. This result can be extended to the case of a Reissner-Nordström black hole with an overall charge Q; Eq. (12.12) has to be modified to include the Coulomb force [1]. In this way one obtains

$$T_{\mathrm{BH}}(M, Q) = T_{\mathrm{BH}}(M, 0) \left\{ \frac{4\sqrt{1 - (Q^2/GM^2)}}{(1 + \sqrt{1 - (Q^2/GM^2)})^2} \right\} \tag{12.15}$$

for the Hawking temperature $T(M, Q)$ as a function of mass M and charge Q. For $Q = 0$, it is seen to reduce to the temperature of the Schwarzschild black hole, while for $Q^2 = GM^2$ (denoted as *extremal* Reissner-Nordström black hole), the temperature vanishes. As a result, we obtain here a "phase diagram" in T and Q^2 which is quite similar to that found in Chap. 7 for strongly interacting matter as function of T and baryochemical potential μ; it is illustrated in Fig. 12.2. For further discussion of this "duality", see [13]; it evidently raises a rather general interesting question: is there a relation between critical behavior and the existence of an event horizon? For some work in this context, see [25].

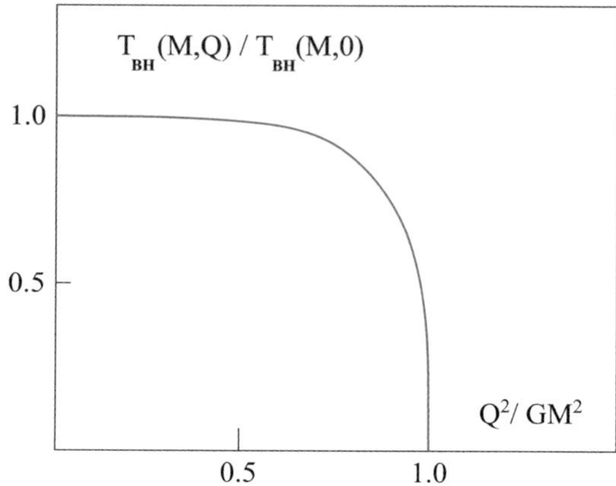

Fig. 12.2 Radiation temperature for a charged black hole

Another example of Unruh radiation is given by the Schwinger mechanism for pair production [16]. Here the basic observation is that in a strong electric field $\mathcal{E}$, the physical vacuum becomes unstable against pair production: when the local energy density becomes larger than the mass of an electron-positron pair, the field energy brings such a pair on-shell, out of the vacuum sea. The probability for this is given by

$$P(m, \mathcal{E}) \sim \exp\{-\pi m^2/e\mathcal{E}\}, \tag{12.16}$$

where e is the charge of the electron. With the equation of motion

$$F = e\mathcal{E} = (m/2)a, \tag{12.17}$$

where $m/2$ is the reduced mass of an electron in the pair, we obtain as Unruh temperature

$$T_U = \frac{a}{2\pi} = \frac{e\mathcal{E}}{\pi m}, \tag{12.18}$$

which with

$$P(m, \mathcal{E}) \sim \exp\{-m/T_U\} \tag{12.19}$$

gives the Schwinger form (12.16). For more elaborate discussion of the Schwinger mechanism as Unruh effect, see [17–21].

In summary: an event horizon does not allow information transfer across it. It can only be passed by quantum tunnelling, leading to stochastic radiation on the other side [26, 27]. We now want to apply this formalism to hadron production in strong interaction.

12.3 Pair Production in e^+e^- Annihilation

As starting point, we consider e^+e^- annihilation into hadrons at cms energy $\sqrt{s}$. In lowest order, the resulting virtual photon couples to a color-singlet quark-antiquark pair,

$$e^+e^- \to \gamma^* \to q\bar{q}; \tag{12.20}$$

see Fig. 12.3. The $q\bar{q}$ pair will subsequently produce the multihadron final state in the form of two hadronic jets.

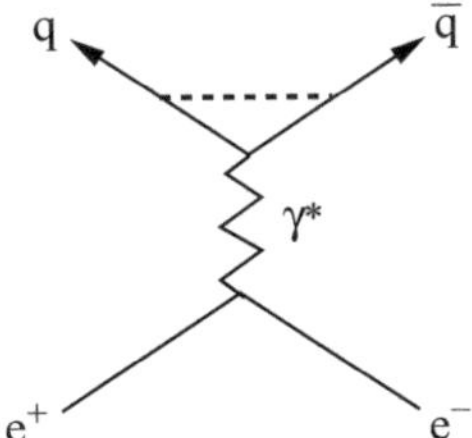

Fig. 12.3 Quark pair production in e^+e^- annihilation

The primary $q\bar{q}$ pair flies apart, but because of color confinement it remains subject to a binding force which increase with separation. Describing this in terms of a classical string, specified by a string tension σ, provides a constant confining force and thus results in hyperbolic motion of the type discussed in the previous section [28]. At $t = 0$, the q and $\bar{q}$ separate with an initial velocity $v_0 = p/p_0$, where p is the momentum and $p_0 = (p^2 + m^2)^{1/2} \simeq \sqrt{s}/2$ the energy of the primary q or $\bar{q}$ in the overall cms; m specifies the effective quark mass. The $q\bar{q}$ pair is bound by the string potential

$$V = \sigma x, \tag{12.21}$$

defined by the string tension σ and the $q\bar{q}$ separation distance x. The classical event horizon can be defined as the value $x = x^*$ for which the initial kinetic energy becomes equal to the potential energy, i.e., when

$$\frac{m}{\sqrt{1 - v_0^2}} = \sigma x^*. \tag{12.22}$$

We thus get

$$x^* = \frac{p_0 - m}{\sigma} \simeq \frac{\sqrt{s}}{2m\sigma},$$

(12.23)

which allows the q and the $\bar{q}$ to separate arbitrarily far, provided the initial energy was high enough; the classical horizon grows with $\sqrt{s}$, as illustrated in Fig. 12.4. This clearly violates color confinement; what went wrong?

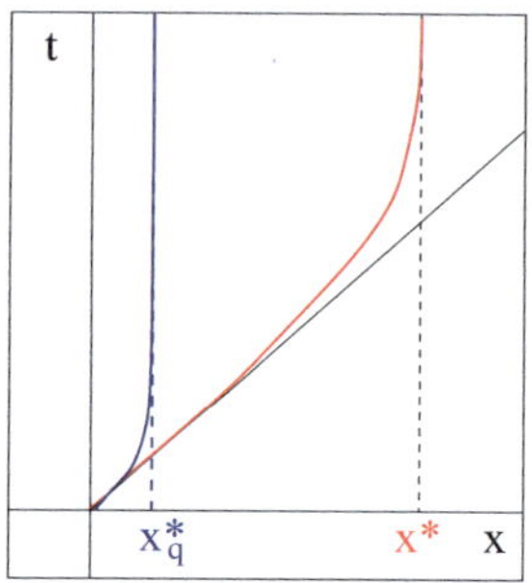

Fig. 12.4 Classical and quantum horizons in $q\bar{q}$ separation

Our mistake was to treat the quantum system $q\bar{q}$ in classical terms; in quantum field theory, the background medium of the $q\bar{q}$ contains virtual $q\bar{q}$ pairs, and hence it is not possible to increase the potential energy of a given $q\bar{q}$ state beyond the threshold value necessary to bring such a virtual $q\bar{q}$ pair on-shell. In QED, as we had just seen in the previous section, the presence of a strong electric field leads to the production of electron-positron pairs. In QCD, we expect a similar effect when the string tension exceeds the pair production limit, i.e., when

$$\sigma x > 2m$$

(12.24)

where m again specifies the effective quark mass. Beyond this point, it becomes energetically favorable to produce a further $q\bar{q}$ pair and start two new string configurations, rather than to continue stretching the primary string (see Fig. 12.3b). This acts like a quantum event horizon $x_q^* \simeq 2m/\sigma$, which becomes operative long before the classical turning point is ever reached; in Fig. 12.4 we compare the two event horizons. Moreover, the allowed separation distance for our $q\bar{q}$ pair, the color confinement radius $r_q = x_q^*/2 \simeq m/\sigma$, now no longer depends on the initial energy of the primary quarks. Let us look at the resulting hadron production cascade for e^+e^- annihilation in more detail.

The hadronization of the jets formed by the primary quark and antiquark is believed to proceed in the form of a self-similar cascade [29, 30]. Initially, we have the separating primary $q\bar{q}$ pair,

$$\gamma \to [q\bar{q}]$$

(12.25)

where the square brackets indicate color neutrality. When the energy of the string connecting the pair becomes large enough, a further color-neutral pair $q_1\bar{q}_1$ is excited from the vacuum by two-gluon exchange (see Fig. 12.5),

$$\gamma \rightarrow [q[\bar{q}_1 q_1]\bar{q}]. \tag{12.26}$$

The $\bar{q}_1$ now screens the primary q from its original partner $\bar{q}$, with an analoguous effect for the q_1 and the primary antiquark. Although the new pair is at rest in the overall cms, each of its constituents has a transverse momentum k_T determined, through the uncertainty relation, by the transverse dimension r_T of the "flux tube" resulting from the connecting string. To estimate the $q\bar{q}$ separation distance at the point of pair production, we need to include this transverse momentum in the effective quark mass in Eq. (12.24). The thickness of the flux tube connecting the $q\bar{q}$ pair is in string theory given by [31]

$$r_T^2 = \frac{2}{\pi\sigma} \sum_{k=0}^{K} \frac{1}{2k+1}, \tag{12.27}$$

where K is the string length in units of an intrinsic vibration measure. Lattice studies [32] show that for strings in the range of 1–2 fm, the first string excitation dominates, so that we have

$$r_T \simeq \sqrt{\frac{2}{\pi\sigma}}. \tag{12.28}$$

Higher excitations lead to a greater thickness and eventually to a divergence (the "roughening" transition). From the uncertainty relation we then have

$$k_T \simeq \sqrt{\frac{\pi\sigma}{2}}. \tag{12.29}$$

When this transverse momentum is included in Eq. (12.24), we obtain for the pair production separation x_Q

$$\sigma x_q = 2\sqrt{m_q^2 + k_T^2} \;\Rightarrow\; x_q \simeq \frac{2}{\sigma}\sqrt{m_q^2 + (\pi\sigma/2)} \simeq \sqrt{\frac{2\pi}{\sigma}} \simeq 1\,\text{fm}, \tag{12.30}$$

with $\sigma = 0.2\,\text{GeV}^2$ and $m_q^2 \ll \sigma$ for the bare quark mass.

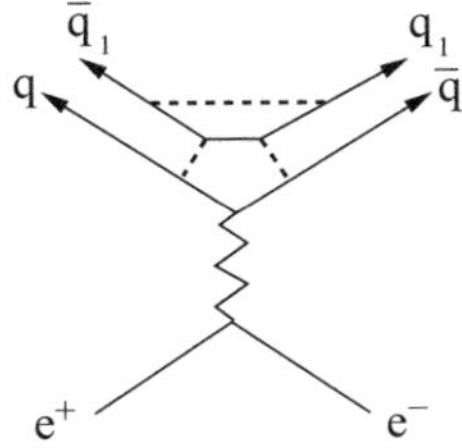

Fig. 12.5 Secondary quark pair production in e^+e^- annihilation

Once the new pair is present, we have a color-neutral system $q\bar{q}_1 q_1\bar{q}$, and in principle it could now separate into to color neutral pairs $q\bar{q}_1$ and $\bar{q}q_1$, flying apart. To form two eventual hadrons $[q\bar{q}_1]$ and $[\bar{q}q_1]$, the force between the q and the $\bar{q}_1$ would have to accelerate the latter from (longitudinal) rest to about half the momentum of the former, and correspondingly for the $[\bar{q}q_1]$. On the left side of Fig. 12.5, we thus have a constantly accelerating quark, the $\bar{q}_1$, and a constantly decelerating quark, the primary q; an analogous situation holds on the right.

To separate the two pairs $q\bar{q}_1$ and $\bar{q}q_1$, the string potential corresponding to the color confinement of the newly formed $q_1\bar{q}_1$ pair has to be overcome – the primary string so far is not yet "broken". To break the binding of the new pair, the q_1 has to tunnel through the barrier of the confining potential provided by the $\bar{q}_1$, and vice versa. Now the longitudinal force of the q on the $\bar{q}_1$, and of the $\bar{q}$ on the q_1, results in a longitudinal acceleration and ordering of q_1 and $\bar{q}_1$. When (see Fig. 12.6)

$$\sigma x(q_1\bar{q}_1) = 2\sqrt{m_q^2 + k_T^2},\tag{12.31}$$

the $\bar{q}_1$ reaches its $q_1\bar{q}_1$ horizon; equivalently, the new string $q\bar{q}_1$ reaches the energy needed to produce a further pair $q_2\bar{q}_2$. The $\bar{q}_2$ screens the primary q from the q_1 and forms a new string $q\bar{q}_2$. At this point, the original string is broken, and the remaining pair $\bar{q}_1 q_2$ forms a color neutral bound state which is emitted as Hawking-Unruh radiation in the form of a hadron, with the relative weights of the different possible states governed by the corresponding Unruh temperature. The resulting pattern is schematically illustrated in Fig. 12.6.

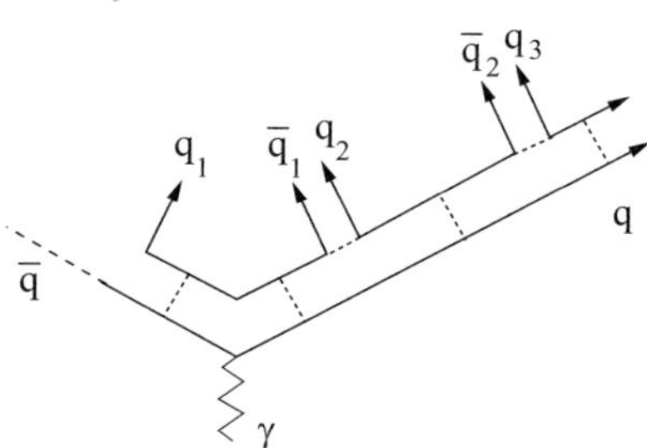

Fig. 12.6 String breaking through $q\bar{q}$ pair production

To determine the temperature of the hadronic Hawking radiation, we note that the produced antiquark $\bar{q}_1$ is accelerated by the primary quark q because of the string tension σ; similarly the $\bar{q}_2$, and so on along the cascade. Hence the antiquark acceleration (as well as the quark acceleration along the primary $\bar{q}$ direction) becomes

$$a_q = \frac{\sigma}{w_q} = \frac{\sigma}{\sqrt{m_q^2 + k_q^2}},\tag{12.32}$$

where $w_q = (m_q^2 + k_q^2)^{1/2}$ is the effective mass of the produced quark or antiquark, with m_q for the bare quark mass and k_q the associated quark momentum due to the spatial uncertainty. Since the string breaks for x_q given by Eq. (12.30), the uncertainty relation $k_q \simeq 1/x_q$ gives

$$w_q = \sqrt{m_q^2 + [\sigma^2/(4m_q^2 + 2\pi\sigma)]} \tag{12.33}$$

for the effective quark mass. The resulting Unruh temperature is thus given by

$$T_q \simeq \frac{\sigma}{2\pi w_q} \simeq \sqrt{\frac{\sigma}{2\pi}}, \tag{12.34}$$

where in the last term we have assumed a negligible bare quark mass. For the canonical string tension range, $\sigma \simeq 0.16\text{--}0.20\,\text{GeV}^2$, this leads to

$$T_q \simeq 160\text{--}180\,\text{MeV} \tag{12.35}$$

for the hadronic Unruh temperature. Note that, for a given string tension, this is a parameter-free prediction of the temperature governing hadron production in e^+e^- annihilation. It agrees well with the hadronization temperature determined in the lattice evaluation of statistical QCD.

A given step in the evolution of the hadronization cascade of a primary quark or antiquark produced in e^+e^- annihilation thus involves several distinct phenomena. The color field created by the separating q and $\bar{q}$ produces a further pair $q_1\bar{q}_1$ and then provides an acceleration of the q_1, increasing its longitudinal momentum. When it reaches the $q_1\bar{q}_1$ confinement horizon, still another pair $q_2\bar{q}_2$ is excited; the state $\bar{q}_1 q_2$ is emitted as a hadron, the $\bar{q}_2$ forms together with the primary q a new flux tube. This pattern thus step by step increases the longitudinal momentum of the "accompanying" $\bar{q}_i$ as well as that of the emitted hadron. This, together with the energy of the produced pairs, causes a corresponding deceleration of the primary quarks q and $\bar{q}$, in order to maintain overall energy conservation. In Fig. 12.7, we show the world line given by the acceleration $\bar{q}_i \to \bar{q}_{i+1}$ ($q_i \to q_{i+1}$) and that of the formation threshold of the hadrons $\bar{q}_i q_{i+1}$ and the corresponding antiparticles.

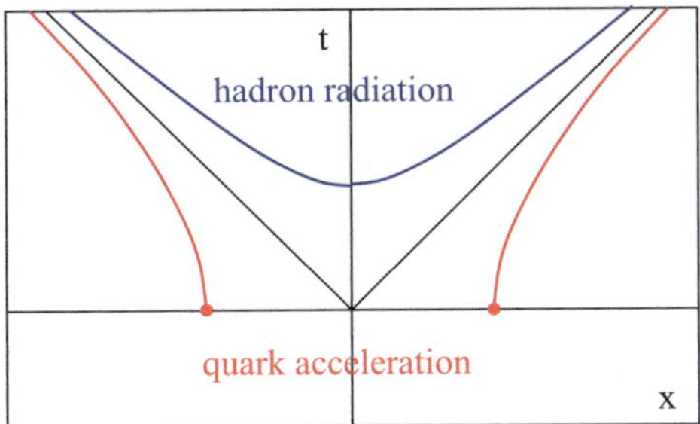

Fig. 12.7 Quark acceleration and hadronization world lines

The energy loss and deceleration of the primary quark q in this self-similar cascade, together with the acceleration of the accompanying partner $\bar{q}_i$ from the successive pairs, brings q and $\bar{q}_i$ closer and closer to each other in momentum, from an initial separation $q\bar{q}_1$ of $\sqrt{s}/2$, until they finally are combined into a hadron and the cascade is ended. The resulting pattern is shown in Fig. 12.8.

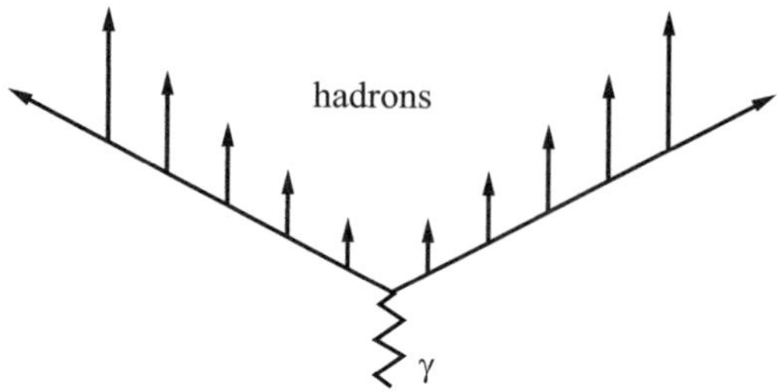

Fig. 12.8 Hadronization in e^+e^- annihilation

The number of emitted hadrons, the multiplicity $v(s)$, follows quite naturally from the picture presented here. The classical string length, in the absence of quantum pair formation, is given by the classical turning point determined in Eq. (12.23). The thickness of a flux tube of such an "overstretched" string is known [31]; from Eq. (12.27) we get

$$R_T^2 = \frac{2}{\pi\sigma} \sum_{k=0}^{K} \frac{1}{2k+1} \simeq \frac{2}{\pi\sigma} \ln 2K, \tag{12.36}$$

where K is the string length. From Eq. (12.23) we thus get

$$R_T^2 \simeq \frac{2}{\pi\sigma} \ln \sqrt{s} \tag{12.37}$$

for the flux tube thickness in the case of the classical string length. In parton language, the logarithmic growth of the transverse hadron size is due to parton random walk ("Gribov diffusion"[33]); this phenomenon is responsible for diffraction cone shrinkage in high–energy hadron scattering.

Because of pair production, the string breaks whenever it is stretched to the length x_q given in Eq. (12.30); its thickness r_T at this point is given by Eq. (12.27). The multiplicity can thus be estimated by the ratio of the corresponding classical to quantum transverse flux tube areas,

$$v(s) \sim \frac{R_T^2}{r_T^2} \sim \ln \sqrt{s}, \tag{12.38}$$

and is found to grow logarithmically with the e^+e^- annihilation energy, as is observed experimentally over a considerable range.

We note here that in our argumentation we have neglected parton evolution, which can cause the emitted radiation (e.g., $\bar{q}_1 q_2$ in Fig. 12.6) to start another cascade of the same type. Such evolution effects result eventually in a stronger increase of the multiplicity. The hadronization mechanism discussed here does not affect the formation of hard processes at early times (e.g., multiple jet production), which is responsible for an additional growth of the measured multiplicity.

A further effect we have not taken into account here is parton saturation. At sufficiently high energy, stronger color fields can lead to gluon saturation and thus to a higher temperature determined by the saturation momentum [34, 35]. The resulting system then first expands and subsequently hadronizes at the universal temperature determined by the string tension.

The hadronic Unruh temperature given by Eq. (12.34) contains a dependence on the bare quark mass; we have so far neglected this, assuming $m_q \ll \sqrt{\sigma} \simeq 400\,\mathrm{MeV}$. Such an assumption is reasonable for u and d quarks, but not really for strange quarks, with a mass of around 100 MeV. If we retain the quark mass dependence, the Unruh radiation of a meson containing one strange (s) and one non-strange (q) quark is determined by the average acceleration

$$\bar{a}_{qs} = \frac{w_q a_q + w_s a_s}{w_q + w_s} = \frac{2\sigma}{w_q + w_s}, \tag{12.39}$$

where w_q and w_s are the corresponding effective masses. From this the Unruh temperature of a strange meson is given by

$$T_{qs} \simeq \frac{\sigma}{\pi(w_q + w_s)}. \tag{12.40}$$

Similarly, the temperature of a ϕ meson, consisting of a $s\bar{s}$ pair, becomes

$$T(ss) \simeq \frac{\sigma}{2\pi w_s}. \tag{12.41}$$

For typical values $\sigma \simeq 0.2\,\mathrm{GeV}^2$ and $m_s \simeq 0.1$ Gev, this leads to temperatures

$$T_{qq} = 178\,\mathrm{MeV},\ T_{qs} = 167\,\mathrm{MeV},\ T_{ss} = 157\,\mathrm{MeV}. \tag{12.42}$$

The relative abundance of a given species is determined by its hadronization temperature (see Chap. 11); e.g., schematically

$$N(qq) \sim \exp\{-m(qq)/T_{qq}\},\ N(ss) \sim \exp\{-m(ss)/T_{ss}\}. \tag{12.43}$$

The decrease of the temperature for mesons containing strange quarks thus corresponds to an effective suppression of such states, compared to those containing only light quarks, and it was shown [36] that this mechanism can in fact account for the strangeness suppression observed in elementary collisions. In a description of hadronization as Unruh radiation, there thus is no longer any need for an ad hoc strangeness suppression factor, introduced in the usual resonance gas analysis (see Chap. 11).

This line of argument can be extended to strange baryons as well, leading to altogether five different Unruh temperatures, depending on the strangeness content of the hadron in question. These temperatures are, however, given in terms of only two "parameters", the string tension σ and the strange quark mass m_s. Since these two are both determined in various other contexts [37–43] and can be taken as given, the resulting description is again parameter-free. The strange quark mass is presently listed as $m_s = 0.095 \pm 0.025$ GeV [37], and the string tension average as $\sigma = 0.19 \pm 0.02\,\text{GeV}^2$ [38–43]. Fits of hadron production data in e^+e^- annihilation covering all presently available energies, from $\sqrt{s} = 14$ to $190\,\text{GeV}$, have shown that these values indeed reproduce very well the observed abundances [36].

12.4   Hadronic Collisions

Up to now, we have considered hadron production in e^+e^- annihilation, in which the virtual photon produces a confined colored $q\bar{q}$ pair as a "colorless hole". Turning now to hadron-hadron collisions, we note that here two incident colorless holes fuse to form a new system of the same kind, as schematically illustrated in Fig. 12.9. Again the resulting string or strong color field produces a sequence of $q\bar{q}$ pairs of increasing cms momentum, leading to the well-known multiperipheral hadroproduction cascade shown in Fig. 12.10. We recall here the comments made above concerning parton evolution and saturation; in hadronic collisions as well, these phenomena will affect the multiplicity, but not the relative abundances.

In the case of nucleus-nucleus collisions, we have a superposition of many such tunnelling cascades, with possible interference. In principle, the dense medium consisting of Hawking-Unruh radiation from the different cascades can lead to rearrangement of quarks, so that a final hadron can consist of a quark from one cascade and an antiquark from another, as illustrated in Fig. 12.11. Such a rearrangement is a quantum-mechanical interference effect, occurring on short space and time scales; hadronic interactions, if they occur, set in when the medium is much more dilute, allowing hadronic scales to become operative.

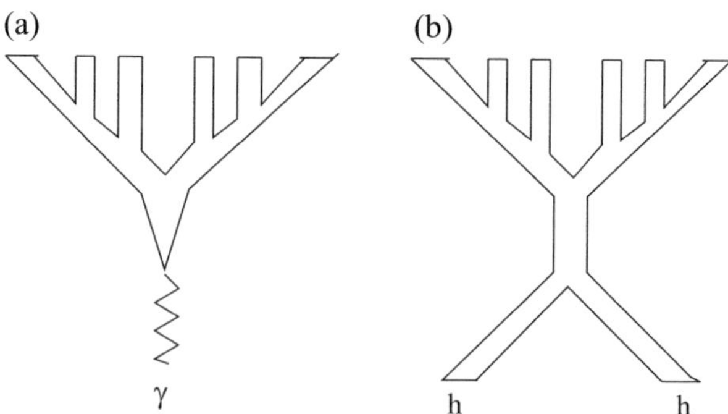

Fig. 12.9 "Colorless hole" structure in e^+e^- annihilation (**a**) and hadronic collisions (**b**)

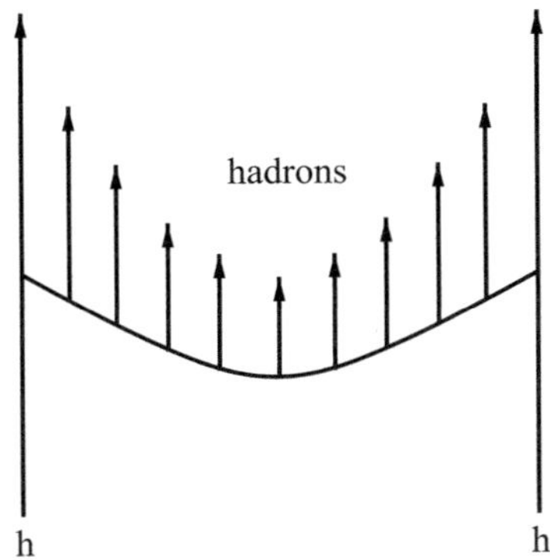

Fig. 12.10 Hadronization in hadron-hadron collisions

12.5 Strange Particle Production

Through multiple interference of this kind, the differences between hadronization temperatures will be removed, leading to one final temperature. As a result, there now is no more suppression of strange hadron production – all hadrons are produced according to phase space at one universal temperature.

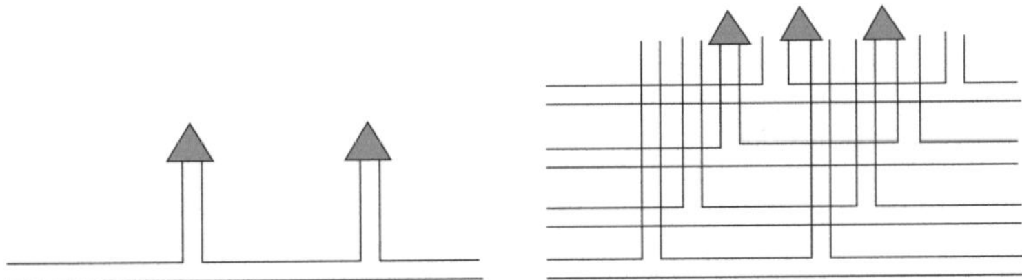

Fig. 12.11 Hadron formation in elementary (left) and in nucleus-nucleus (right) collisions

Since the axes of the different cascades will be slightly shifted in a random way by the interactions, this will also lead to a broadening of the transverse momentum distributions of the emitted hadrons [44]; this effect is illustrated in Fig. 12.12.

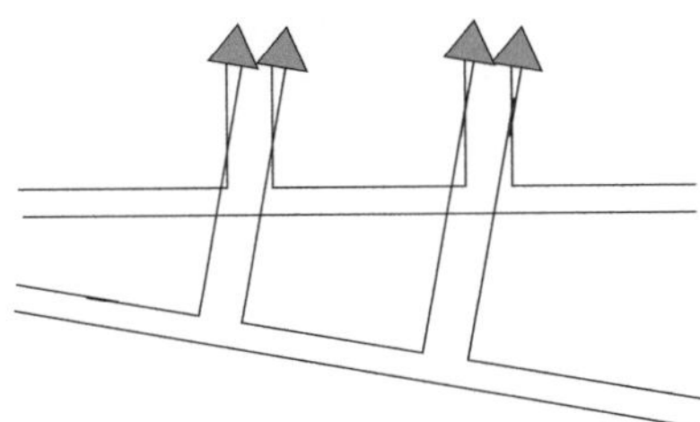

Fig. 12.12 Transverse momentum broadening in nuclear collisions

12.6   Stochastic Thermalization

In statistical mechanics, a basic topic is the evolution of a system of many degrees of freedom from non-equilibrium to equilibrium. Starting from a non-equilibrium initial state of low entropy, the system is assumed to evolve as a function of time through collisions to a time-independent equilibrium state of maximum entropy. In other words, the system loses the information about its initial state through a sequence of collisions and thus becomes thermalized. In this sense, thermalization in nuclear collisions was studied as the transition from an initial state of two colliding beams of "parallel" partons to a final state in which these partons have locally isotropic distributions. This "kinetic" thermalization requires a sufficient density of constituents, sufficiently large interaction cross sections, and an amount of time sufficient for equilibration.

From such a point of view, the observation of thermal hadron production in high energy collisions, in particular in e^+e^- and pp interactions, is a puzzle: how could these systems ever "have reached" thermalization? Already Hagedorn [45, 46] had therefore concluded that the emitted hadrons were "born in equilibrium". Even given an exponentially increasing resonance mass spectrum, it remained totally unclear why collisions should result in a thermal system.

Hawking-Unruh radiation provides a stochastic rather than kinetic approach to equilibrium, with a randomization essentially provided by the quantum physics of the Einstein-Podolsky-Rosen effect. The barrier to information transfer presented by the event horizon requires that the resulting radiation states excited from the vacuum are distributed according to maximum entropy, with a temperature determined by the strength of the "confining" field. In other words, the produced state is selected at random from the set of all states corresponding to this temperature. The ensemble of all produced hadrons then results in the same equilibrium distribution as would be obtained in hadronic matter by kinetic equilibration.

We thus encounter here something like an ergodic equivalence principle. In general relativity, the effects of gravitation and acceleration become equivalent; a given observer cannot distinguish a gravitational attraction from an accelerating reference frame. Here, in a similar way, the observer cannot tell whether the observed equilibrium was attained by stochastic selection or by kinetic thermalization.

In the case of a very high energy collision with a high average multiplicity, already one event can provide such equilibrium; because of the interruption of information transfer at each of the successive quantum color horizons, there is no phase relation between two successive production steps in a given event. The destruction of memory, which in kinetic equilibration is achieved through sufficiently many successive collisions, is here automatically provided by the tunnelling process.

So the thermal hadronic final state in high energy collisions is not reached through a kinetic process; it is rather provided by successively throwing dice. This origin of thermal hadron production moreover also provides an explanation for the observation that the hadronization temperature in collision processes is the same as that found for the hadronization of a quark-gluon plasma in statistical QCD. In such a medium as well, hadronization occurs when cooling and expansion "threaten" to separate a given quark from any possible antiquark partner by more than the canonical hadronic distance. Through dilution all the colored quarks thus are forced to reach their event horizons, leading to hadronic Unruh radiation. In other words, what we observe in high energy collisions is simply a specific case of the general phenomenon causing the hadronization of a colored medium.

12.7 Conclusions

We have shown that quantum tunnelling through the color confinement horizon leads to thermal hadron production in the form of Hawking-Unruh radiation. In particular, this implies:

- The radiation temperature T_q is determined by the force of color confinement, giving (for light quarks)

$$T_q \simeq \sqrt{\frac{\sigma}{2\pi}}, \tag{12.44}$$

  in terms of the string tension σ. For strange quarks, the larger quark mass m_s leads to a slightly reduced temperature; this accounts for the strangeness suppression observed in elementary collisions. In nuclear collisions, these temperature differences are reduced by interference effects between different radiation cascades.
- The multiplicity $v(s)$ of the produced hadrons is approximately given by the increase of the flux tube thickness with string length, leading to

$$v(s) \simeq \ln \sqrt{s}, \tag{12.45}$$

  where $\sqrt{s}$ denotes the center-of-mass collision energy. Parton evolution and gluon saturation will, however, increase this multiplicity, as will early hard production. The universality of the resulting abundances is, however, not affected.
- In statistical QCD, thermal equilibrium is reached kinetically from an initial nonequilibrium state, with memory destruction through successive interactions of the constituents. In high energy collisions, tunnelling prohibits information transfer and hence leads to stochastic production, so that we have a thermal distribution from the outset.

References

1. See e.g., L.Z. Fang, R. Ruffini, *Basic Concepts in Relativistic Astrophysics* (World Scientific, Singapore, 1983)
2. E. Recami, P. Castorina, Lett. Nuovo Cim. **15**, 347 (1976)
3. See e.g., R. Alkofer, C.S. Fisher, F.J. Llanes-Estrada, Phys. Lett. B **611**, 279 (2005)
4. A. Salam, J. Strathdee, Phys. Rev. **D18**, 4596 (1978)

5. The *Clay Mathematics Institute*, Cambridge, MA, has posted a list of *Millenium Problems*, whose solutions will in each case be awarded with a million dollar prize; problem no. 7 is the proof on color confinement in Yang-Mills Theory
6. M. Novello et al., Phys. Rev. D **61**, 045001 (2000)
7. T.D. Lee, in *Statistical Mechanics of Quarks and Hadrons*, ed. by H. Satz (North Holland Publishing Co., Amsterdam, 1980)
8. H. Pagels, E.T. Tomboulis, Nucl. Phys. B **143**, 453 (1978)
9. L.B. Abbott, Nucl. Phys. B **185**, 189 (1981)
10. L. Maiani et al., Nucl. Phys. B **273**, 275 (1986)
11. P. Castorina, M. Consoli, Phys. Rev. D **35**, 3249 (1987)
12. D. Kharzeev, J. Raufeisen, nucl-th/0206073
13. P. Castorina, D. Kharzeev, H. Satz, Eur. Phys. J. C **52**, 187 (2007)
14. S.W. Hawking, Commun. Math. Phys. **43**, 199 (1975)
15. W.G. Unruh, Phys. Rev. D **14**, 870 (1976)
16. J. Schwinger, Phys. Rev. **82**, 664 (1951)
17. R. Brout, R. Parentani, Ph. Spindel, Nucl. Phys. B **353**, 209 (1991)
18. P. Parentani, S. Massar, Phys. Rev. D **55**, 3603 (1997)
19. K. Srinivasan, T. Padmanabhan, Phys. Rev. D **60**, 024007 (1999)
20. D. Kharzeev, K. Tuchin, Nucl. Phys. A **753**, 316 (2005)
21. Sang Pyo Kim, arXiv:0709.4313 [hep-th] 2007
22. See e.g. W. Pauli, *Relativitätstheorie*, in *Enzyklopdie der mathematischen Wissenschaften* (Teubner Verlag, Leipzig, 1921); English version *Theory of Relativity* (Pergamon Press, New York, 1958)
23. A. Einstein, B. Podolsky, N. Rosen, Phys. Rev. **47**, 777 (1935)
24. J.S. Bell, Physics **1**, 195 (1964)
25. G. Chapline et al., Int. J. Mod. Phys. A **18**, 3587 (2003)
26. T.D. Lee, Nucl. Phys. B **264**, 437 (1986)
27. M.K. Parikh, F. Wilczek, Phys. Rev. Lett. **85**, 5042 (2000)
28. A. Hosoya, Progr. Theoret. Phys. **61**, 280 (1979)
29. J.D. Bjorken, Lect. Notes Phys. (Springer) **56**, 93 (1976)
30. A. Casher, H. Neuberger, S. Nussinov, Phys. Rev. D **20**, 179 (1979)
31. M. Lüscher, G. Münster, P. Weisz, Nucl. Phys. B **180**, 1 (1981)
32. G. Bali, K. Schilling, C. Schlichter, Phys. Rev. D **51**, 5165 (1995)
33. V.N. Gribov, arXiv:hep-ph/0006158
34. D. Kharzeev, K. Tuchin, Nucl. Phys. A **753**, 316 (2005)
35. D. Kharzeev, Nucl. Phys. A **774**, 315 (2006)
36. F. Becattini et al., Eur. Phys. J. C **56**, 493 (2008)
37. W.-M. Yao et al. (2006 Review of Particle Physics), J. Phys. G **33**, 1 (2006)
38. S. Jacobs, M.G. Olsson, C. Suchyta, Phys. Rev. D **33**, 3338 (1986)
39. F.J. Yndurain, *Theory of Quark and Gluon Interactions* (Springer, Berlin, 1999)
40. N. Brambilla et al., CERN Yellow Report CERN-2005-005
41. C. Aubin et al. (MILC Collaboration), Phys. Rev. D **70**, 094505 (2004)
42. A. Gray et al., Phys. Rev. D **72**, 0894507 (2005)
43. M. Cheng et al., arXiv:hep-lat/0608013
44. A. Leonidov, H. Satz, Z. Phys. C **74**, 535 (1997)
45. R. Hagedorn, *Thermodynamics of Strong Interactions*, CERN 71-'-12, 1971
46. see also R. Stock, Phys. Lett. B **456**, 277 (1999) and arXiv:nucl-th/0703050

Chapter 13
Fluids and Flow

ΠANTA PEI
[Everything flows]

Heraclitus of Ephesus (535–475 B.C.)

Abstract In this chapter, we want to consider some phenomena which arise when the medium being studied is not in overall thermal equilibrium – i.e., if local variations in energy density, pressure or other variables lead to expansion and flow effects.

13.1 Introduction

Up to now, we have considered matter as a many-body state in overall thermal equilibrium: time did not enter. Whatever changes of state were considered, they were assumed to occur adiabatically, slow enough so that at any point in the evolution the system as a whole could be taken to be in thermal equilibrium. As an illustration of what this means, let's consider the well-known Joule experiment (Fig. 13.1).

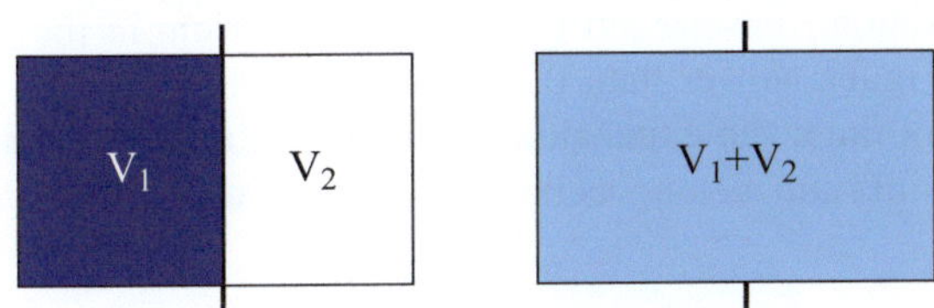

Fig. 13.1 The Joule experiment

The initial state is a volume consisting of two partitions V_1 and V_2, of which one (V_1) is filled with a gas at thermal equilibrium at temperature T, the other (V_2) is empty. Upon removing the divider, the gas flows into the empty partition and eventually equilibrium is restored in a volume of size $V_1 + V_2$ and, since no work

© Springer International Publishing AG 2018
H. Satz, *Extreme States of Matter in Strong Interaction Physics*,
Lecture Notes in Physics 945, https://doi.org/10.1007/978-3-319-71894-1_13

is done, again at temperature T. In the intermediate time interval between the two equilibrium states, while the gas is "flowing" from the full into the empty volume, the system is evidently not in equilibrium. The momenta of the constituents entering volume V_2 are not isotropically oriented: the constituents are predominantly flying in the direction towards the far wall of the empty partition. In the initial state, the entropy was maximal in the filled partition, $S_1 \sim kV_1T^3$, and in the final state it is once more so in the now larger volume, $S_{1+2} \sim k(V_1 + V_2)T^3$. In the intermediate "equilibration stage", in the "relaxation time", the entropy is increasing from S_1 to S_{1+2} (Fig. 13.2).

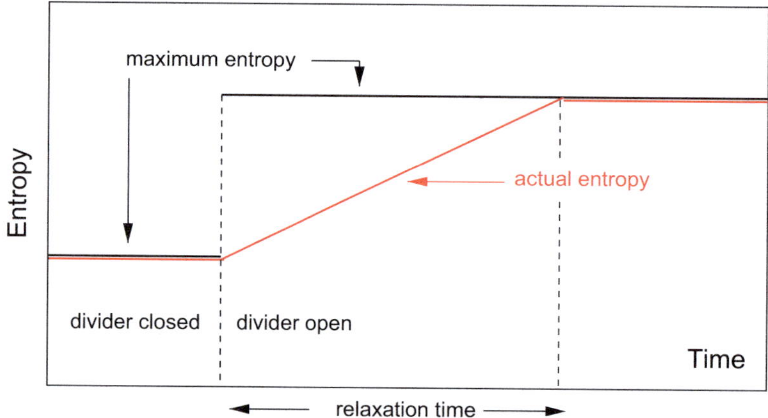

Fig. 13.2 The temporal evolution of the Joule experiment

We now consider a continuous version of such an experiment, with a piston being pulled out at a constant rate to increase the volume available to the gas (Fig. 13.3). To keep things well-defined, we assume a very heavy piston, so that it really has to be pulled out; the gas cannot do work to push it out. In such an experiment, there is evidently a competition between two time scales, the rate of expansion of the volume and relaxation time needed to restore equilibrium in the given gas. If the expansion rate is much slower than the equilibration rate, the gas is effectively in equilibrium at any time: the expansion is adiabatic. In the opposite case, the gas never reaches equilibrium as long as the volume is being increased.

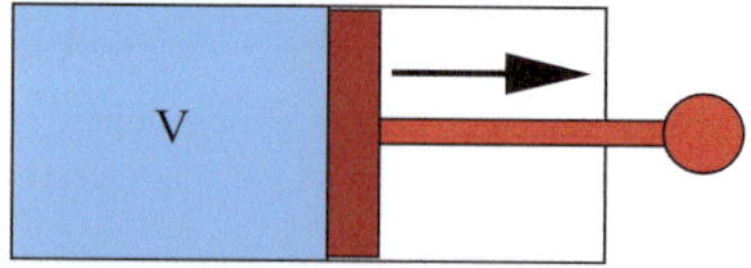

Fig. 13.3 A continuous version of the Joule experiment

Let us now apply these considerations to two cases: the expansion of the early universe and the expansion of the fireball produced in high energy nuclear collisions.

The expansion rate of the universe is determined through the solution of the Einstein equations in general relativity. The relevant result for us is the so-called Friedmann form, obtained by the Russian theorist Alexander Friedmann not long after Einstein's pioneering work. The crucial quantity is the scale factor $a(t)$, which specifies the scale value for the expanding universe at time t. It is determined by the Friedmann equation

$$\left(\frac{1}{a}\frac{da}{dt}\right)^2 \equiv H^2(t) = \frac{8\pi}{3c^2}G\,\epsilon(t), \tag{13.1}$$

where G denotes the Newtonian constant of gravity and $\epsilon(t)$ the energy density at time t; as usual, c is the speed of light (for dimensional reasons we do not set it equal to unity here). The expansion rate is commonly denoted as the Hubble parameter $H(t)$, which we had already encountered in Sect. 9.1. The velocity $v(t)$ with which galaxies at a distance $d(t)$ away recede from us is given by the Hubble equation

$$v = H\,d. \tag{13.2}$$

With $G \simeq 6.7 \times 10^{-39}\ (\mathrm{GeV})^{-2}$ and (see Eq. (4.15))

$$\epsilon = \frac{37\pi^2}{30}T^4 \tag{13.3}$$

for an ideal two-flavor color SU(3) quark-gluon plasma, we have in the transition region around $T \simeq 200\,\mathrm{MeV}$ an expansion rate

$$H \simeq 2 \times 10^{-20}\ \mathrm{GeV} \simeq 3 \times 10^4\ s^{-1}. \tag{13.4}$$

To estimate the relaxation rate, we recall that the constituent density of the QGP (see Eq. (4.18)),

$$n(T) = \frac{37\pi^2}{90}T^3, \tag{13.5}$$

gives for constituents at $T = 200\,\mathrm{MeV}$ an average separation distance $d = n^{-1/3} = 0.63/T = 0.63\,\mathrm{fm}$, as a reasonable estimate of the mean free path. If the constituents move at the speed of light, the average travel time between constituents is d^{-1}, giving

$$R_{\mathrm{coll}} = c\,n^{1/3} \simeq 0.32\,\mathrm{GeV} \simeq 4.8 \times 10^{23}\ s^{-1}. \tag{13.6}$$

as estimate for the collision rate. In other words, the rate for equilibration is by a factor of 10^{19} faster than the expansion rate (13.4): the system is at all times effectively in full equilibrium. For a description of the early universe around the hadronization transition, equilibrium thermodynamics thus appears fully adequate.

The fireball produced in heavy ion collisions is expected to expand with the speed of sound c_s in the medium at the given temperature. In the QGP a little above the hadronization transition it is found to approach $c_s^2 = (1/3)c^2$; see Sect. 8.7. We just had seen that the average separation between constituents in the QGP in the transition regime is of the order of $d \simeq 0.63/T$. Using the Hubble relation Eq. (13.2) between expansion velocity v and separation distance d, this leads to the expansion rate

$$H = \sqrt{1/3}\, 1.6\, T \simeq 0.18\, \text{GeV}. \tag{13.7}$$

Above we had found that the reaction rate in the plasma is $R_{\text{coll}} \simeq 0.32\,\text{GeV}$; hence here the reaction rate is only slightly greater than the expansion rate. It thus seems quite likely that non-equilibrium effects will still play a significant role in high energy nuclear collisions. At best we can assume that the medium is *locally* in thermal equilibrium, but distinct volume regions are expected to differ in energy density or baryon density. As a result, they will move relative to each other to equalize these differences: we can expect the QGP to flow with time.

There are two ways to address this problem. On a microscopic level, we can study the behavior near equilibrium in kinetic theory, based on the Boltzmann equation. Its solution would give us the modifications to the Boltzmann distribution $\exp -(\sqrt{p^2 + m^2}/T)$ due to collisions with other constituents and due to eventual external forces. This approach is essentially the extension of statistical mechanics to include non-equilibrium effects, and it has so far been carried out only for specific models. In a macroscopic picture, one can invoke fluid dynamics as a phenomenological approach: this is the extension of thermodynamics to include effects of flow both in an ideal (non-dissipative) form and in case of transport phenomena (dissipative fluid dynamics). The applicability of fluid dynamics, just as that of thermodynamics, requires the mean free path of the constituents to be much smaller than the dimensions of the system. With the value $d \simeq 0.63\,\text{fm}$ obtained above, nuclear collisions with an overall volume of a diameter $D \simeq 12\,\text{fm}$ this may well be the case. Assuming local equilibrium, i.e., using the equation of state of strongly interacting matter for a smaller local volume, one invokes the equations of relativistic fluid dynamics to determine the resulting ideal flow pattern, eventually adding dissipative effects such as viscosity and heat conductivity. We want to summarize here this second approach in a very conceptual way; for more details, see e.g. [1–4].

At this point, a comment on different possible motivations seems appropriate. Our aim here is the study of strongly interacting matter, and one widely used experimental tool for this aim is given by high energy collisions of heavy nuclei, as introduced in Chap. 9. In the present chapter, we want to address the question of how the observed final state in such collisions can provide us with information about strongly interacting matter under different conditions. A complementary but very distinct approach is the theoretical or phenomenological description of the detailed behavior observed in heavy ion collisions per se, based on specific models with perhaps only incidental reference to QCD equilibrium thermodynamics. These latter issues are dealt with in various approaches, from color glass condensate and glasma studies to relativistic molecular dynamics models. For further references, see e.g. [1].

13.2 Liquids, Gases, Fluids

We recall two different ways to classify states of matter. The scheme introduced more than two thousand years ago by Empedocles has *earth, water, air* and *fire*, in todays form *solids, liquids, gases* and *plasmas*. Under changes of temperature and pressure, solids maintain a fixed volume and shape. Liquids maintain a fixed volume, but adapt their shape to that of the container. For gases, both volume and shape are determined by the temperature and pressure. A plasma is a gas exhibiting charge conductivity. Generally speaking then, gases are compressible; liquids and solids are not, they have fixed, approximately invariant densities. Since strongly interacting matter, confined as well as deconfined, exists over a wide range of pressures and densities, it is evidently a gas and not a liquid in the conventional sense of the word.

A second type of classification is obtained by subjecting matter to shear stress, leading to two forms: *solids* and *fluids*. Solids remain (essentially) unaffected by such stress, while fluids are continuously deformed: they flow. Hence both liquids and gases are fluids. In colloquial use, liquid and fluid are sometimes used as synonymous. We just saw that here this leads to problems. Strongly interacting matter is an interacting gas and as such exhibits fluid properties, to be addressed by fluid dynamics; this discipline contains aerodynamics and hydrodynamics as subfields for the study of gases and liquids, respectively. Nevertheless, for historical reasons one very often still speaks of hydrodynamics even when dealing with gaseous media.

The basic measure of fluidity is the viscosity, which specifies the resistance of the fluid to deformation. We restrict ourselves here to *dynamic* or *shear viscosity*, which arises when two layers of fluid move relative to each other. It is due to the interaction between constituents moving at different speeds. In addition, there can be *bulk viscosity*, which describes a kind of internal friction resisting uniform expansion or compression of a fluid. This generally becomes significant in the case of very rapidly expanding or contracting fluids, and we shall not deal with it here.

Let us then look in more detail at shear viscosity. Consider a uniform fluid between two parallel plates of area A each, one stationary, the other moving at a constant velocity u through application of a force F (see Fig. 13.4). The (local) shear viscosity $\eta(y)$ is then defined as

$$\frac{F}{A} = \eta(y)\left(\frac{du}{dy}\right), \tag{13.8}$$

where y measures the distance between the plates and $u(y)$ the velocity with which the layer at level y moves relative to the bottom plate. Due to interactions between the constituents of the fluid with those in the upper plate, the molecules in the uppermost layer move effectively with velocity u; layer by layer the average velocity then decreases, until the molecules in the bottom layer remain effectively at rest. The role of the two plates is symbolical: the crucial feature is the transfer of momentum from the top fluid layer on down. Thus η is in a sense a measure of the friction between two layers of the medium moving relative to each other. In the limit in which this friction can be neglected we speak of a *perfect* or *ideal fluid*. We shall see below that quantum effects prevent any actual fluid from achieving this limit.

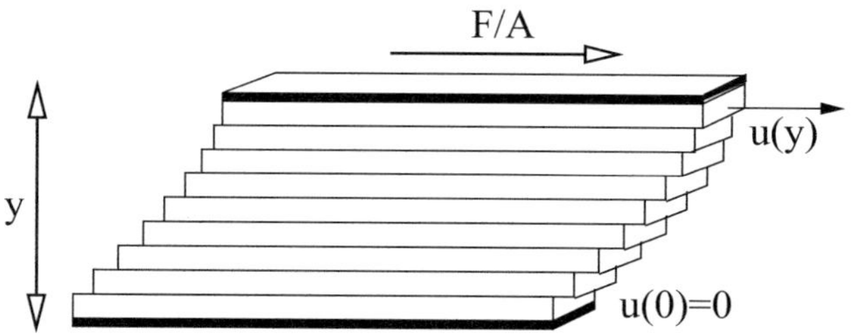

Fig. 13.4 Schematic viscosity measurement

Kinetic theory arguments going back to Maxwell relate the viscosity of a gas to its density n, the mean free path $\lambda = 1/n\sigma$ of its constituents, and to their average momenta p,

$$\eta = \frac{1}{3}np\lambda = \frac{1}{3}\left(\frac{p}{\sigma}\right);\qquad(13.9)$$

here σ denotes the interaction cross section between the constituents. The dependence on the density thus cancels out, initially even to Maxwell's surprise. But it becomes evident when one realizes that it is not the number of potential scattering centers that matters but the number of those on which scattering actually occurs. The silent majority does not matter here.

Relation (13.9) implies that strongly/weakly interacting media with large/small σ have a small/large viscosity. The strong interaction limit is quite evident: the larger the interaction cross section, the greater the transfer of the momentum change due the stress force to the next layer. It moves along faster with the top layer and hence results in smaller viscosity. The opposite limit, towards a weakly interacting gas, leads to increasing viscosity: the small cross-section inhibits momentum transfer between adjacent layers and thus makes it difficult for the upper layer to pull along the lower one. Strictly speaking, however, fluid dynamics requires a mean free path (much) shorter than the size of medium, and that as well as viscosity itself presupposes a certain interaction strength between the constituents. Nevertheless, if one considers the behavior for decreasing interaction cross sections, the viscosity of an interacting gas continues to increase, and that is what we should expect in the high temperature limit of the quark-gluon plasma.

It seems worthwhile to note here that the value of the viscosity alone tells us something about the relative fluidity of the medium only if we compare media of similar structure. Water is more fluid than oil and in fact does have a smaller viscosity. Similarly, hydrogen has a smaller viscosity than oxygen. But the typical gas viscosity is by two orders of magnitude smaller than that of liquids, and it is not completely clear what that means. One would like to measure the fraction of the work of the shear force dissipated in friction, and for that, a reference frame appears necessary. Since in some sense η measures the fraction of entropy density created by the flow friction, its ratio to the total entropy density s seems meaningful. For this reason, the ratio η/s has become a commonly employed measure. For many fluids, the entropy density is effectively given by the density of the number of constituents, $s \sim k_B n$, with k_B denoting the Boltzmann constant. As an example, an ideal relativistic Bose gas has $s/n \simeq 3.6\,k_B$, giving us [5]

$$\frac{\eta}{s} \simeq \frac{p\lambda}{10.8\,k_B}. \tag{13.10}$$

For quantum Bose fluids, the uncertainty relation $p\lambda \geq \hbar$ thus limits η/s by the bound

$$\frac{\eta}{s} \gtrsim \frac{1}{10.8}\frac{\hbar}{k_B}. \tag{13.11}$$

In other words, the viscosity of such a Bose gas cannot vanish; note however, that the actual values for most dilute Bose gases lie above this limit by an order of magnitude or more. In recent years, many studies have tried to establish a *universal* lower limit for η/s; we shall return to this issue later on, after first looking at some predictions of ideal fluid dynamics for the behavior of the medium formed in high energy nuclear collisions.

13.3 Ideal Relativistic Fluid Dynamics

The equations of relativistic fluid dynamics express the conservation of energy and momentum as well as that for specific charges (baryon number, electric charge, etc.) for a flowing medium consisting of spatial elements, a volume densely filled with *bubbles*. Within each bubble, thermal equilibrium is assumed. The size of the bubbles is assumed to be much smaller than the overall fluid volume, yet much larger than the separation between the microscopic constituents of the fluid. In a given reference frame, the momenta of the constituents of a bubble will thus be a superposition of their momenta in the rest-frame of the bubble and the overall flow momentum of the bubble. In the general case, dissipative effects (viscosity, conductivity) are in addition taken into account; we consider here for the time an ideal fluid without any such transport effects.

The fundamental variable of the resulting fluid dynamics is the stress or energy-momentum tensor,

$$T^{\mu\nu} = (\epsilon + P)u^{\mu}u^{\nu} - Pg^{\mu\nu}, \tag{13.12}$$

where $\epsilon(x)$ and $P(x)$ denote energy density and pressure in the local rest frame at spacetime point x; the flow four velocity is given by u^{μ}, normalized with $u_{\mu}u^{\mu} = 1$, and $g^{\mu\nu}$ is the Minkowski metric tensor. With this form of $T^{\mu\nu}$, we assume for the moment an ideal fluid, neglecting transport coefficients accounting for viscous effects and heat conductivity. In the rest frame of the fluid, i.e., for $u = (1, 0, 0, 0)$, the energy-momentum tensor has the diagonal form

$$T_0^{\mu\nu} = \begin{bmatrix} \epsilon & 0 & 0 & 0 \\ 0 & P & 0 & 0 \\ 0 & 0 & P & 0 \\ 0 & 0 & 0 & P \end{bmatrix}. \tag{13.13}$$

Thus the fluid has an energy density ϵ and exerts an isotropic pressure P on its surface. For simplicity, we shall assume here and in the following that charge densities, in particular the baryon density, are negligible; this means that the results we obtain are relevant only for the central rapidity region of high energy collisions. For the more general case, see [2]. The four equations

$$\partial_{\mu}T^{\mu\nu} = 0, \tag{13.14}$$

with $\partial_{\mu} = \partial/\partial x^{\mu} = (\partial/\partial t, \partial/\partial x_1, \partial/\partial x_2, \partial/\partial x_3)$, then correspond to the conservation of energy and momentum in flowing matter. They involve five variables, ϵ, P and three fluid velocity variables for the normalized flow u^{μ}; one thus has one more variable than equations. For an ideal fluid, this problem is solved by assuming that locally energy density and pressure are related by the equation of state.

In the rest frame of the fluid, for $u = (1, 0, 0, 0)$, Eq. (13.14) becomes very simple:

$$\frac{\partial \epsilon}{\partial t} = 0 \tag{13.15}$$

keeps the energy density constant in time, while

$$\frac{\partial P}{\partial x_i} = 0, \; i = 1, 2, 3 \tag{13.16}$$

implies that no pressure gradients exist, the pressure is constant throughout space.

One basic input for ideal relativistic fluid dynamics is then the *equation of state (EoS)* of strongly interacting matter, i.e., the energy density $\epsilon(T)$ and its relation to the pressure $P(T)$ for temperature T at an arbitrary point x in space time. Once we assume local thermal equilibrium, the temperature dependence and the relation between $\epsilon(T)$ and $P(T)$ are given by the strong interaction thermodynamics presented in the previous chapters. The specific values of these quantities at some given starting point x, however, the *initial conditions*, have to be introduced in some way from the outside: from what point on is local equilibrium a reasonable assumption for a given reaction? Given the EoS and the initial conditions, fluid dynamics then provides a scheme to calculate the evolution of the thermal parameters of the system in space and time.

Let us consider the fluid dynamic equations in some more detail, retaining as mentioned the case of vanishing baryon number (and other charges). Contracting the energy-momentum conservation equations (13.14) with the fluid velocity,

$$u_\nu \partial_\mu T^{\mu\nu} = 0, \tag{13.17}$$

and inserting Eq. (13.12) leads to

$$u_\nu \partial_\mu [(\epsilon + P)u^\mu u^\nu - g_{\mu\nu}P] = 0. \tag{13.18}$$

Using the normalization condition $u_\mu u^\mu = 1$, this gives the so-called first fluid dynamic equation

$$u^\mu \partial_\mu \epsilon + (\epsilon + P)\partial_\mu u^\mu = 0. \tag{13.19}$$

With the help of the thermodynamic relations obtained in Chap. 4,

$$\epsilon + P = Ts \tag{13.20}$$

and

$$\left(\frac{\partial \epsilon}{\partial s}\right) = T \tag{13.21}$$

where s denotes the entropy density, this equation simplifies to the entropy conservation law

$$\partial_\mu (u^\mu s) = 0. \tag{13.22}$$

Using the defining equation for the speed of sound c_s^2,

$$\frac{\partial P}{\partial \epsilon} = \frac{s}{T}\frac{\partial T}{\partial s} = c_s^2, \tag{13.23}$$

the entropy equation obtained from (13.22)

$$\partial_\mu u^\mu + u^\mu \partial_\mu \ln s = 0. \tag{13.24}$$

then becomes an equation for the temperature,

$$\partial_\mu u^\mu + \frac{1}{c_s^2} u^\mu \partial_\mu \ln T = 0. \tag{13.25}$$

A second set of fluid dynamic equations is obtained by the projection

$$(g_{\mu\nu} - u_\mu u_\nu)\partial_\sigma T^{\mu\nu} = 0, \tag{13.26}$$

which becomes

$$(\epsilon + P)u^\mu \partial_\mu u_\nu - \partial_\nu P + u_\nu u^\mu \partial_\mu P = 0. \tag{13.27}$$

As above, this can be converted into temperature equations, using the thermodynamic relations (13.20) and (13.21), leading to

$$u^\mu \partial_\mu (u_\nu T) - \partial_\nu T = 0. \tag{13.28}$$

The basic fluid dynamic equations in terms of energy density and pressure, Eqs. (13.25) and (13.28), thus become (13.25) and (13.28) in terms of temperature.

13.4 Longitudinal Expansion

Let us next discuss the solution of the fluid dynamical equations, using some physics input to simplify the situation. For a *central* nucleus-nucleus collision, the system has an intrinsic rotational symmetry around the collision axis, so that only two velocity components enter, v_z for longitudinal and v_r for radial motion, along and orthogonal to the collision axis, respectively. We thus have

$$u^\mu = \frac{1}{\sqrt{1 - v_z^2 - v_r^2}}(1, v_z, v_r, 0), \tag{13.29}$$

with no dependence on the azimuthal angle; note that this no longer holds for non-central collisions, which result in elliptic flow and hence a dependence on the azimuthal angle. In high energy nuclear collisions, the radial flow is much weaker than that in the longitudinal direction; we thus use $v_r \simeq 0$ as starting approximation. The velocity can then be put into the simpler form

$$u^\mu = (\cosh \theta, \sinh \theta, 0, 0), \tag{13.30}$$

where the fluid velocity variable $\theta(t, z)$ depends on both time t and (longitudinal) space z. In terms of this variable, Eq. (13.19) becomes

$$\cosh\theta(\partial\epsilon/\partial t) + \sinh\theta(\partial\epsilon/\partial z) + (\epsilon + P)[\sinh\theta(\partial\theta/\partial t) + \cosh\theta(\partial\theta/\partial z)] = 0, \tag{13.31}$$

while Eq. (13.27) leads to

$$\sinh\theta(\partial P/\partial t) + \cosh\theta(\partial P/\partial z) + (\epsilon + P)[\cosh\theta(\partial\theta/\partial t) + \sinh\theta(\partial\theta/\partial z)] = 0. \tag{13.32}$$

These two equations specify the longitudinal expansion of the fluid in terms of the energy density ϵ, the pressure P and the fluid velocity θ. For a solution we thus still need a third relation – as indicated, the equation of state connecting ϵ and P can play that role.

To proceed with the solution, it is convenient to change the space-time variables x, t to space-time rapidity η_s and proper time τ,

$$\eta_s = \frac{1}{2}\ln\frac{t+z}{t-z}, \quad \tau = \sqrt{t^2 - z^2} \tag{13.33}$$

For a Lorentz boost along the z-axis, η_s changes by an additive constant, while τ remains invariant. The inversion of Eqs. (13.33) is given by

$$t = \cosh\eta_s, \quad z = \sinh\eta_s. \tag{13.34}$$

In terms of $y\eta_s$ and τ, the fluid dynamic equations become

$$\tau(\partial\epsilon/\partial\tau) + \tanh(\theta - \eta_s)(\partial\epsilon/\partial\eta_s) + (\epsilon + P)[(\partial\theta/\partial\eta_s) + \tanh(\theta - \eta_s)\tau(\partial\theta/\partial\tau) = 0 \tag{13.35}$$

and

$$(\partial P/\partial\eta_s) + \tanh(\theta - y)\tau(\partial P/\partial\tau) + (\epsilon + P)[\tau(\partial\theta/\partial\tau) + \tanh(\theta - \eta_s)(\partial\theta/\partial\eta_s)] = 0. \tag{13.36}$$

The fluid velocity thus is now a function of τ and η_s, $\theta(\tau, \eta_s)$.

To arrive at a solution of Eqs. (13.35) and (13.36), we have to specify the initial conditions of the system at the time when a fluid dynamic description starts being applicable, i.e., after the equilibration time. The first proposal, by Landau [6], assumed that the two colliding nuclei would completely stop each other, depositing the incident energy in a small volume Lorentz-contracted along the collision axis. For high energy collisions, this scenario was eventually ruled out on both theoretical and experimental grounds. Causality implied that disjoint regions of the nuclei could not be stopped simultaneously, resulting in the concept of *nuclear transparency*. The observation of fast leading baryons supported this empirically: the incident nuclei continued largely unhindered out of the collision region. In proton-proton collisions this moreover led to the observation of a plateau in the multiplicity distribution of secondaries as function of rapidity, as illustrated in Fig. 13.5. It led to the Bjorken scenario [7], to which we return shortly.

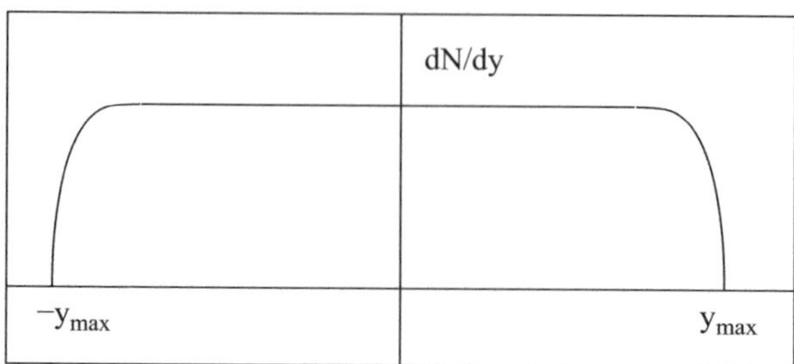

Fig. 13.5 Schematic rapidity distribution of secondaries in high energy proton-proton collisions, as function of the center-of-mass rapidity y

Here y is the momentum space rapidity,

$$y = \frac{1}{2}\log\left(\frac{p_y + E}{p_z - E}\right) \simeq \eta_s, \tag{13.37}$$

where p_z and E denote the longitudinal momentum and E the energy of the secondary. In the Bjorken scheme, y is taken to be equal to the space-time fluid rapidity η_s [8].

To obtain such a plateau behaviour, Eqs. (13.35) and (13.36) should be independent under translations in rapidity y, which evidently requires

$$\theta(y, \tau_0) = y, \tag{13.38}$$

where τ_0 is the initial time for the fluid dynamic evolution. Such a scaling initial condition was proposed by Bjoerken [7], and although nuclear collisions show some deviations, the resulting picture provides the basis of fluid dynamic studies. The condition (13.38) leads to a greatly simplified form of the fluid dynamic equations, with

$$\tau(\partial\epsilon/\partial\tau) + \epsilon + P = 0 \tag{13.39}$$

and

$$(\partial P/\partial y) = 0, \tag{13.40}$$

since the condition (13.38) also holds at all later times τ. The entropy relation (13.25) takes on a similarly simplified form,

$$\tau(\partial s/\partial\tau) + s = 0 \tag{13.41}$$

These equations are now readily solvable. From Eq. (13.41) we have

$$s/s_0 = \tau_0/\tau \tag{13.42}$$

for the evolution of the entropy density. Since a volume element is given by

$$d^3x = dz\,d^2x_r = \tau\,dy\,d^2x_r, \tag{13.43}$$

with x_r denoting the (constant) radial components, Eq. (13.42) implies that the entropy S, with $s = S/d^3x$, in a given rapidity interval dy remains constant during the evolution: the expansion is *isentropic*.

From the temperature form of the entropy equation (13.25) we obtain

$$T/T_0 = (\tau_0/\tau)^{c_s^2}, \tag{13.44}$$

Since $c_s^2 < 1$, this means that the temperature drops more slowly than the corresponding entropy density. If we now invoke the equation of state for an ideal gas of massless particles, $\epsilon = 3P$, we can use the thermodynamic relation (13.20) to obtain from Eq. (13.39)

$$\partial\epsilon/\partial\tau + 4\epsilon/3\tau = 0, \tag{13.45}$$

with the solution

$$\epsilon_0 = \epsilon\left(\frac{\tau}{\tau_0}\right)^{4/3}. \tag{13.46}$$

In contrast, free streaming emission without fluid dynamic flow implies $P = 0$ and hence gives

$$\epsilon_0 = \epsilon\left(\frac{\tau}{\tau_0}\right). \tag{13.47}$$

If we now measure the value ϵ of the energy density at the end of the collision evolution, the presence of fluid dynamic flow implies that the initial energy density ϵ_0 must have been higher than it would have been for free streaming emission, by a factor $(\tau/\tau_0)^{1/3}$. In this case, all initial energy goes into free particle production, while in the presence of flow, part is converted into collective motion.

To connect these considerations directly to measured quantities, we assume that in a given rapidity interval dy around $y = 0$ one observes the production of dN/dy secondaries, each of which has an average transverse momentum m_T. The final energy density at mid-rapidity then is

$$\epsilon = \frac{m_T}{\pi R_A^2 \tau}\left(\frac{dN}{dy}\right)_{y=0} \tag{13.48}$$

where πR_A^2 specifies the transverse nuclear area; recall that so far we assumed no transverse expansion. As a result, we deduce for the case of one-dimensional fluid dynamic flow [7] an initial energy density

$$\epsilon_0 = \frac{m_T}{\pi R_A^2 \tau_0} \left(\frac{dN}{dy}\right)_{y=0} \left(\frac{\tau}{\tau_0}\right)^{1/3},$$

(13.49)

while freely flowing secondaries give with

$$\epsilon_0 = \frac{m_T}{\pi R_A^2 \tau_0} \left(\frac{dN}{dy}\right)_{y=0}$$

(13.50)

the approximate form we had used in Sect. 9.2 above. To estimate the uncertainty involved, we recall from above that the energy density at the quark-hadron transition is in the range 0.5–1.0 GeV/fm^3. For initial energy densities below 10 GeV/fm^3, the factor $(\tau/\tau_0)^{1/3}$ is in the range 1–2. This provides some *a posteriori* justification for the frequent use of expression (13.50) for the initial energy density in nuclear collisions. Moreover, we conclude from these considerations that in this energy density regime, longitudinal expansion alone does not really provide us with a direct reflection of flowing matter.

13.5　Transverse Flow

The full 3-d solution of the fluid dynamical equations even in their ideal form is a rather complicated task and can only be carried out numerically. That is why we proceed here in a rather qualitative way. We have seen that the restriction to only longitudinal flow and expansion allows an analytic and conceptually understandable solution. In this section, we now want to discuss the role of transverse flow, and in the following section then turn to modifications due to dissipative transport phenomena.

Since the scaling solution is invariant under boosts in rapidity, we can consider the radial behavior for a given value, say $y = 0$, and then Lorentz-boost the result in the longitudinal direction. For the moment, we still remain with central collisions. The secondaries emitted here come from a cylindrical piece of fluid at rest in the longitudinal direction, expanding with time in the radial direction, again according to the relevant fluid dynamic equations. The resulting radial expansion provides the temperature profiles as function of time and radial distance. We shall not consider here the fluid dynamical calculations leading to the solution; for details see [2, 3]. Instead we want look for possible observable effects.

If the secondaries come from a single thermodynamic source at rest, the transverse momentum distributions have the Boltzmann form (see Sect. 11.1)

$$\left(\frac{d^2N}{dydm_T^2}\right)_{y=0} \sim \left(\frac{dN}{dy}\right)_{y=0} \exp-(m_T/T_f)$$

(13.51)

where $m_T = \sqrt{(m^2 + p_T^2)}$ defines the transverse mass of the secondary, while T_f denotes the universal freeze-out temperature. Equation (13.51) is obviously of schematic nature – we have assumed one species of secondary only and ignored possible resonance feed-down modifications in the real world. We now want to consider the effect on transverse hadron spectra due to a superposition of radially moving fireballs. It can be shown through the cited detailed calculations [1] that transverse fluid flow leads to the form

$$\left(\frac{d^2N}{dydm_T^2}\right)_{y=0} \sim \left(\frac{dN}{dy}\right)_{y=0} K_1(z_1)I_0(z_2),$$
(13.52)

where

$$z_1 = \frac{m_T \cosh y_r}{T_f} \quad z_2 = \frac{p_T \sinh y_r}{T_f}$$
(13.53)

are transverse rapidity variables and K_1 and I_0 denote modified Bessel function of the second and first kind, respectively. The transverse rapidity y_r is specified by

$$y_r = \frac{1}{2} \ln\left(\frac{1 + v_r}{1 - v_r}\right),$$
(13.54)

with v_r denoting the radial flow velocity. For $v_r \to 0$, we have $y_r \to 0$ and hence $z_1 \to m_T/T_f$ and $z_2 \to 0$. In this limit, we get $I_0(0) = 1$, so that for large m_T/T_f with $K_1(m_T/T_f) \to \exp-(m_T/T_f)$ we recover the single fireball form (13.51). In the following we want to consider how the presence of fluid flow modifies this generic transverse momentum distribution, and how flow results in different behavior for secondaries of different masses.

In the case when $m_T \sim p_T \gg T$, both Bessel functions in Eq. (13.51) lead to exponential behavior, with $K_1(x) \sim \exp(-x)$ and $I_0(x) \sim \exp(+x)$. As a result, we obtain an exponential form

$$\left(\frac{d^2N}{dydm_T^2}\right)_{y=0} \sim \left(\frac{dN}{dy}\right)_{y=0} \exp-(m_T/T_f^{\text{eff}})$$
(13.55)

with a higher effective temperature

$$T_f^{\text{eff}} \simeq T_f \sqrt{\frac{1 + v_r}{1 - v_r}}.$$
(13.56)

This means that due to flow the average value of the transverse mass is shifted to higher values: the flow motion gives the emitted hadron a kick in the outward direction, depleting low and enhancing high transverse momenta. The superposition thus leads to a shift of the schematic form illustrated in Fig. 13.6. The transverse momentum distribution of the secondaries becomes broadened as a result of radial flow.

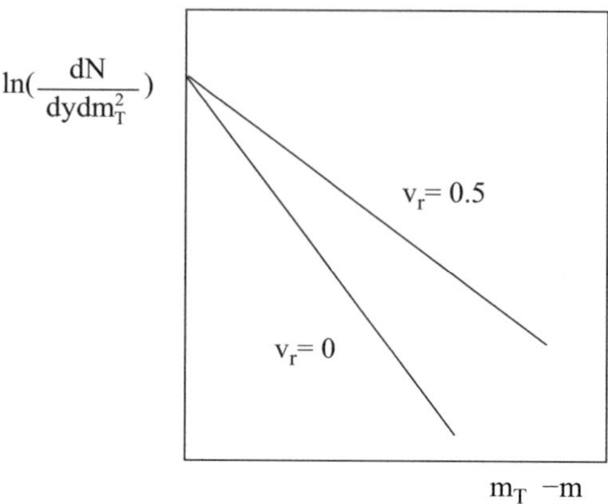

Fig. 13.6 Transverse momentum broadening through radial flow

The role of the specific mass of the secondary becomes evident by considering another limit, with $m \gg p_T$, $m/T \gg 1$ and $T/m \gg v_r^2$. In this case, one finds [1] as effective temperature

$$T_f^{\text{eff}} = T_f + \frac{1}{2}mv_r^2. \tag{13.57}$$

Since $\Delta T = T_f^{\text{eff}} - T_f = mv_r^2/2$ is a measure of the flow-induced momentum broadening shown by the emitted secondary, it is clear that heavier particles flow more strongly. If we compare the inverse slope parameter T_f^{eff} for the transverse mass distributions of different hadronic secondaries produced in nuclear collisions, we expect to find the behavior schematically illustrated in Fig. 13.7. Note that in contrast proton-proton collisions lead to a universal value for all hadron species (see Sect. 11.1).

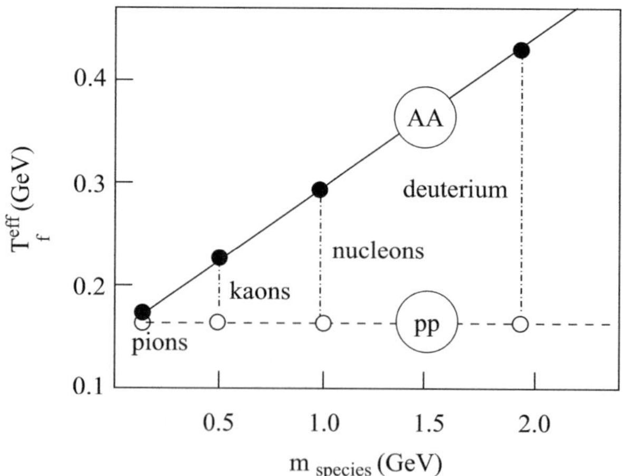

Fig. 13.7 Schematic form of the inverse transverse slope parameter for different hadron species in nuclear (closed circles) and proton-proton collisions (open circles)

A pattern of the form shown in Fig. 13.7 is indeed observed when comparing transverse mass spectra for pions, kaons, nucleons and deuterium from proton-proton and nuclear collisions. This mass dependence of T_f^{eff} in nuclear collisions therefore is one of the crucial pieces of evidence for the presence of radial flow of the produced medium. We have here presented the case in a somewhat simplified form; in the determination of T_f^{eff} from actual data, various other effects (resonance feed-down, hard production at high p_T, etc.) have to be taken into account [9].

So far, we have considered central collisions. A very specific flow pattern emerges in the case of non-central interactions. Consider the transverse geometric profile of such a collision (see Fig. 13.8), for two equal nuclei of radius R colliding at an impact parameter b. Here we label the collision axis as z, the axis aligned with the impact parameter as x, and the remaining one as y. The plane $x - z$ is generally denoted as the reaction plane. Due to the inherent asymmetry around the collision axis now present, we here expect the particle production to depend on the azimuthal angle ϕ, in contrast to the case of central collisions ($b = 0$, no ϕ-dependence).

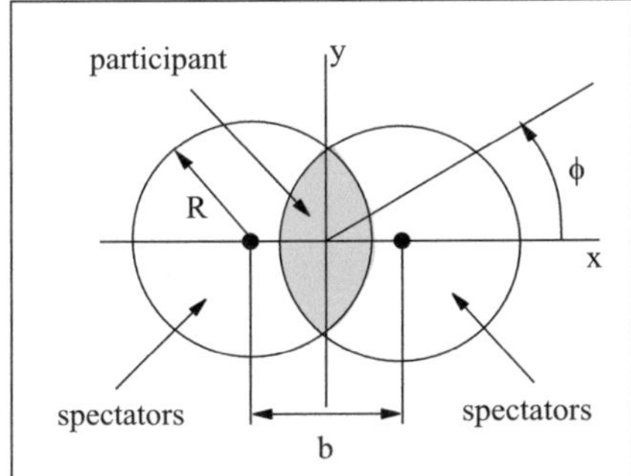

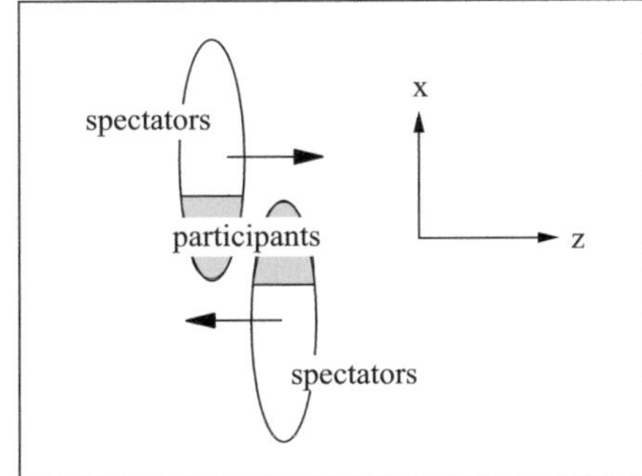

Fig. 13.8 Schematic form of non-central collisions of two equal nuclei. Left: the x-axis is aligned with the impact parameter b; the collision axis z is into the page. Right: the profile in the reaction plane x, z; the nuclei are Lorentz-contracted along the collision axis z

Let us look at such a non-central collision at high energy. Along the collision axis, the nuclei are very flat through Lorentz-contraction, so in a very short time, the spectator sections of the nuclei have left the interaction region. What is left behind, at rest in the center of mass, is the interacting medium of the participants; in coordinate space, it has an almond-like shape, shortest along the x-axis. Both density and pressure gradients in this volume are non-uniform, attaining larger values (more compression) along the x-axis than along the y-axis. As a result, the expansion and the flow will be stronger in the reaction plane (along x) than out of it (along y); hence in momentum space, the almond is flipped to have its longest axis in the x direction. The detailed flow analysis is carried out using the form

$$\frac{dN}{dy\,d^2 p_T} = \frac{dN}{2\pi\,dp_T^2\,dy}\left[1 + \sum_{k=1}^{\infty} 2v_k \cos(k\phi)\right] \tag{13.58}$$

with the angle ϕ as defined above (see Fig. 13.8). The azimuthal asymmetry of the production pattern is thus determined by the flow coefficients v_k; they depend in general on the rapidity y and the transverse momentum p_T of the secondary, $v_k(y, p_T)$. For symmetry reasons, for equal nuclei we have $v_1(y) = -v_1(-y)$: the result at backward rapidity must be the reflection of that at forward rapidity. Hence the coefficient v_1 of what is called *directed flow* must vanish for $y = 0$, $v_1(y=0)=0$, and to make use of this simplification, we consider production at mid-rapidity. The next coefficient, $v_2(y=0, p_T)$ characterizes what is known as *elliptic flow*; it has been extensively studied both experimentally and theoretically. The origin of the name becomes evident if we consider as illustration the square bracket in Eq. (13.58) for the case $v_2 = 1/4$ and all other $v_k = 0$. The resulting distribution is illustrated in Fig. 13.9; in the actual case, $v_2(y=0, p_T)$ will vary with p_T, and it will moreover depend on the mass of the secondary hadron in question.

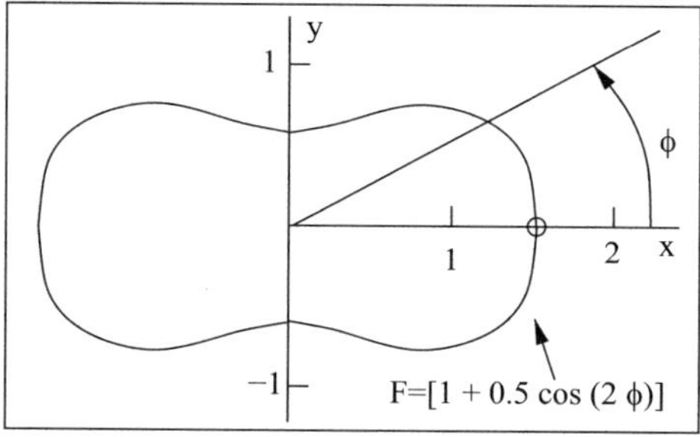

Fig. 13.9 Schematic form of the azimuthal distribution function $F = 1 + 2v_2 \cos(2\phi)$, for illustration with $v_2 = 1/4$; the collision axis z goes orthogonally into the page

The mass dependence of v_2 is a consequence of the stronger transverse flow for heavier hadrons mentioned above (see Eq. (13.57)). As a result, a given value of v_2 corresponds to a higher p_T for nucleons than for pions. The resulting pattern for $v_2(y = 0, p_T)$ for different hadrons is schematically illustrated in Fig. 13.10, and such a mass ordering is indeed observed in high energy nuclear collisions.

The observation of elliptic flow constitutes perhaps the strongest evidence for the occurrence of collective behavior in high energy nuclear collisions: for non-central collisions, the initial azimuthal asymmetry in coordinate space is through flow translated into an azimuthal momentum space asymmetry. If the produced medium would not exhibit collective behavior, i.e., if the secondaries would be emitted independently by the individual nucleon-nucleon interactions, no such asymmetry would arise. We note that for much higher transverse momenta, the produced particles arise from individual hard interactions, so that above some 2–3 GeV, $v_2(p_T)$ begins to decrease with p_T and eventually vanishes. Furthermore, we note that the comparison with fluid dynamic calculations was based on a ideal fluid, without any viscosity. We now have to check how the reduction of flow arising in the presence of viscosity affects these results.

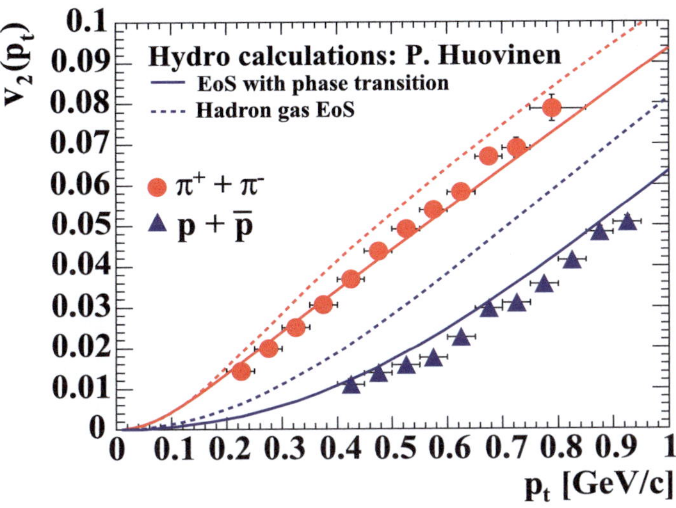

Fig. 13.10 Elliptic flow coefficient $v_2(p_T)$ based on ideal hydrodynamic calculations [10] for pions and protons (solid lines); the data points indicate the results from Au-Au collisions at 130 GeV [11]

13.6 Viscous Flow

So far, we have considered the effect of ideal (non-viscous) flow on the longitudinal and transverse production of secondaries in high energy nuclear collisions. In the absence of any collective effects, such collisions would produce hadrons in the form of a simple superposition of individual nucleon-nucleon collisions. Deviations from such a simple picture thus indicate the presence of some form of collective behavior. Fluid dynamics assumes that through a thermalization process early in the collision the total incident energy is repartitioned such that a certain fraction goes into collective motion of clusters of secondaries from different collisions. Ideally, these clusters move relative to each other without friction, so that the total entropy is preserved during the entire evolution. The next step would be to hinder the relative motion of the clusters through friction, producing heat: the reaction then becomes dissipative, increasing the entropy.

The presence of dissipation through shear viscosity modifies the stress tensor as given for the ideal fluid in Eq. 13.12); it acquires an additional term for shear viscosity η,

$$T^{\mu\nu} = (\epsilon + P)u^\mu u^\nu - Pg^{\mu\nu} + \eta S^{\mu\nu}, \tag{13.59}$$

where $S^{\mu\nu}$ is a traceless tensor constructed from the fluid velocity u, its derivatives and the metric tensor $g^{\mu\nu}$; for more details, see [12]. Such dissipation has essentially two effects. In the longitudinal direction, it reduces flow; the viscosity acts like friction between the flowing layers. In the transverse direction, on other hand, the produced heat enhances the outward flow beyond what it would be for non-viscous longitudinal flow. That means that the pressure in the stress tensor (13.13) is no longer isotropic; in the longitudinal direction it is reduced by the viscosity,

$$P_z = P - \frac{4}{3}\left(\frac{\eta}{\tau}\right),$$ (13.60)

while in the radial direction it becomes enhanced,

$$P_r = P + \frac{2}{3}\left(\frac{\eta}{\tau}\right).$$ (13.61)

Here P denotes the isotropic pressure in the non-viscous case, while τ specifies the proper time. This evidently affects both longitudinal and transverse flow; its consequence on elliptic flow is most striking and has attracted the greatest attention. The increase of v_2 with p_T is significantly damped by even small values of η/s, as shown in Fig. 13.11 [13]. The data taken at the Relativistic Heavy Ion Collider RHIC in Brookhaven, included in this figure, therefore established that the interacting medium produced in high energy nuclear collisions must have an extremely low viscosity: the quark-gluon plasma acts as an almost ideal fluid and hence must experience very strong interactions in the regime just above deconfinement. We note that this confirms nicely the prediction of lattice QCD, showing that the trace anomaly as interaction measure exhibits a pronounced peak in the temperature range between T_c and $2\,T_c$ (see Sect. 8.3).

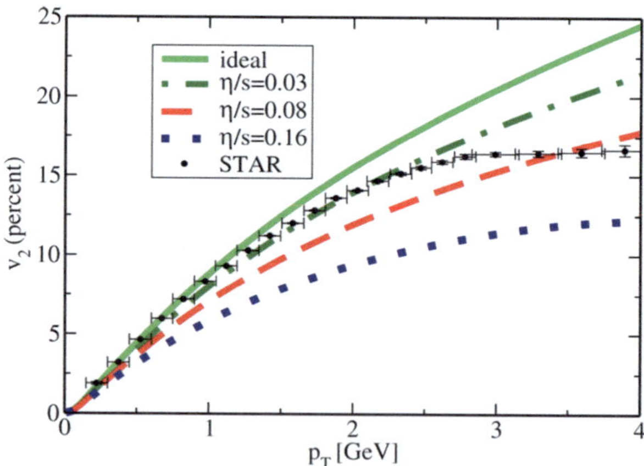

Fig. 13.11 Elliptic flow vs. p_T for different values of η/s, together with Au-Au data at $\sqrt{s} = 200\,\text{GeV}$ from RHIC [13]

The observation of such low viscosity triggered a renewed interest in the issue of a possible universal lower bound to η/s. A comparison with conventional fluids showed that liquids have values orders of magnitude higher, and even most gases lead to an η/s around one, well above the range required for the RHIC data. The lowest values, for superfluid liquid helium for a temperature near absolute zero seem still too high for the behavior obtained in the mentioned elliptic flow studies.

13.7 Gravity Dual Limits

Viscosity limits are difficult to establish in lattice QCD, and perturbation theory addresses the weakly interacting regime. Therefore a first and still speculative answer came from an unexpected side: string theory. Before we turn to the gauge/gravity duality leading to it, we briefly pause to discuss viscosity in the context of black hole physics.

In Chap. 12 we had recalled the crucial features of black holes. In particular, we had seen that the production of virtual pairs near the event horizon leads to Hawking-Unruh radiation, emerging from the black hole surface. This and some other features have been combined in the so-called *membrane paradigm*, in which the surface of the black hole is pictured as a thin, hot membrane, which emits the Hawking radiation into outer space. This membrane exhibits fluid properties, and it is found [14] to have a shear viscosity

$$\eta_{\mathrm{bh}} = \frac{1}{16\pi}. \tag{13.62}$$

On the other hand, the entropy of a black hole is given by it surface area in Planck length units, $S = k_B(A/4)(c^3/\hbar)$, with $G = 1$ for the gravitational constant. Hence the entropy density becomes

$$s_{\mathrm{bh}} = \frac{1}{4}k_B\left[\frac{c^3}{\hbar}\right], \tag{13.63}$$

so that we obtain

$$\left(\frac{\eta}{s}\right)_{\mathrm{bh}} = \frac{1}{4\pi}\left[\frac{\hbar}{k_B}\right] \tag{13.64}$$

for the ratio viscosity to entropy density of the black hole membrane. It is slightly smaller than the Bose value given in Eq. (13.10), and, as we shall see, it has in fact been conjectured to be the universal lower bound of η/s for all fluids [15].

In our present context, the question is less general: what does the shear viscosity of the black hole surface membrane have to do with that of the quark-gluon plasma? There is no definite answer, but some very interesting speculations exist, perhaps best labelled *gauge/gravity duality*. We can discuss this topic here only very qualitatively; for details, see [16].

Consider two quite different worlds. On one hand, a gauge field theory in the usual $3+1$ space-time dimensions. On the other hand, a general theory of gravitation-like interactions in $4+1$ dimensions. The idea now is that the surface of a black hole in this gravitation world is equivalent, dual, to the mentioned gauge field theory. We can use one or the other description, depending on which is more suitable for our given purpose. In particular, strongly interacting phenomena in the gauge world correspond to weakly interacting features in the gravity world. So ideally the strong interaction physics so difficult to address in QCD could be easier to deal with in a dual gravity theory. That is the dream – which actually started the other way around, as a hope to deal with difficult features in quantum gravity by looking at them in a dual gauge theory.

A decisive step towards the realization of this dream was made by J. Maldacena [17], who showed that such a duality is in fact possible. However, the gauge theory he used was not QCD, but a conformal field theory without confinement, chiral symmetry breaking and asymptotic freedom. He showed that a specific conformal gauge theory (CFT), supersymmetric Yang-Mills theory, is equivalent to a surface in a higher dimensional Anti-de-Sitter (AdS) space. This AdS/CFT correspondence thus was a successful application of the so-called holographic principle [18–20], which states that the information contained in a volume of space can be encoded on a boundary, a surface of space. This principle had solved the entropy puzzle in black hole physics: if you throw a complex structure into a black hole, what happens to its entropy? The holographic answer is that all information about what went into the black hole is preserved in the entropy of its surface.

The calculations of features of the resulting black hole membranes in general gravity theories correspond to the strong coupling limit of the associated dual gauge theory. And it was shown [15] that all theories with gravity duals lead for η/s to the value (13.64) that we had obtained above for "normal" black holes. That is why it was then conjectured that this might in fact be the universal lower bound for the viscosity to entropy ratio – a limit also expected for QCD.

Experimentally, this limit appears to be obeyed for all fluids studied. Theoretically, the situation appears less clear, but so far it does not seem possible to construct models with arbitrarily small η/s, in accord with the discussion about Eqs. (13.10) and (13.11).

In closing this section, we conclude that the observation of flow, in particular of elliptic flow, has established that high energy nuclear collisions do lead to collective behavior: they produce a medium which we can indeed call matter. This medium reaches the highest temperatures ever produced in the laboratory, and its viscosity in the regime of strong interactions makes it perhaps the closest to an ideal fluid ever created.

References

1. K. Yagi, T. Hatsuda, Y. Miake, *Quark-Gluon Plasma* (Cambridge University Press, Cambridge, 2005)
2. R. Vogt, *Ultrarelativistic Heavy-Ion Collisions* (Elsevier, Amsterdam/London, 2007)
3. W. Florkowski, *Phenomenology of Ultra-Relativistic Heavy-Ion Collisions* (World Scientific, Singapore, 2010)
4. T. Hirano, N. van der Kolk, A. Bilandzic, Hydrodynamics and flow, in *The Physics of the Quark-Gluon Plasma*, ed. by S. Sarkar, H. Satz, B. Sinha. Springer Lecture Notes in Physics, vol. 785 (Springer, Berlin, 2010)
5. T. Schaefer, D. Teaney, Rept. Prog. Phys. **72**, 126001 (2009)
6. L.D. Landau, Izv. Akad. Nauk SSSR **17**, 51 (1953)
7. J.D. Bjorken, Phys. Rev. D **27**, 140 (1983)
8. D. Teaney, arXiv:0905.2433; published in *Quark-Gluon Plasma 4*, ed. by R. Hwa, X.-N. Wang (World Scientific, Singapore)
9. N. Borghini, P.M. Dinh, J.-Y. Ollitrault, Phys. Rev. C **62**, 034902 (2000)
10. P. Huovinen, P.V. Ruuskanen, Ann. Rev. Nucl. Part. Sci. **56**, 163 (2006)
11. R. Snellings, Eur. Phys. J. C **49**, 87 (2007)
12. E. Shuryak, Prog. Part. Nucl. Phys. **53**, 273 (2004)
13. P. Romatschke, U. Romatschke, Phys. Rev. Lett. **99**, 172301 (2007)
14. K.S. Thorne, R.H. Price, D.A. Macdonald, *Black Holes: The Membrane Paradigm* (Yale University Press, New Haven/London, 1986)
15. P. Kovtun, D.T. Son, A.O. Starinets, Phys. Rev. Lett. **94**, 4204 (2005)
16. J. Casalderrey et al., *Gauge String Duality, Hot QCD and Heavy Ion Collisions* (Cambridge University Press, Cambridge, 2014)
17. J.M. Maldacena, Int. J. Theor. Phys. **38**, 1113 (1999)
18. G. t'Hooft, *Dimensional Reduction in Quantum Gravity*. arXiv:gr-qc/9310006
19. G. t'Hooft, *The Holographic Principle*. arXiv:hep-th/0003004
20. L. Susskind, J. Math. Phys. **36**, 6377 (1995)

Chapter 14
Outlook

Et le géographe, ayant ouvert son registre, tailla son crayon. On note d'abord au crayon les récrits des explorateurs. On attend, pour noter a l'encre, que l'explorateur ait fourni des preuves

Antoine de Saint-Exupéry, *Le Petit Prince*

(And the geographer opened his register and sharpened his pencil. One first writes down in pencil the stories of the explorers. Before writing them in ink, one waits until the explorer has furnished proofs.)

(Antoine de Saint-Exupéry, *The Little Prince*)

Abstract We summarize the status of the search for the quark-gluon plasma and the deconfinement transition in both theory and experiment. Lattice studies of finite temperature QCD provide clear predictions for this, and high energy nuclear collision experiments have shown promising indications for the laboratory production of the primordial matter of the early universe.

Strong interaction thermodynamics, based on quantum chromodynamics as fundamental interaction theory, predicts the quark-gluon plasma as a new state of matter. This prediction is in nature similar to that of the existence of Bose-Einstein condensates or of black holes. Given QCD as basic theory, the existence of the QGP follows. To theory, the challenge is to determine as many features of this new state as possible. This is quite a particular challenge, since we are dealing with a relativistic quantum field theory in a regime where, clearly, the interaction cannot be addressed by perturbative techniques. So far, the computer simulation of the lattice formulation appears to be the only viable approach for ab initio calculations in the temperature range of interest to possible experimental applications. Fortunately, the development of ever more powerful and more efficient computing facilities makes the continuum extrapolation even for full QCD only a matter of time. Complementary progress in analytic studies should then lead to a meeting of numerical with perturbative

© Springer International Publishing AG 2018
H. Satz, *Extreme States of Matter in Strong Interaction Physics*,
Lecture Notes in Physics 945, https://doi.org/10.1007/978-3-319-71894-1_14

approaches. In any case, for the interpretation of the results, accompanying model or "effective" theory studies have so far been indispensible, and they may well remain so.

Experimentally, the obvious challenge is to produce and study the predicted new state of matter in the laboratory. Also here we have quite a novel challenge. What exactly is the quark-gluon plasma from an empirical point of view? Given the definition and the specification of essential properties by statistical QCD: what can and must be measured in order to claim that the QGP was indeed formed in the nuclear collisions studied? And moreover, experiments can observe unexpected features which statistical QCD then a posteriori finds to be specific properties of the QGP. So the search for the QGP in the laboratory obviously requires intensive complementary theoretical and experimental efforts. It should be kept in mind, however, that the problem cannot be solved by defining the results of high energy nuclear collision experiments per se as QGP production, however novel they might be.

The observable phenomena discussed in this book, hadronic and electromagnetic radiation, quarkonium dissociation and jet quenching, can, in principle, provide crucial information on the nature of the QGP; but they might also arise from more mundane initial state as well as pre-thermal or confined final state sources. So the identification of measured features as due to QGP formation is clearly a difficult task. It is perhaps of some consolation that the Bose-Einstein condensate was predicted in 1924 and first observed experimentally in 1995. We may well need more patience.

The straight-forward "canonical" approach would be to measure, if possible, specific features which have been calculated in statistical QCD. Quantum chromodynamics itself was established in this way, by showing that hard scattering processes and the associated running coupling agree with the calculations of perturbative QCD. A conceivable example of this type was discussed in Chap. 10: the dissociation of quarkonia. The dissociation temperatures and energy densities of up to ten different charmonium and bottomonium states can, in principle, be determined theoretically, although so far, this determination is far from conclusive. Through the dependence of quarkonium production on centrality and on collision energy, it seems feasible to determine the energy densities at which the different states are dissolved. Do they match with the theoretical dissociation points? Obviously, it is a primary task to obtain for other probes similar relations between experiment and QCD calculations.

Undoubtedly it can be said today that the QGP search by means of nuclear collision experiments has led to the observation of various novel features and the creation of an environment hotter and denser than ever before. A variety of phenomenological considerations suggest that, indeed, QGP formation provides a common explanation. But except for one observation, we have so far not yet reached the point of directly matching data with QCD calculations. The forthcoming experiments at the CERN-LHC are thus accompanied by great hopes and expectations.

The one feature which has emerged over the years in high energy hadron interactions is the existence of a universal hadronization temperature, which in fact coincides with the confinement/deconfinement temperature calculated in statistical QCD. The relative abundances of the hadron species produced in *all* high energy collisions are in agreement with a resonance-gas temperature of about 160 MeV. As mentioned in Chap. 12, the observation of this fact in elementary collisions cannot be easily accommodated in a kinetic thermalization scenario. On the other hand, it may well point to a universal hadronization mechanism, which becomes operative whenever quarks are forced to separate beyond a confinement distance. In elementary collisions, this occurs dynamically, in parton cascades; in a hot QGP, it occurs thermally, when the medium expands, cools off and the constituents are forced to separate in a way to retain color neutrality. Nucleus-nucleus collisions could, in principle, follow either or both of the two routes. In any case, at present the quantitative empirical information we can compare to QCD calculations concerns the end of the new state of matter. It specifies the parameter at which the unknown pre-hadronic world turns into our conventional world of color-neutral hadrons in a physical vacuum.

And so for the time being, the challenge remains, concerning an unambiguous experimental determination and study of the quark-gluon plasma itself. We have numerous promising and exciting indications pointing "in the right direction", but we still have to identify that measurable phenomenon which can also be calculated in and compared to statistical QCD studies. Besides this, there are at least two further, more theoretical challenges.

At low or vanishing baryon density, lattice studies give us the temperature-dependence of the energy density, the specific heat, and other thermodynamic observables above deconfinement. It is not clear, however, what the origin of the observed behavior is. Eventually some form of effective field theory should hold, but in the critical region, up to $2\,T_c$ or more, they must fail. What mechanism dominates there? In the critical regime below T_c, resonance formation appears to provide a reasonable conceptual picture of what is happening. Above T_c, we don't yet have a corresponding scenario.

On the "other side", at high baryon density and low temperatures, all descriptions remain speculative. Do deconfinement and chiral symmetry restoration still coincide there, or do we have a plasma of massive colored quarks? And a major possible state, the color superconductor in its various forms, has not been addressed here at all. So the high temperature part of the world for which we do have quite a bit of insight – that may be just the tip of iceberg, with all the cold dense "underwater" matter not accessible to our present standard tools. Is it really impossible to formulate an ab initio scheme for non-perturbative QCD calculations at low temperature and high baryon density?

So we can conclude that this field of physics, addressing complex systems in the realm of strong interactions, combines the most interesting features of statistical, nuclear and particle physics – and it still presents to the ingenuity and skill of the coming generations a large and conceptually important set of open questions.

Index